BIRKHÄUSER

Birkhäuser
Advanced Texts

Edited by
Herbert Amann, Zürich University
Steven G. Krantz, Washington University, St. Louis
Shrawan Kumar, University of North Carolina at Chapel Hill
Jan Nekovář, Université Pierre et Marie Curie, Paris

Vladimir Azarin

Growth Theory of Subharmonic Functions

Birkhäuser
Basel · Boston · Berlin

Author:

Vladimir Azarin
Department of Mathematics
Bar-Ilan University
Ramat-Gan 52900
Israel
e-mail: azarin@macs.biu.ac.il

2000 Mathematics Subject Classification: Primary 31B05; Secondary 30D20

Library of Congress Control Number: 2008930978

Bibliographic information published by Die Deutsche Bibliothek
Die Deutsche Bibliothek lists this publication in the Deutsche Nationalbibliografie; detailed bibliographic data is available in the Internet at <http://dnb.ddb.de>.

ISBN 978-3-7643-8885-0 Birkhäuser Verlag AG, Basel · Boston · Berlin

Basel · Boston · Berlin
P.O. Box 133, CH-4010 Basel, Switzerland
Part of Springer Science+Business Media
Printed on acid-free paper produced of chlorine-free pulp. TCF ∞
Printed in Germany

ISBN 978-3-7643-8885-0 e-ISBN 978-3-7643-8886-7

9 8 7 6 5 4 3 2 1 www.birkhauser.ch

Contents

Chapter 1

Preface

This book aims to convert the noble art of constructing an entire function with prescribed asymptotic behavior to a handicraft.

For this you should just consider the limit set that describes the asymptotic behavior of the entire function, i.e., you should consider the set $U[\rho, \sigma]$ of subharmonic functions (that is, $\{v \text{ is subharmonic} : v(re^{i\phi}) \leq \sigma r^{\rho}\}$) and pick out the subset U which characterizes its asymptotic properties.

How to do it? The properties of limit sets are listed in Section 3. All the standard growth characteristics are expressed in terms of limit sets in Sections 3.2, 3.3, 5.7. Examples of construction are to be found in Sections 5.4–6.3. So you can use this book as a reference book for construction of entire functions.

Of course, you need some terms. All the terms that we use in this book are listed on pages 249–253.

If you want to study the theory, I recommend that you solve the exercises that are in the text. Most of them are trivial. However, I recommend that you do all of them by the moment that they appear trivial to you.

A few words about the history of this book. It arose from a course of lectures that I gave at Kharkov University in 1977. After some time, under pressure and with active help of Prof. I.V. Ostrovskii, a rotaprint edition (Edition of KhGU)of this course appeared: the first part in 1978, the second one in 1982. Mathematical Reviews did not notice this fact.

Since that time lots of new and important results have been obtained. Some of them were presented in Chapter 3 of the review [GLO].

In 1994, when I started to work in the Bar-Ilan University and obtained a personal computer, my first wish was to study typing on it in English. This was the first impulse for translating this course into English (there are no more than five copies of this book in the world, I believe, one of them being mine). I continued this project while working in Bar-Ilan (1994–2006) but there was not much time for this. And now I have finished.

Acknowledgements

I am indebted to many people. I start from Prof. I.V. Ostrovskii, who supported this idea for many years, and Prof. A.A. Gol′dberg, who stimulated my mathematical activity all my life by his letters and conversations.

I am indebted to Prof's A. Eremenko and M. Sodin, who, not being my "aspirants," solved a lot of problems connected to limit sets, and also to Dr.'s V. Giner, L. Podoshev and E. Fainberg who worked with me to develop the theory.

I am indebted to Prof's L. Hörmander and R. Sigurdsson who have sent me the preprints of their papers that were not yet published. I am indebted to Prof. I.F. Krasichkov–Ternovskii, who explained to me many years ago the connection between the multiplicator problem and completeness of the exponent system in a convex domain.

I am indebted to Prof's. M.I. Kadec and V.P. Fonf for proving Theorem 4.1.5.2, which is rather far off my speciality.

I am indebted to my coauthors Prof's D. Drasin and P. Poggi–Corradini; I have exploited the results of our joint paper in Section 6.2.

Of course, I am indebted to my late teacher Prof. B.Ya. Levin, who taught me entire and subharmonic functions and gave me the first problems in this area. Actually, the theory of limit sets is a generalization of the theory of functions of completely regular growth.

I am also indebted very much to my grandson Sasha Sodin, who transformed "my English" into English.

Chapter 2

Auxiliary Information. Subharmonic Functions

2.1 Semicontinuous functions

2.1.1. Let $x \in \mathbb{R}^m$ be a point in an m-dimensional Euclidean space, E a Borel set and $f(x)$ a function on E such that $f(x) \neq \infty$.

Set

$$M(f, x, \varepsilon) := \sup\{f(x') : |x - x'| < \varepsilon,\ x' \in E\} \tag{2.1.1.1}$$

The function

$$f^*(x) := \lim_{\varepsilon \to 0} M(f, x, \varepsilon)$$

is called the *upper semicontinuous regularization* of the function $f(x)$.

In the case of a finite jump, the regularization "raises" the values of the function. However, there is no influence on $f^*(x)$, if $f(x)$ tends to $-\infty$ "continuously".

Proposition 2.1.1.1 (Regularization Properties) *The following properties hold:*

(rg1) $$f(x) \leq f^*(x);$$

(rg2) $$(\alpha f)^*(x) = \alpha f^*(x);$$

(rg3) $$(f^*)^*(x) = f^*(x);$$

(rg4) $$\begin{aligned}(f_1 + f_2)^*(x) &\leq f_1^*(x) + f_2^*(x);\\ (\max(f_1, f_2))^*(x) &\leq \max(f_1^*, f_2^*)(x);\\ (\min(f_1, f_2))^*(x) &\leq \min(f_1^*, f_2^*)(x).\end{aligned}$$

These properties are obvious corollaries of the definition of $f^*(x)$.

Exercise 2.1.1.1 Prove them.

2.1.2 The function $f(x)$ is called *upper semicontinuous* at a point x if $f^*(x) = f(x)$. We denote the class of upper semicontinuous functions on E by $C^+(E)$. The function $f(x)$ is called *lower semicontinuous* if $-f(x)$ is upper semicontinuous (notation $f \in C^-(E)$).

Examples of semicontinuous functions are given by

Proposition 2.1.2.1 (Semicontinuity of Characteristic Functions of Sets) *Let $G \subset \mathbb{R}^m$ be an open set. Then its characteristic function χ_G is lower semicontinuous in $\mathbb{R}^m$. Let F be a closed set, then χ_F is upper semicontinuous.*

The proof is obvious.

Exercise 2.1.2.1 Prove this.

Proposition 2.1.2.2 (Connection with Continuity) *If $f \in C^+ \cap C^-$, then f is continuous.*

The assertion follows from the equalities

$$f^*(x) = \limsup_{\varepsilon \to 0}\{f(x') : |x - x'| < \varepsilon\}; -(-f)^*(x) = \liminf_{\varepsilon \to 0}\{f(x') : |x - x'| < \varepsilon\}.$$

Proposition 2.1.2.3 (C^+-Properties) *The following holds:*

$$(C^+\ 1) \qquad f \in C^+(E) \Rightarrow \alpha f \in C^+(E), \ \text{for } \alpha \geq 0$$

$$(C^+\ 2) \qquad f_1, f_2 \in C^+ \Rightarrow f_1 + f_2, \max(f_1, f_2), \min(f_1, f_2) \ \in C^+.$$

These properties follow from the properties of regularization (Proposition 2.1.1.1).

Exercise 2.1.2.2 Prove them.

Let G be an open set. Set $G_A := \{x \in G : f(x) < A\}$.

Theorem 2.1.2.4 (First Criterion of Semicontinuity) *One has $f \in C^+$ if and only if G_A is open for all $A \in \mathbb{R}$.*

Proof. Let $f(x) = f^*(x)$, $x \in G$. Then $\{f(x) < A\} \Longrightarrow \{f^*(x) < A\} \Longrightarrow \{M(f, x, \varepsilon) < A\}$ for all sufficiently small ε. Thus the neighborhood of x $V_{\varepsilon,x} := \{x' : |x - x'| < \varepsilon\}$ is contained in G_A.

Conversely, since the set G_A is open for $A = f(x_0) + \delta$, we have $f^*(x_0) \leq f(x_0) + \delta$ for any $\delta > 0$, hence for $\delta = 0$. With property (rg1) of Proposition 2.1.1.1 this gives $f^*(x_0) = f(x_0)$. □

Let F be a closed set. Set $F^A := \{x \in F : f(x) \geq A\}$. An obvious corollary of the previous theorem is

Corollary 2.1.2.5 *One has $f \in C^+$ if and only if F^A is closed for all A.*

Exercise 2.1.2.3 Prove the corollary.

We denote compact sets by K. Set $M(f, K) = \sup\{f(x) : x \in K\}$.

Theorem 2.1.2.6 (Weierstrass) *Let $K \subset \mathbb{R}^m$ be a compact set and $f \in C^+(K)$. Then there exists $x_0 \in K$ such that $f(x_0) = M(f, K)$.*

I.e., f attains its supremum on any compact set.

Proof. Set $K_n := \{x \in K : f(x) \geq M(f, K) - 1/n\}$.

The K_n are closed by Corollary 2.1.2.5, nonempty by definition of $M(f, K)$. Their intersection is nonempty and is equal to the set

$$K_{\max} := \{x \in K : f(x) \geq M(f, K)\}.$$

It means that there exists x_0 in K such that $f(x_0) \geq M(f, K)$.

The opposite inequality holds for any x in K. □

Exercise 2.1.2.4 Why?

The following theorem shows that the functional $M(f, K)$ is continuous with respect to monotonic convergence of semicontinuous functions.

Proposition 2.1.2.7 (Continuity from the right of $M(f, K)$) *Let $f_n \in C^+(K)$, $f_n \downarrow f$, $n = 1, 2, 3 \ldots$.*

Then $M(f_n, K) \downarrow M(f, K)$.

Proof. It is clear that $\lim\limits_{n \to \infty} M(f_n, K) := M$ exists.

Set $K_n := \{x \in K : f_n(x) \geq M\}$. The intersection of the closed nonempty sets K_n is nonempty and has the following form: $\bigcap\limits_n K_n = \{x : f(x) \geq M\}$. So $M(f, K) \geq M$.

The opposite inequality is obvious. □

Exercise 2.1.2.5 Why?

In the same way one proves

Proposition 2.1.2.8 (Commutativity of inf and $M(\cdot)$) *Let*

$$\{f_\alpha \in C_+(K),\ \alpha \in (0; \infty)\}$$

be an arbitrarily decreasing family of semicontinuous functions. Then

$$\inf_\alpha M(f_\alpha, K) = M(\inf_\alpha f_\alpha, K).$$

Exercise 2.1.2.6 Prove this proposition.

Theorem 2.1.2.9 (Second Criterion of Semicontinuity) *$f \in C^+(K)$ iff there exists a sequence f_n of continuous functions such that $f_n \downarrow f$.*

Sufficiency. Let $f_n \in C^+(K)$, $f_n \downarrow f$.Set $K_n^A := \{x \in K : f_n(x) \geq A\}$. This is a sequence of nonempty closed sets. If the set $K^A := \{x : f(x) \geq A\}$ is nonempty, then K^A is closed because $\bigcap_n K_n^A = K^A$. Hence $f \in C^+(K)$ by Corollary 2.1.2.5.

Necessity. Set $f_n(x, y) := f(y) - n|x - y|$.

This sequence of functions has the following properties:

a) it decreases monotonically in n and

$$\lim_{n\to\infty} f_n(x, y) = \begin{cases} f(x), & \text{for } x = y; \\ -\infty, & \text{for } x \neq y; \end{cases}$$

b) for any fixed n the functions f_n are continuous in x uniformly with respect to y, because $|f_n(x, y) - f_n(x', y)| \leq n|x - x'|$;

c) f_n are upper semicontinuous in y.

Proposition 2.1.2.7 and c) imply that

$$\lim_{n\to\infty} M_y(f_n(x, y), K) = M_y(\lim_{n\to\infty} f_n(x, y), K).$$

b) implies that the functions $f_n(x) := M_y(f_n(x, y), K)$ are continuous, and a) implies that they decrease monotonically to $f(x)$. □

2.1.3 We will consider a family of upper semicontinuous functions:$\{f_t : \ t \in T \subset (0, \infty)\}$. It is easy to prove

Proposition 2.1.3.1 $f_t \in C^+ \Longrightarrow \inf_{t\in T} f_t(x) \in C^+$.

Exercise 2.1.3.1 Prove this proposition.

Set $f_T(x) := \sup_{t\in T} f_t(x)$. The function f_T is not, generally speaking, upper semicontinuous even if T is countable and f_t are continuous. It is not possible to replace $\sup_{t\in T}$ in the definition of f_T by $\sup_{t\in T_0}$, where T_0 is a countable set. However, the following theorem holds:

Theorem 2.1.3.2 (Choquet's Lemma) *There exists a countable set $T_0 \subset T$ such that*

$$(\sup_{t\in T_0} f_t)^*(x) = (\sup_{t\in T} f_t)^*(x).$$

Proof. Let $\{x_n\}$ be a countable set that is dense in $\mathbb{R}^m$ and $\varepsilon_j \downarrow 0$. Then the balls

$$K_{n,j} := \{x : |x - x_n| < \varepsilon_j\}$$

cover every point $x \in \mathbb{R}^m$ infinitely many times.

Renumbering we obtain a sequence $\{K_l : l \in \mathbb{N}\}$. For any l there exists, by definition of $\sup_{K_l}$, such a point $x_0 \in K_l$ that

$$\sup_{K_l} f_T(x) \leq f_T(x_0) + 1/2l. \tag{2.1.3.1}$$

By definition of $\sup_T$ there exists t_l such that

$$f_T(x_0) < f_{t_l}(x_0) + 1/2l.$$

Thus

$$f_T(x_0) < \sup\{f_{t_l}(x) : x \in K_l\} + 1/2l. \tag{2.1.3.2}$$

The inequalities (2.1.3.1) and (2.1.3.2) imply that for any l there exists t_l such that

$$\sup\{f_T(x) : x \in K_l\} \leq \sup\{f_{t_l}(x) : x \in K_l\} + 1/l. \tag{2.1.3.3}$$

Now set $T_0 = \{t_l\}$. Evidently, $f_{T_0}(x) \leq f_T(x)$ and thus

$$f^*_{T_0}(x) \leq f^*_T(x). \tag{2.1.3.4}$$

Let us prove the opposite inequality.

Let $x \in \mathbb{R}^m$. Choose a subsequence $\{K_{l_j}\}$ that tends to x. From (2.1.3.3) we obtain

$$\begin{aligned} f^*_T(x) &\leq \limsup_{j\to\infty} \sup_{x'\in K_{l_j}} f_T(x') \\ &\leq \limsup_{j\to\infty} \sup_{x'\in K_{l_j}} f_{t_{l_j}}(x') \\ &\leq \limsup_{j\to\infty} \sup_{x'\in K_{l_j}} f_{T_0}(x') = f^*_{T_0}(x). \end{aligned} \tag{2.1.3.5}$$

(2.1.3.4) and (2.1.3.5) imply the assertion of the theorem. □

2.2 Measures and integrals

2.2.1 Let G be an open set in $\mathbb{R}^m$ and $\sigma(G)$ a σ-algebra of Borel sets containing all the compact sets $K \subset G$.

Let μ be a countably additive nonnegative function on $\sigma(G)$, which is finite on all $K \subset G$. We will call it a *measure* or a *mass distribution*.

Let $G_0(\mu)$ be the largest open set for which μ is zero. (It is the union of all the open sets G' such that $\mu(G') = 0$.)

The set $\operatorname{supp}\mu := G\backslash G_0(\mu)$ is called the *support* of μ. It is closed in G.

We say that μ is *concentrated* on $E \in \sigma(G)$ if $\mu(G\backslash E) = 0$.

Theorem 2.2.1.1 (Support) *The support of a measure μ is the smallest closed set on which the measure μ is concentrated.*

Exercise 2.2.1.1 Prove this.

A measure μ can be concentrated on a non-closed set E and then $E \Subset \operatorname{supp}\ \mu$.

Example 2.2.1.1 Let E be a countable set dense in G. Then supp $\mu = G$ and, of course, $E \neq G$.

The set of all measures on G will be denoted by $\mathcal{M}(G)$.

The measure $\mu_F(E) := \mu(E \cap F)$ is called the *restriction* of μ onto $F \in \sigma(G)$. It is easy to see that μ_F is concentrated on F and $\operatorname{supp} \mu \subset \overline{F}$.

A countably additive function ν on $\sigma(G)$ that is finite for all $K \subset G$ is called a *charge*. We consider only real charges.

Example 2.2.1.2 $\nu := \mu_1 - \mu_2, \quad \mu_1, \mu_2 \in \mathcal{M}(G)$.

The set of all charges will be denoted $\mathcal{M}^d$.

Theorem 2.2.1.2 (Jordan decomposition) *Let $\nu \in \mathcal{M}^d(G)$. Then there exist two sets G^+, G^- such that*

a) $G = G^+ \cup G^-, \ G^+ \cap G^- = \varnothing$;
b) $\nu(E) \geq 0$ *for* $E \subset G^+$; $\nu(E) \leq 0$ *for* $E \subset G^-$.

One can find the proof in [Ha, Ch. VI Sec. 29]

The measures $\nu^+ := \nu_{G^+}$ and $\nu^- := \nu_{G^-}$, where ν_{G^+}, ν_{G^-} are restrictions of ν to G^+, G^-, are called the *positive* and *negative*, respectively, *variations* of ν. The measure $|\nu| := \nu_+ + \nu_-$ is called the *full variation* of ν or just a *variation.*

Theorem 2.2.1.3 (Variations) *The following holds:*

$$\nu^+(E) = \sup_{E' \subset E} \nu(E'); \ \nu^-(E) = \inf_{E' \subset E} \nu(E'); \ \nu = \nu^+ + \nu^-.$$

The proof is easy enough.

Exercise 2.2.1.2 Prove this.

Example 2.2.1.3 Let $\psi(x)$ be a locally summable function with respect to the Lebesgue measure. Set $\nu(E) := \int_E \psi(x)dx$. Then

$$\nu^+(E) = \int_E \psi^+(x)dx, \ \nu^-(E) = \int_E \psi^-(x)dx; \ |\nu|(E) = \int_E |\psi|(x)dx,$$

where

$$\psi^+(x) = \max(0, \psi(x)); \ \psi^-(x) = -\min(0, \psi(x)). \tag{2.2.1.1}$$

2.2.2 The function $f(x)$, $x \in G$ is called a *Borel function* if the set $E^A := \{f(x) > A\}$ belongs to $\sigma(G)$ for any $A \in \mathbb{R}$.

Let $K \Subset G$ be a compact set and f a Borel function. Then the Lebesgue-Stieltjes integrals of the form $\int_K f^+ d\mu$, $\int_K f^- d\mu$ with respect to a measure $\mu \in \mathcal{M}(G)$ are defined, and $\int_K f d\mu := \int_K f^+ d\mu - \int_K f^- d\mu$ is defined if at least one of the terms is finite.

We say that a property holds *μ-almost everywhere on E* if the set E_0 of x for which it does not hold satisfies the condition $\mu(E_0) = 0$.

We will denote all the compact sets in G as K (sometimes with indexes). The following theorems hold:

Theorem 2.2.2.1 (Lebesgue) *Let $\{f_n,\ n \in \mathbb{N}\}$ be a sequence of Borel functions on K and $g(x) \geq 0$ a function on K that is summable with respect to μ (i.e., its integral is finite), $|f_n(x)| \leq g(x)$ for x in K, and $f_n \to f$ when $n \to \infty$.*

Then $\lim\limits_{n\to\infty} \int_K f_n d\mu = \int_K f d\mu$.

Theorem 2.2.2.2 (B. Levy) *Let $f_n \downarrow f$ when $n \to \infty$, and f be a summable function on K.*

Then $\lim\limits_{n\to\infty} \int_K f_n d\mu = \int_K f d\mu$.

Theorem 2.2.2.3 (Fatou's Lemma) *Let $f_n(x) \leq \mathrm{const} < \infty$ for x in K.*

Then $\limsup\limits_{n\to\infty} \int_K f_n d\mu \leq \int_K \limsup\limits_{n\to\infty} f_n d\mu$.

The proofs can be found in [Ha, Ch. V, Sec. 27].

Let $L(\mu)$ be the space of functions that are summable with respect to μ. We say that $f_n \to f$ in $L(\mu)$ if $f_n, f \in L(\mu)$ and

$$\|f_n - f\| := \int |f_n - f|(x) d\mu \to 0$$

Theorem 2.2.2.4 (Uniqueness in $L(\mu)$) *Let $f_n \to f$ in $L(\mu)$ and*

$$\int f_n \psi d\mu \to \int g \psi d\mu$$

for any ψ continuous on $\operatorname{supp} \mu$. *Then* $\|g - f\| = 0$.

For the proof see, e.g., [Hö, Th. 1.2.5].

2.2.3 Let $\phi(x)$ be a Borel function on G. The set $\operatorname{supp}\ \phi := \overline{\{x : \phi(x) \neq 0\}}$ is called the *support of $\phi(x)$*. A function ϕ is called *finite in* G if $\operatorname{supp}\ \phi \Subset G$.

We say that a sequence $\mu_n \in \mathcal{M}$ converges *weakly* to $\mu \in \mathcal{M}$ if the condition $\int \phi d\mu_n \to \int \phi d\mu$ holds for any continuous function ϕ.

We will not show the integration domain, because it is always $\operatorname{supp} \phi$.

The weak (it is called also C^*-) convergence will be denoted as $\xrightarrow{*}$.

Theorem 2.2.3.1 (C^*-limits) *If $\mu_n \xrightarrow{*} \mu$, then for $E \in \sigma(G)$ the following assertions hold:*

$$\limsup_{n\to\infty} \mu_n(\overline{E}) \leq \mu(\overline{E});$$

$$\liminf_{n\to\infty} \mu_n(\overset{\circ}{E}) \geq \mu(\overset{\circ}{E});$$

where $\overset{\circ}{E}$ is the interior of E, $\overline{E}$ is the closure of E.

Proof. Let $\chi_{\overline{E}}$ be the characteristic function of the set $\overline{E}$. It is upper semicontinuous. Thus there exists a decreasing sequence φ_m of continuous functions finite in G that converges to $\chi_{\overline{E}}$ as $m \to \infty$. Then we have

$$\mu_n(\overline{E}) = \int \chi_{\overline{E}} d\mu_n \leq \int \varphi_m d\mu_n.$$

Passing to the limit as $n \to \infty$ we obtain

$$\limsup_{n\to\infty} \mu_n(\overline{E}) \leq \int \varphi_m d\mu.$$

Passing to the limit as $m \to \infty$ we obtain by Theorem 2.2.2.2

$$\limsup_{n\to\infty} \mu_n(\overline{E}) \leq \int \chi_{\overline{E}} d\mu = \mu(\overline{E}).$$

The proof for $\overset{\circ}{E}$ is analogous. □

Theorem 2.2.3.2 (Helly) *Let $\{\mu_\alpha : \alpha \in A\}$ be a family of measures uniformly bounded on any compact set $K \subset G$, i.e., $\exists C = C(K) : \mu_\alpha(K) \leq C(K)$, for $K \Subset G$. Then this family is weakly compact, i.e., there exists a sequence $\{\alpha_j : \alpha_j \in A\}$ and a measure μ such that $\mu_{\alpha_j} \xrightarrow{*} \mu$.*

The proof can be found in [Ha].

A set E is called *squarable* with respect to measure μ (μ-squarable) if $\mu(\partial E) = 0$.

Theorem 2.2.3.3. (Squarable Ring) *The following holds:*

sqr1) *if E_1, E_2 are μ-squarable, the sets $E_1 \cap E_2, E_1 \cup E_2, E_1 \setminus E_2$ are μ-squarable;*

sqr2) *for any couple: an open set G and a compact set $K \subset G$ there exists a μ-squarable set E such that $K \subset E \subset G$.*

Proof. The assertion sqr1) follows from

$$\partial(E_1 \cup E_2) \bigcup \partial(E_1 \cap E_2) \bigcup \partial(E_1 \setminus E_2) \subset \partial E_1 \cup \partial E_2.$$

Let us prove sqr2). Let $K_t := \{x : \exists y \in K : |x - y| < t\}$ be a t-neighborhood of the K. It is clear that for all the small t we have $K \Subset K_t \Subset G$. The function $a(t) := \mu(K_t)$ is monotonic on t and thus has no more than a countable set of jumps.

Let t be a point of continuity of $a(t)$. Then

$$\mu(\partial K_t) \leq \lim_{\epsilon \to 0}[\mu(K_{t+\epsilon}) - \mu(K_{t-\epsilon})] = 0.$$

Thus it is possible to set $E := K_t$ for this t. □

A family Φ of sets is called a *dense ring* if the following conditions hold:

dr1) $\forall F_1, F_2 \in \Phi \Longrightarrow F_1 \cup F_2, F_1 \cap F_2 \in \Phi$;

dr2) $\forall K, G : K \Subset G \; \exists F \in \Phi : K \subset F \subset G$.

The previous theorem shows that the class of μ-squarable sets is a dense ring. The following theorem shows how one can extend a measure from a dense ring to the Borel algebra.

Let Φ be a dense ring and $\Delta(F)$, $F \in \Phi$ a function of a set which satisfies the conditions:

$\Delta 1)$ *monotonicity* on Φ: $F_1 \subset F_2 \Longrightarrow \Delta(F_1) \leq \Delta(F_2)$;

$\Delta 2)$ *additivity* on Φ: $\Delta(F_1 \cup F_2) \leq \Delta(F_1) + \Delta(F_2)$ and $\Delta(F_1 \cup F_2) = \Delta(F_1) + \Delta(F_2)$ if $F_1 \cap F_2 = \varnothing$

$\Delta 3)$ *continuity* on Φ: $\forall F \in \Phi$ and $\epsilon > 0$ there exists a compact set K and an open set $G \supset K$ such that $\forall F' \in \Phi : K \subset F' \subset G$ the inequality $|\Delta(F) - \Delta(F')| < \epsilon$ holds.

Theorem 2.2.3.4 (N. Bourbaki) *There exists a measure μ such that*

$$\mu(F) = \Delta(F), \; \forall F \in \Phi$$

iff the conditions $\Delta 1)$–$\Delta 3)$ hold.

Theorem 2.2.3.5 (Uniqueness of Measure) *Under the conditions $\Delta 1)$–$\Delta 3)$ the measure is defined uniquely by the formulae:*

$$\mu(K) = \inf\{\Delta(F) : F \in \Phi, \; F \supset K\}; \tag{2.2.3.1}$$

$$\mu(G) = \sup\{\Delta(F) : F \in \Phi, \; F \subset G\}; \tag{2.2.3.2}$$

$$\mu(E) = \sup\{\mu(K) : K \subset E\} = \inf\{\mu(G) : G \supset E\}, \tag{2.2.3.3}$$

and every $F \in \Phi$ is μ-squarable.

For the proof see [Bo, Ch. 4, Sec 3, it. 10]. The squarability follows from (2.2.3.3).

The following theorem connects the convergence of measures on any dense ring and on the ring of sets squarable with respect to the limit measure.

Theorem 2.2.3.6 (Set-convergences) *If $\mu_n(F) \to \mu(F)$ for all F in a dense ring Φ, then $\mu_n(E) \to \mu(E)$ for any μ-squarable set E.*

Proof. Suppose $\overset{\circ}{E} \neq \varnothing$.

Let $\epsilon > 0$. By (2.2.3.3) one can find a compact set K such that

$$\mu(K) + \epsilon \geq \mu(\overset{\circ}{E}) = \mu(E). \tag{2.2.3.4}$$

One can also find an open set G such that

$$\mu(G) - \epsilon \leq \mu(\overline{E}) = \mu(E). \tag{2.2.3.5}$$

By property dr2) of a dense ring one can find $F, F' \in \Phi$ such that

$$K \subset F \subset \overset{\circ}{E} \subset E \subset \overline{E} \subset F' \subset G.$$

Thus $\mu_n(F) \leq \mu_n(E) \leq \mu_n(F')$ and hence

$$\mu(F) \leq \varliminf_{n\to\infty} \mu_n(E) \leq \varlimsup_{n\to\infty} \mu_n(E) \leq \mu(F'). \tag{2.2.3.6}$$

From (2.2.3.4) and (2.2.3.5) we obtain $0 \leq \mu(F') - \mu(F) \leq \mu(G) - \mu(K) \leq 2\epsilon$ for arbitrarily small ϵ. Thus from (2.2.3.6) we obtain

$$\varliminf_{n\to\infty} \mu_n(E) = \varlimsup_{n\to\infty} \mu_n(E) = \mu(E). \tag{2.2.3.7}$$

That is to say that $\mu_n(E) \to \mu(E)$.

If $\overset{\circ}{E} = \varnothing$, then $\mu(\overline{E}) = 0$ by the definition of a squarable set. One can show in the same way that $\mu_n(E) \to 0$. □

Now we connect the weak convergence to the convergence on squarable sets.

Theorem 2.2.3.7 (Set- and C*-convergences) *The conditions*

$$\mu_n \overset{*}{\to} \mu \tag{2.2.3.8}$$

and $\mu_n(E) \to \mu(E)$ on μ-squarable sets E are equivalent.

Proof. Sufficiency of (2.2.3.8) follows from Theorem 2.2.3.1.

Exercise 2.2.3.1 Prove this.

Let us prove necessity.

For any compact set one can find a μ-squarable E such that $K \subset E$. Hence $\mu_n(K) \leq \mu(E) + 1 := C(K)$ when n is big enough.

By Helly's theorem (Theorem 2.2.3.2) there exists a measure μ' and a subsequence $\mu_{n_j} \overset{*}{\to} \mu'$. By the proved sufficiency, $\mu'(E) = \mu(E)$ on a dense ring of the squarable sets. Thus $\mu' = \mu$ by Uniqueness Theorem 2.2.3.5. And thus $\mu_n \overset{*}{\to} \mu$. □

Denote by

$$\mu_E(G) := \begin{cases} \mu(G \cap E) & \text{if } G \cap E \neq \varnothing, \\ 0 & \text{if } G \cap E = \varnothing \end{cases}$$

the *restriction* of μ on the set E.

Corollary 2.2.3.8 *Let $\mu_n \xrightarrow{*} \mu$ and E be a squarable set for μ. Then $(\mu_n)_E \xrightarrow{*} (\mu)_E$.*

Indeed, if E is a squarable set for μ it is a squarable set for μ_E. So Theorem 2.2.3.7 implies the corollary.

2.2.4 Let $\sigma(\mathbb{R}^{m_1} \times \mathbb{R}^{m_2})$ be the σ-algebra of all the Borel sets, $\Phi_i \subset \sigma(\mathbb{R}^{m_i})$, $i = 1, 2$, be dense rings, $\Phi := \Phi_1 \otimes \Phi_2 \subset \sigma(\mathbb{R}^{m_1} \times \mathbb{R}^{m_2})$ be a ring generated by all the sets of form $F_1 \times F_2$, $F_i \in \Phi_i$.

Theorem 2.2.4.1 (Product of Rings) *If Φ_i, $i = 1, 2$ are dense rings, then $\Phi_1 \otimes \Phi_2$ is a dense ring; if they consist of squarable sets, then Φ consists of squarable sets.*

Proof. Let $K \subset G \subset \mathbb{R}^{m_1} \times \mathbb{R}^{m_2}$. For every point $x \in K$ one can (evidently) find $F_1 \times F_2$ such that $x \subset F_1 \times F_2 \subset G$. One can find a finite covering and obtain a finite union F of sets of such form. Thus $F \in \Phi_1 \otimes \Phi_2$ and $F \subset G$.

The second assertion follows from the formula

$$\partial(F_1 \times F_2) = (\partial F_1 \times F_2) \cup (F_1 \times \partial F_2). \qquad \square$$

Let μ_i be a measure on $\sigma(\mathbb{R}^{m_i})$, $i = 1, 2$, and $\mu := \mu_1 \otimes \mu_2$ the *product of measures*, i.e., a measure on $\sigma(\mathbb{R}^{m_1} \times \mathbb{R}^{m_2})$ such that $\mu(E_1 \times E_2) = \mu_1(E_1)\mu_2(E_2)$ for all $E_i \in \sigma(\mathbb{R}^{m_i})$, $i = 1, 2$.

Theorem 2.2.4.2 (Product of Measures) *A measure $\mu_1 \otimes \mu_2$ is uniquely defined by its values on $\Phi_1 \otimes \Phi_2$.*

The assertion follows from Theorem 2.2.4.1 and Uniqueness Theorem 2.2.3.5.

Theorem 2.2.4.3 (Fubini) *Let $f(x_1, x_2)$ be a Borel function on $\mathbb{R}^{m_1} \times \mathbb{R}^{m_2}$. Then*

$$\begin{aligned} \int\limits_{\mathbb{R}^{m_1} \times \mathbb{R}^{m_2}} f(x_1, x_2)(\mu_1 \otimes \mu_2)(dx_1 dx_2) &= \int\limits_{\mathbb{R}^{m_1}} \mu_1(dx_1) \int\limits_{\mathbb{R}^{m_2}} f(x_1, x_2)\mu_2(dx_2) \\ &= \int\limits_{\mathbb{R}^{m_2}} \mu_2(dx_2) \int\limits_{\mathbb{R}^{m_1}} f(x_1, x_2)\mu_1(dx_1), \end{aligned} \tag{2.2.4.1}$$

if at least one of the parts of (2.2.4.1) *is well defined.*

The proof can be found in [Ha, Ch. VII, Sec. 36].

2.3 Distributions

2.3.1 Let us consider the set $\mathcal{D}(G)$ of all infinitely differentiable functions $\varphi(x),\ x \in G \subset \mathbb{R}^m$.

It is a *linear space* because for any constants c_1, c_2,

$$\varphi_1, \varphi_2 \in \mathcal{D}(G) \Longrightarrow c_1\varphi_1 + c_2\varphi_2 \in \mathcal{D}(G). \tag{$D1$}$$

It is a *topological* space with convergence defined by

$$\varphi_n \xrightarrow{\mathcal{D}} \varphi : \begin{cases} \text{a) } \operatorname{supp}\varphi_n \subset K \Subset \mathbb{R}^m \\ \quad \text{for some compact } K \\ \text{and} \\ \text{b) } \varphi_n \to \varphi \text{ uniformly on } K \\ \quad \text{with all their derivatives.} \end{cases} \tag{$D2$}$$

We consider some examples of functions $\varphi \in \mathcal{D}$. Set

$$\alpha(t) = \begin{cases} Ce^{-\frac{1}{1-t^2}}, & \text{for } t \in (-1;1), \\ 0, & \text{for } t \,\overline{\in}\, (-1;1). \end{cases} \tag{2.3.1.1}$$

Evidently $\alpha(|x|) \in \mathcal{D}(\mathbb{R}^m)$ and $\operatorname{supp}\alpha \subset \{x : |x| \le 1\}$.

Exercise 2.3.1.1 Check this.

Let us find C such that

$$\int \alpha(|x|)dx = \sigma_m \int_0^1 \alpha(t)t^{m-1}dt = 1 \tag{2.3.1.2}$$

where σ_m is the area of the unit sphere $\{|x| = 1\}$. Set

$$\alpha_\varepsilon(x) := \varepsilon^{-m}\alpha\left(\frac{|x|}{\varepsilon}\right). \tag{2.3.1.3}$$

For any ε we have $\alpha_\varepsilon \in \mathcal{D}$ and $\operatorname{supp}\alpha_\varepsilon \subset \{x : |x| \le \varepsilon\}$.

Let $\psi(y), y \in K \subset G$ be a Lebesgue summable function. Consider the function

$$\psi_\varepsilon(x) := \int_K \psi(y)\alpha_\varepsilon(x-y)dy. \tag{2.3.1.4}$$

The function belongs to $\mathcal{D}(G)$ for ε small enough and its support is contained in the ε-neighborhood of K.

2.3.2 Let $f(x),\ x \in G \subset \mathbb{R}^m$ be a locally summable function in G. The formula

$$\langle f, \varphi\rangle := \int f(y)\varphi(y)dy,\ \varphi \in \mathcal{D}(G) \tag{2.3.2.1}$$

defines a *linear continuous functional* on $\mathcal{D}$, i.e., one that satisfies the conditions

$$\langle f, c_1\varphi_1 + c_2\varphi_2\rangle = c_1\langle f, \varphi_1\rangle + c_2\langle f, \varphi_2\rangle; \tag{D'1}$$

$$(\varphi_n \xrightarrow{\mathcal{D}} \varphi) \Longrightarrow \langle f, \varphi_n\rangle \to \langle f, \varphi\rangle. \tag{D'2}$$

However, (2.3.2.1) does not exhaust all the linear continuous functionals as we will see further. An arbitrarily linear continuous functional on $\mathcal{D}(G)$ is called a Schwartz *distribution* and the linear topological space of the functionals is denoted as $\mathcal{D}'(G)$.

Following are some examples of functionals that do not have the form of (2.3.2.1):

$$\langle \delta_x, \varphi\rangle := \varphi(x). \tag{2.3.2.2}$$

This distribution is called the *Dirac delta-function.* Further,

$$\langle \delta_x^{(n)}, \varphi\rangle := (-1)^n \varphi^{(n)}(x). \tag{2.3.2.3}$$

This distribution is called *the nth derivative* of the Dirac delta-function.

Exercise 2.3.2.1 Check that the functionals (2.3.2.2) and (2.3.2.3) are both distributions.

A distribution of the form (2.3.2.1) is called *regular.*

Theorem 2.3.2.1 (Du Bois Reymond) *If two locally summable functions f_1 and f_2 define the same distribution, then they coincide almost everywhere.*

For the proof see, e.g., [Hö, Thm. 2.1.6].

Note that the converse assertion is obvious.

A distribution μ is called *positive* if $\langle \mu, \varphi\rangle \geq 0$ for any $\varphi \in \mathcal{D}(G)$ such that $\varphi(x) \geq 0$ for all $x \in \mathbb{R}^m$. We shall write this as $\mu > 0$ in $\mathcal{D}'$.

Example 2.3.2.1 Let $\mu(E)$ be a measure in G. Then the distribution

$$\langle \mu, \varphi\rangle := \int \varphi(x)\mu(dx) \tag{2.3.2.4}$$

is positive.

This formula represents all the positive distributions as one can see from

Theorem 2.3.2.2 (Positive Distributions) *Let $\mu > 0$ in $\mathcal{D}(G)$. Then there exists a unique measure $\mu(E)$ such that the distribution μ is given by* (2.3.2.4).

For the proof see, e.g., [Hö, Thm. 2.1.7].

2.3.3 Let us consider operations on distributions.

A *product* of a distribution f by an *infinitely differentiable* function $\alpha(x)$ is defined by

$$\langle \alpha f, \varphi\rangle := \langle f, \alpha\varphi\rangle. \tag{2.3.3.1}$$

It is well defined because $\alpha\varphi \in \mathcal{D}$ too.

A *sum* of distributions f_1 and f_2 is defined by

$$\langle f_1 + f_2, \varphi\rangle := \langle f_1, \varphi\rangle + \langle f_2, \varphi\rangle, \tag{2.3.3.2}$$

and the *partial derivative* $\frac{\partial}{\partial x_k}$ is defined by the equality

$$\langle \frac{\partial}{\partial x_k} f, \varphi\rangle := \langle f, -\frac{\partial}{\partial x_k}\varphi\rangle. \tag{2.3.3.3}$$

These definitions look reasonable because of the following

Theorem 2.3.3.1 (Operations on Distributions) *The sum of regular distributions corresponds to the sum of the functions; the product of a regular distribution by an infinitely differentiable function corresponds to the product of the functions; the derivative of a regular distribution that is generated by a differentiable function corresponds to the derivative of that function.*

Proof. We have, for example,

$$\langle \alpha \cdot (f), \varphi\rangle := \int f(x)[\alpha(x)\varphi(x)]dx = \int [\alpha(x)f(x)]\varphi(x)dx := \langle (\alpha f), \varphi\rangle.$$

For the sum we have

$$\begin{aligned}\langle (f_1) + (f_2), \varphi\rangle := \langle f_1, \varphi\rangle + \langle f_2, \varphi\rangle &= \int f_1(x)\varphi(x)dx + \int f_2(x)\varphi(x)dx \\ &= \int [f_1(x) + f_2(x)]\varphi(x)dx = \langle (f_1 + f_2), \varphi\rangle.\end{aligned}$$

Let $f(x)$ have the derivative $\frac{\partial}{\partial x_k} f$. Then

$$\begin{aligned}\langle \frac{\partial}{\partial x_k} f, \varphi\rangle :&= \langle f, -\frac{\partial}{\partial x_k}\varphi\rangle \\ &= \int f(x_1, x_2, \dots, x_m)\left[-\frac{\partial}{\partial x_k}\varphi(x_1, x_2, \dots, x_m)\right] dx_1 dx_2, \dots, dx_m \\ &= \int dx_1, \dots, dx_{k-1}dx_{k+1}, \dots, dx_m \\ &\quad \times \int f(x_1, x_2, \dots, x_m)\left[-\frac{\partial}{\partial x_k}\varphi(x_1, x_2, \dots, x_m)\right] dx_k.\end{aligned}$$

Now we shall integrate by parts and all the substitution will vanish, because φ is finite. So we obtain

$$\langle \frac{\partial}{\partial x_k} f, \varphi\rangle = \int \frac{\partial}{\partial x_k} f(x)\varphi(x)dx.$$

That is to say the derivative of the distribution corresponds to the function derivative. □

2.3.4 We say that a sequence of distributions f_n *converges* to a distribution f if

$$\langle f_n, \varphi\rangle \to \langle f, \varphi\rangle \ \forall \varphi \in \mathcal{D}(G). \tag{2.3.4.1}$$

Theorem 2.3.4.1 (Completeness of $\mathcal{D}'$) *If the sequence of numbers $\langle f_n, \varphi\rangle$ has a limit for every $\varphi \in \mathcal{D}(G)$, then this functional is a linear continuous functional on $\mathcal{D}(G)$, i.e., a distribution.*

For the proof see, e.g., [Hö, Thm. 2.1.8].

Differentiation is continuous with respect to convergence of distributions.

Theorem 2.3.4.2 (Continuity of Differential Operators) *If $f_n \to f$ in $\mathcal{D}(G)$, then $\frac{\partial}{\partial x_k} f_n \to \frac{\partial}{\partial x_k} f$.*

Proof. Set in (2.3.4.1) $\varphi := -\frac{\partial}{\partial x_k}\varphi$. Then

$$\left\langle \frac{\partial}{\partial x_k} f_n, \varphi \right\rangle = \left\langle f_n, -\frac{\partial}{\partial x_k}\varphi \right\rangle \to \left\langle f, -\frac{\partial}{\partial x_k}\varphi \right\rangle = \left\langle \frac{\partial}{\partial x_k} f, \varphi \right\rangle. \qquad \square$$

The following theorem shows that the $\mathcal{D}'$-convergence is the weakest of the convergences considered earlier.

Theorem 2.3.4.3 (Connection between Convergences) *Let f_n be a sequence of Lebesgue summable functions on domain G such that at least one of the following conditions holds:*

Cnvr1) *$f_n \to f$ uniformly on any compact set $K \Subset G$ and f is a locally summable function;*

Cnvr2) *$f_n \to f$ on any $K \Subset G$, satisfying the conditions of the Lebesgue theorem (Theorem 2.2.2.1);*

Cnvr3) *$f_n \downarrow f$ monotonically and f is a locally summable function.*

Then $f_n \to f$ in $\mathcal{D}'(G)$.

Proof. All the assertions are corollaries of Section 2.2.2 on passing to the limit under an integral.

Let us prove, for example, Cnvr3). Let $f_n \downarrow f$. Then

$$\langle f_n, \varphi\rangle = \int f_n(x)\varphi(x)dx = \int f_n(x)\varphi^+(x)dx - \int f_n(x)\varphi^-(x)dx \tag{2.3.4.2}$$

where φ^+ and φ^- are defined in (2.2.1.1).

Both last integrals in (2.3.4.2) have a limit by the B. Levy theorem (Theorem 2.2.2.2), and thus

$$\lim_{n\to\infty} \langle f_n, \varphi\rangle = \int f(x)\varphi^+(x)dx - \int f(x)\varphi^-(x)dx = \int f(x)\varphi(x)dx = \langle f, \varphi\rangle. \tag{2.3.4.3}$$

(2.3.4.3) means that $f_n \to f$ in $\mathcal{D}'$. $\square$

Exercise 2.3.4.1 Prove Cnvr 1) and 2).

Theorem 2.3.4.4 ($\mathcal{D}'$ and C^* convergences) *Let μ_n, μ be measures in G. The conditions $\mu_n \to \mu$ in $\mathcal{D}'(G)$ and $\mu_n \overset{*}{\to} \mu$ are equivalent.*

It is clear that the first condition is necessary for the second one. The sufficiency holds, because every continuous function can be approximated with functions that belong to $\mathcal{D}$. For more details see, e.g., [Hö, Thm. 2.1.9].

Let $\alpha_\epsilon(x)$ be defined as in (2.3.1.3).For any $f \in \mathcal{D}'(D)$ we can consider the function $f_\epsilon(x) := \langle f, \alpha_\epsilon(x + \bullet)\rangle$. It is called a *regularization* of the distribution f.

Theorem 2.3.4.5.(Properties of Regularizations) *The following holds:*

reg1) *$f_\epsilon(x)$ is an infinitely differentiable function in any $K \Subset D$ for sufficiently small ϵ;*

reg2) *$f_\epsilon(x) \to f$ in $\mathcal{D}'(D)$ as $\epsilon \downarrow 0$;*

reg3) *if $f_n \to f$ in $\mathcal{D}'(D)$, $(f_n)_\epsilon \to f_\epsilon$ uniformly with all its derivatives on any compact set in D.*

The property reg1) follows from the formula

$$\frac{\partial}{\partial x_j} f_\epsilon = \left\langle f, \frac{\partial}{\partial x_j}\alpha_\epsilon(x + \bullet)\right\rangle .$$

The property reg2) follows from the assertion

$$\phi_\epsilon(x) := \int \phi(y)\alpha_\epsilon(x + y)dy \to \phi(x) \text{ in } \mathcal{D}(D)$$

as $\epsilon \downarrow 0$.

For the proof of reg3) see [Hö, Theorems 2.1.8, 4.1.5].

Let us note the following assertion;

Theorem 2.3.4.6 (Continuity $\langle\bullet, \bullet\rangle$) *The function*

$$\langle f, \phi\rangle : \mathcal{D}'(G) \times \mathcal{D}(G) \mapsto \mathbb{R}$$

is continuous in the appropriate topology.

I.e., $f_n \to f$ in $\mathcal{D}'(G)$ and $\phi_j \to \phi$ in $\mathcal{D}(G)$ imply $\langle f_n, \phi_j\rangle \to \langle f, \phi\rangle$.

For the proof see [Hö, Theorem 2.1.8].

2.3.5 Let $G_1 \subset G$. Then $\mathcal{D}'(G) \subset \mathcal{D}'(G_1)$, because every functional on $\mathcal{D}(G)$ can be considered as a functional on $\mathcal{D}(G_1)$.

A distribution $f \in \mathcal{D}'(G)$ considered as a distribution in $\mathcal{D}'(G_1)$ is called the *restriction* of f to G_1 and is denoted $f\,|_{G_1}$.

Theorem 2.3.5.1 (Sewing Theorem) *Let $G_\alpha \subset \mathbb{R}^m$ be a family of domains and in every of them let there be a distribution $f_\alpha \in \mathcal{D}(G_\alpha)$, such that:*

If $G_{\alpha_1} \cap G_{\alpha_2} \neq \varnothing$, the equality

$$f_{\alpha_1} |_{G_{\alpha_1} \cap G_{\alpha_2}} = f_{\alpha_2} |_{G_{\alpha_1} \cap G_{\alpha_2}} \tag{2.3.5.1}$$

holds. Then there exists one and only one distribution $f \in \mathcal{D}'(G)$ where $G = \bigcup\limits_\alpha G_\alpha$ such that $f |_{G_\alpha} = f_\alpha$.

In particular, it means that every distribution is defined uniquely by its restriction to a neighborhood of every point.

Let $\mathcal{D}(S_R)$ be a space of infinitely differentiable functions on the sphere $S_R := \{x : |x| = R\}$. The corresponding distribution space is denoted as $\mathcal{D}'(S_R)$. The sewing theorem holds for this space in the following form:

Theorem 2.3.5.2 ($\mathcal{D}'$ on Sphere) *Let a family of domains Ω_α cover S_R and in every of them let there be a distribution $f_\alpha \in \mathcal{D}(\Omega_\alpha)$, such that:*

If $\Omega_{\alpha_1} \cap \Omega_{\alpha_2} \neq \varnothing$, the equality

$$f_{\alpha_1} |_{\Omega_{\alpha_1} \cap \Omega_{\alpha_2}} = f_{\alpha_2} |_{\Omega_{\alpha_1} \cap \Omega_{\alpha_2}} \tag{2.3.5.2}$$

holds. Then there exists one and only one distribution $f \in \mathcal{D}'(S_R)$ such that $f |_{\Omega_\alpha} = f_\alpha$.

2.3.6 Let

$$L := \sum_{i,j} \frac{\partial}{\partial x_i} a_{i,j}(x) \frac{\partial}{\partial x_j} + q(x) \tag{2.3.6.1}$$

be a differential operator of second order with infinitely differentiable coefficients $a_{i,j}, q$.

We will consider only three types of differential operators: a one-dimensional operator with constant coefficients,the Laplace operator and the so-called spherical operator (see Section 2.4).

For all these operators we have the following assertion which follows from the general theory (see, e.g., [Hö, Theorem 11.1.1]):

Theorem 2.3.6.1 (Regularity of Generalized Solution) *If the equation $Lu = 0$ has a solution $u \in \mathcal{D}'(G)$, then u is a regular distribution and can be realized as an infinitely differentiable function.*

A distribution that satisfies the equation

$$Lu = \delta_y \ \ in \ \mathcal{D}'(G), \tag{2.3.6.2}$$

where δ_y is a Dirac delta function (see (2.3.2.2)), is called a *fundamental solution* of L at the point y.

Every differential operator that we are going to consider has a fundamental solution (see, e.g., [Hö, Theorem 10.2.1]).

A restriction of the equation (2.3.6.2) to the domain $G_y := G\backslash y$ is a homogeneous equation $Lu = 0$ in $\mathcal{D}'(G_y)$. Thus we have

Theorem 2.3.6.2 (Regularity of Fundamental Solution) *The fundamental solution is an infinitely differentiable function outside the point y.*

2.3.7 We will need further also the Fourier coefficients for the distribution on the circle.

Let $\mathcal{D}(S^1)$ be a set of all infinitely differentiable functions on the unit circle S^1. The set of all linear continuous functionals over $\mathcal{D}(S^1)$ with the corresponding topology (see 2.3.2) is the corresponding space of distributions $\mathcal{D}'(S^1)$ for which all the previous properties of distributions holds.

The functions $\{e^{ik\phi},\ k = 0, \pm 1, \pm 2, \dots\}$ belong to $\mathcal{D}(S^1)$. The Fourier coefficients of $\nu \in \mathcal{D}'(S^1)$ are defined by

$$\hat{\nu}(k) := \langle \nu, e^{-ik\phi} \rangle. \tag{2.3.7.1}$$

The inverse operator is defined by

$$\langle \nu, g \rangle = \frac{1}{2\pi} \sum_{k=-\infty}^{\infty} \hat{\nu}(k) \langle g, e^{ik\phi} \rangle, \tag{2.3.7.2}$$

and the series converges, in any case, for those ν that are finite derivatives of summable functions, because Fourier coefficients of g decrease faster then every power of x.

The convolution of distribution $\nu \in \mathcal{D}'(S^1)$ and $g \in \mathcal{D}(S^1)$ is defined by

$$\nu * g(\phi) = \langle \nu, g(\phi - \bullet) \rangle. \qquad (2.3.7.3.)$$

This is a function from $\mathcal{D}(S^1)$.

The convolution of distributions $\nu_1,\ \nu_2 \in \mathcal{D}'(S^1)$ is defined by

$$\langle \nu_1 * \nu_2, g \rangle = \nu_1 * (\nu_2 * g). \tag{2.3.7.4}$$

In spite of the view it is commutative and

$$\widehat{\nu_1 * \nu_2}(k) = \hat{\nu}_1(k) \cdot \hat{\nu}_2(k).$$

Exercise 2.3.7.1 Count the Fourier coefficients of the functions

$$G(re^{i\phi}) = \log |1 - re^{i\phi}| \tag{2.3.7.5}$$

for $r > 1,\ r = 1, r < 1$; the function defined by

$$\widetilde{\cos \rho}(\phi) := \cos \rho\phi,\ -\pi < \phi < \pi,\ \rho \in (0, \infty) \tag{2.3.7.6}$$

and 2π-periodically extended; the function

$$\tilde{\phi} \sin p\phi, \; p \in \mathbb{N} \tag{2.3.7.7}$$

where $\tilde{\phi}$ is the 2π-periodical extension of the function $f(\phi) = \phi, \; \phi \in [0, 2\pi)$.

Exercise 2.3.7.2 Set

$$P_{p-1}(re^{i\phi}) := \Re\left\{\sum_{k=1}^{p-1} \frac{r^k e^{ik\phi}}{k}\right\}, \quad p \in \mathbb{N}. \tag{2.3.7.8}$$

Prove that for every distribution ν :

$$(\widehat{P_{p-1}(re^{i\bullet}) * \nu})(p) = 0. \tag{2.3.7.9}$$

The same for the function

$$G_p(re^{i\phi}) := G(re^{i\phi}) + P_p(re^{i\phi})$$

for $r < 1$.

2.4 Harmonic functions

2.4.1 We will denote as Δ the Laplace operator in $\mathbb{R}^m$:

$$\Delta := \frac{\partial^2}{\partial x_1^2} + \cdots + \frac{\partial^2}{\partial x_m^2}.$$

We introduce in $\mathbb{R}^m$ the spherical coordinate system by the formulae:

$$\begin{aligned}
x_1 &= r \sin \phi_0 \sin \phi_1 \ldots \sin \phi_{m-2};\\
x_2 &= r \cos \phi_0 \sin \phi_1 \ldots \sin \phi_{m-2};\\
x_3 &= r \cos \phi_1 \sin \phi_2 \ldots \sin \phi_{m-2};\\
&\ldots\ldots\ldots\ldots\\
x_k &= r \cos \phi_{k-2} \sin \phi_{k-1} \ldots \sin \phi_{m-2};\\
&\ldots\ldots\ldots\ldots\\
x_m &= r \cos \phi_{m-2},
\end{aligned}$$

where

$$0 < \phi_0 \leq 2\pi; \;\; 0 \leq \phi_j < \pi, \; j = \overline{1, m-2}; \;\; 0 < r < \infty.$$

Passing to the coordinates $(r, \phi_0, \phi_1, \ldots, \phi_{m-2})$ in the Laplace operator we obtain

$$\Delta = \frac{1}{r^{m-1}} \frac{\partial}{\partial r} r^{m-1} \frac{\partial}{\partial r} + \frac{1}{r^2} \Delta_{\boldsymbol{x}^0}.$$

The operator $\Delta_{\boldsymbol{x}^0}$ is called *spherical*, and has the form

$$\Delta_{\boldsymbol{x}^0} := \sum_{i=0}^{m-2} \frac{1}{\Pi}\frac{\partial}{\partial \phi_i}\frac{\Pi}{\Pi_i}\frac{\partial}{\partial \phi_i},$$

where

$$\Pi := \prod_{j=1}^{m-2} \sin^j \phi_j; \ \Pi_i := \prod_{j=i+1}^{m-2} \sin^2 \phi_j; \ \Pi_{m-2} := 1.$$

In particular, for $m = 2$, i.e., for the polar coordinates,

$$\Delta = \frac{1}{r}\frac{\partial}{\partial r}r\frac{\partial}{\partial r} + \frac{1}{r^2}\frac{\partial^2}{\partial \phi^2}.$$

A distribution $H \in \mathcal{D}'(G)$ is called *harmonic* if it satisfies the equation $\Delta H = 0$.

The next theorem follows from Theorem 2.3.6.1.

Theorem 2.4.1.1 (Smoothness of harmonic functions) *Any harmonic distribution is equivalent to an infinitely differentiable function.*

This function, of course, satisfies the same equation and is a *harmonic function* in the ordinary sense. A direct proof can be found, e.g., in [Ro, Ch. 1, § 2 (1.2.5), p. 60].

Let $f(z)$, $z = x + \imath y$ be a holomorphic function in a domain $G \subset \mathbb{C}$. Then the functions $u(x, y) := \Re f(z)$ and $v(x, y) := \Im f(z)$ are harmonic in G. In particular, the functions $r^n \cos n\varphi$ and $r^n \sin n\varphi$ where $r = |z|$, $\varphi = \arg z$ are harmonic.

Set

$$\mathcal{E}_m(x) := \begin{cases} -|x|^{2-m}, & \text{for } m \geq 3, \\ \log |z|, & \text{for } m = 2. \end{cases} \tag{2.4.1.1}$$

(We will often denote points of the plane as z.)

It is easy to check that $\mathcal{E}_m(x)$ is a harmonic function for $|x| \neq 0$.

Set

$$\theta_m := \begin{cases} (m-2)\sigma_m, & \text{for } m \geq 3; \\ 2\pi, & \text{for } m = 2, \end{cases}$$

where σ_m is the surface area of the unit sphere in $\mathbb{R}^m$.

Theorem 2.4.1.2 (Fundamental Solution) *The function $\mathcal{E}_m(x - y)$ satisfies in $\mathcal{D}'(\mathbb{R}^m)$ the equation*[1]

$$\Delta_x \mathcal{E}_m(x - y) = \theta_m \delta(x - y), \tag{2.4.1.2}$$

where $\delta(x)$ is the Dirac δ-function (*see* 2.3.2).

---

[1] $\mathcal{E}_m$ is slightly different from the fundamental solution (see, (2.3.6.2)), but this is traditional in Potential Theory

Proof. Let us prove the equality (2.4.1.2) for $y = 0$. Suppose $\phi \in \mathcal{D}(\mathbb{R}^m)$ and $\operatorname{supp}\phi \subset K \Subset \mathbb{R}^m$. We have

$$\langle \Delta\mathcal{E}_m, \phi\rangle := \int \mathcal{E}_m(x)\Delta\phi(x)dx = \lim_{\epsilon\to 0} \int\limits_{|x|\geq\epsilon} \mathcal{E}_m(x)\Delta\phi(x)dx.$$

Transforming this integral by the Green formula and using the fact that ϕ is finite we obtain

$$\int\limits_{|x|\geq\epsilon} \mathcal{E}_m(x)\Delta\phi(x)dx = \int\limits_{|x|\geq\epsilon} \Delta\mathcal{E}_m(x)\phi(x)dx + \int\limits_{|x|=\epsilon} \mathcal{E}_m\frac{\partial\phi}{\partial n}ds - \int\limits_{|x|=\epsilon} \phi\frac{\partial\mathcal{E}_m}{\partial n}ds,$$

where ds is an element of surface area and $\frac{\partial}{\partial n}$ is the differentiation in the direction of the external normal.

Use the harmonicity of $\mathcal{E}_m$. Then the first integral is equal to zero. Further we have

$$\int\limits_{|x|=\epsilon} \mathcal{E}_m\frac{\partial\phi}{\partial n}ds = \epsilon\Bigg(\int\limits_{|x^0|=1} \frac{\partial}{\partial r}\phi(rx^0)ds\Bigg)\Bigg|_{r=\epsilon} = O(\epsilon), \text{ for } \epsilon\to 0.$$

For the third term we have

$$\int\limits_{|x|=\epsilon} \phi\frac{\partial\mathcal{E}_m}{\partial n}ds = \frac{m-2}{\epsilon^{m-1}}\epsilon^{m-1} \int\limits_{|x^0|=1} \phi(rx^0)ds = [\phi(0)+o(1)](m-2)\sigma_m.$$

Thus we obtain $\langle \Delta\mathcal{E}_m, \phi\rangle = \phi(0)\theta_m$, and this proves (2.4.1.2) for $y = 0$.

It is clear that by changing $\phi(x)$ for $\phi(x+y)$ we obtain (2.4.1.2) in the general case. □

We will consider now a domain Ω with a *Lipschitz* boundary (*Lipschitz domain*). It means that every part of $\partial\Omega$ can be represented in some local coordinates (x, x'), $x \in \mathbb{R}$, $x' \in \mathbb{R}^{m-1}$ in the form $x = f(x')$, where f is a Lipschitz function, i.e.,

$$|f(x_1') - f(x_2')| \leq M_{\partial\Omega}|x_1' - x_2'|$$

where M depends only on the whole $\partial\Omega$ and does not depend on this local part.

Let $G(x, y, \Omega)$ be the Green function of a Lipschitz domain Ω.

It is known (see, e.g., [Vl, Ch. V, §28]) that the Green function has the following properties:

$$G(x,y,\Omega) < 0, \text{ for } (x,y)\in\Omega\times\Omega;\ G(x,y,\Omega) = 0 \text{ for } (x,y)\in\Omega\times\partial\Omega; \tag{g1}$$

$$G(x,y,\bullet) = G(y,x,\bullet); \tag{g2}$$

$$G(x,y,\bullet) - \mathcal{E}_m(x-y) = H(x,y), \tag{g3}$$

where H is harmonic on x and on y within Ω;

$$-G(x,y,\Omega_1) \leq -G(x,y,\Omega_2) \text{ for } \Omega_1\subset\Omega_2. \tag{g4}$$

From (g3) follows

Theorem 2.4.1.3 (Green Function) *The equality*

$$\Delta_x G(x,y,\Omega) = \theta_m \delta(x-y), \tag{2.4.1.3}$$

holds in $\mathcal{D}'(\Omega)$.

Let $f(x)$ be a continuous function on $\partial\Omega$. It is known (see, e.g., [Vl, Ch. V, § 29]) that the function

$$H(x,f) := \int\limits_{\partial\Omega} f(y)\frac{\partial}{\partial n} G(x,y,\Omega)ds_y \tag{2.4.1.4}$$

is the only harmonic function that coincides with f on $\partial\Omega$.

The unique solution of the Poisson equation

$$\Delta u = p, \ u|_{\partial\Omega} = f$$

for a continuous function p is given by the formula

$$u(x,f,p) := \int\limits_{\partial\Omega} f(y)\frac{\partial}{\partial n_y} G(x,y,\Omega)ds_y + \theta_m^{-1}\int\limits_{\Omega} G(x,y,\Omega)p(y)dy. \tag{2.4.1.5}$$

Let D be an *arbitrarily* open domain.We can define a $G(x,y,D)$ in the following way. Consider a sequence Ω_n of a Lipschitz domain such that $\Omega_n \uparrow D$. The sequence of the corresponding Green functions $G(x,y,\Omega_n)$ monotonically decreases. If it is bounded from below in some point, it is bounded everywhere while $x \neq y$ (as it follows from Theorem 2.4.1.7). It can be shown that the limit exists for any domain, the boundary of which has positive *capacity* (see 2.5 and references there). We will mainly use the Green function for the Lipschitz domains.

Let $G(x,y,K_{a,R})$ be the Green function of the ball $K_{a,R} := \{|x-a| < R\}$.

Theorem 2.4.1.4 (Green Function for a Ball)

$$G_(x,y,K_{a,R}) = \begin{cases} -|x-y|^{2-m} - (\frac{|y-a||x-y^*_{a,R}|}{R})^{2-m}, & \textit{for } m \geq 3, \\ \log\frac{|\zeta-z|R}{|\zeta-a||z-\zeta^*_{a.R}|} & \textit{for } m = 2, \end{cases}$$

where $y^*_{a.R} := a + (y-a)\left(R^2/\,|y-a|^2\right)$ *is the inversion of* y *relative to the sphere* $\{|x-a| = R\}$.

For the proof see, e.g., [Br, Ch. 6, § 3].

Theorem 2.4.1.5 (Poisson Integral) *Let* H *be a harmonic function in* $K_{a,R}$ *and continuous in its closure. Then*

$$H(x) = \frac{1}{\sigma_m R}\int\limits_{|x-a|=R} H(y)\frac{R^2-|x-a|^2}{|x-y|^m}ds_y, \ x \in K(a,R). \tag{2.4.1.6}$$

In particular, for $m = 2$,

$$H(a + re^{\imath\phi}) = \frac{1}{2\pi}\int_0^{2\pi} H(a + Re^{\imath\psi})\frac{R^2 - r^2}{R^2 - 2Rr\cos(\phi - \psi) + r^2}d\psi.$$

This theorem follows from (2.4.1.4).

Theorem 2.4.1.6 (Mean Value) *Let H be harmonic in $G \subset \mathbb{R}^m$. Then*

$$H(x) = \frac{1}{\sigma_m R^{m-1}} \int_{|x-a|=R} H(y)ds_y, \tag{2.4.1.7}$$

where $x \in G$ and R is taken such that $K(x, R) \Subset G$.

We must only set $a := x$ in (2.4.1.6). We can rewrite (2.4.1.7) in the form

$$H(x) = \frac{1}{\sigma_m} \int_{|y|=1} H(x + Ry)ds_y.$$

Theorem 2.4.1.7 (Harnack) *Suppose the family $\{H_\alpha), \alpha \in A\}$ of harmonic functions in G satisfies the conditions*

$$H_\alpha(x) \leq C(K), \text{ for } x \in K; \tag{Har1}$$

$$H_\alpha(x_0) \geq B > -\infty, \text{ for } x_0 \in K \tag{Har2}$$

for every compact $K \Subset G$ and $C(K), B$ are constants not depending on α.

Then the family is precompact in the uniform topology, i.e., there exists such a sequence H_{α_n}, and a function H harmonic in the interior of K and continuous in K such that $H_{\alpha_n} \to H$ uniformly in every K.

One can prove by using (2.4.1.6) that $|\operatorname{grad} H_\alpha|$ are bounded on every compact set by a constant not depending on α. Thus the family is uniformly continuous and thus it is precompact by the Ascoli theorem.

For details see, e.g., [Br, Supplement, § 7].

Theorem 2.4.1.8 (Uniform and $\mathcal{D}'$-convergences) *Suppose the sequence H_n satisfies the conditions of the Harnack theorem and converges to a function H in $\mathcal{D}'(G)$. Then H_n converges to H uniformly on every compact set $K \Subset G$.*

Of course, H is harmonic in G.

Proof. By the Harnack theorem the family is precompact. Thus we must only prove the uniqueness of H. Suppose there exist two subsequences such that $H_k^1 \to H^1$ and $H_k^2 \to H^2$ uniformly on every compact $K \Subset G$.

By Connection between Convergences (Theorem 2.3.4.3) $H_k^1 \to H^1$ and $H_k^2 \to H^2$ in $\mathcal{D}'$. Hence, $H^1 = H^2$ in $\mathcal{D}'(G)$. By the De Bois Raimond theorem (Theorem 2.3.2.1) $H^1 = H^2$ almost everywhere and hence everywhere because these functions are continuous. □

Let D be a domain with a smooth boundary ∂D and let $F \subset \partial D$. Set

$$\omega(x, F, D) := \int_F \frac{\partial G}{\partial n_y}(x, y) ds_y.$$

It is called a *harmonic measure* of F with respect to D. A harmonic measure can be defined for an arbitrary domain D by a limit process similar to the one we had for the Green function. In this case the formula (2.4.1.4) has the form

$$H(x, f) := \int_{\partial D} f(y) d\omega(x, y, D).$$

However we can not assert that $H(x, f)$ coincides with f in any point $x \in \partial D$. We can only consider it as an operator that maps a function defined on ∂D to a harmonic function in D.

By (2.4.1.3) we obtain

Theorem 2.4.1.9 (Two Constants Theorem) *Let H be harmonic in D and satisfy the conditions*

$$H(x) \leq A_1 \text{ for } x \in F; H(x) \leq A_2 \text{ for } x \in \partial D \backslash F$$

where A_1 and A_2 are constants. Then

$$H(x) \leq A_1 \omega(x, F, D) + A_2 \omega(x, \partial D \backslash F, D) \text{ for } x \in D.$$

Let $y^*_{a,R}$ be the inversion from Green Function for a Ball (Theorem 2.4.1.4). Set $y^* := y^*_{0,1}$, i.e., the inversion relative to a unit sphere with the center in the origin. Let $G^* := \{y^* : y \in G\}$ be the inversion of a domain G.

Theorem 2.4.1.10 (Kelvin's Transformation) *If H is harmonic in G, then*

$$H^*(y) := |y|^{2-m} H(y^*) \tag{2.4.1.8}$$

is harmonic in G^*.

For the proof you must honestly compute Laplacian of H^*. "The computation is straightforward but tedious" ([He, Thm. 2.24]). It is not so tedious if you use the spherical coordinate system.

Exercise 2.4.1.1 Do this.

2.4.2 Denote as $S_1 := \{x^0 : |x^0| = 1\}$ the unit sphere with center in the origin. A function $Y_\rho(x^0)$, $x^0 \in \Omega \subset S_1$ is called a *spherical function* of *degree* ρ if it satisfies the equation

$$\Delta_{x^0} Y + \rho(\rho + m - 2) Y = 0. \tag{2.4.2.1}$$

For $m = 2$, (2.4.2.1) gets the form

$$Y''(\theta) + \rho^2 Y(\theta) = 0, \quad \text{i.e.,} \quad Y(\theta) = a \cos \rho\theta + b \sin \rho\theta.$$

Spherical functions are obtained if we solve the equation $\Delta H = 0$ by the change $H(x) = |x|^\rho Y(x^0)$.

Theorem 2.4.2.1 (Sphericality and Harmonicity) *The function $Y_\rho(x^0)$ is spherical in a domain $\Omega \subset S_1$ if and only if the functions $H(x) = |x|^\rho Y_\rho(x^0)$ and $H^*(x) = |x|^{-\rho-m+2} Y_\rho(x^0)$ are harmonic in the cone*

$$\mathrm{Con}(\Omega) := \{x = rx^0 : x^0 \in \Omega,\ 0 < r < \infty\}. \tag{2.4.2.2}$$

If $\rho = k,\ k \geq 0,\ k \in \mathbb{Z}$, and only in this case, $Y_k(x^0)$ is spherical on the whole S_1, $H(x)$ is a homogeneous harmonic polynomial of degree k and H^* is harmonic in $\mathbb{R}^m \backslash 0$.

For the proof see, e.g., [Ax, Ch. 5]

The spherical functions of an integer degree k form a finite-dimension space of dimension

$$\dim(m,k) = \frac{(2k+m-2)(k+m-3)!}{(m-2)!k!}.$$

In particular, $d(2,k) = 2$ for any k.

For different k the spherical functions $Y_k(x^0)$ are orthogonal on S_1. In particular, for $m = 2$, it means the orthogonality of the trigonometric functions system.

Theorem 2.4.2.2 (Expansion of a Harmonic Function) *Let $H(x)$ be a harmonic function in the ball $K_R := \{|x| < R\}$. There exists an orthonormal system of spherical functions $Y_k(x^0),\ k = \overline{0, \infty}$, depending on H such that*

$$H(x) = \sum_{k=0}^{\infty} c_k Y_k(x^0)|x|^k, \text{ for } |x| < R. \tag{2.4.2.3}$$

For any such system we have

$$c_k = \frac{1}{R^k} \int_{S_1} H(Rx^0) Y_k(x^0) ds_{x^0}. \tag{2.4.2.4}$$

For the proof see, e.g., [Ax, Ch. 10], [TT, Ch. 4, § 10].

Theorem 2.4.2.3 (Liouville) *Let H be harmonic in $\mathbb{R}^m$ and suppose*

$$\liminf_{R\to\infty} R^{-\rho} \max_{|x|=R} H(x) < \infty \tag{2.4.2.5}$$

holds. Then H is a polynomial of a degree $q \leq \rho$.

Proof. We can suppose $H(0) = 0$ because $H(x) - H(0)$ is harmonic and also satisfies (2.4.2.5). Let $R_n \to \infty$ be a sequence for which

$$R_n^{-\rho} \max_{|x|=R_n} H(x) \leq \text{const} < \infty. \tag{2.4.2.6}$$

From (2.4.2.4) we obtain

$$|c_k| \leq A_k R^{-k} \int\limits_{S_1} |H(Rx^0)| ds_{x_0}, \tag{2.4.2.7}$$

where $A_k = \max_{S_1} |Y_k(x^0)|$.

From the mean value property (Theorem 2.4.1.6)

$$\int\limits_{S_1} H(Rx^0) ds_{x_0} = H(0)\sigma_m = 0.$$

Thus

$$\int\limits_{S_1} |H(Rx^0)| ds_{x_0} = 2 \int\limits_{S_1} H^+(Rx^0) ds_{x_0} \leq 2\sigma_m \max_{|x|=R} H(x). \tag{2.4.2.8}$$

From (2.4.2.8) and (2.4.2.7) we have

$$|c_k| \leq 2A_k R^{-k} \sigma_m \max_{|x|=R} H(x). \tag{2.4.2.9}$$

Set $R := R_n$ and $k > \rho$. Passing to the limit when $n \to \infty$, we obtain $c_k = 0$ for $k > \rho$. Then (2.4.2.3.) implies that H is a harmonic polynomial of degree $q \leq \rho$. □

2.5 Potentials and capacities

2.5.1 Let $G(x, y.D)$ be the Green function of a Lipschitz domain D. We will suppose it is extended as zero outside of D.

$$\Pi(x, \mu, D) := -\int G(x, y, D)\mu(dy)$$

is called the *Green potential* of μ relative to D. The domain of integration will always be $\mathbb{R}^m$.

Theorem 2.5.1.1 (Green Potential Properties) *The following holds:*

GPo1) *$\Pi(x, \mu, D)$ is lower semicontinuous;*

GPo2) *it is summable over any $(m-1)$-dimensional hyperplane or smooth hypersurface;*

GPo3) *$\Delta\Pi(\bullet, \mu, D) = -\theta_m \mu$ in $\mathcal{D}'(D)$;*

GPo4) *the* **reciprocity law** *holds:*

$$\int \Pi(x, \mu_1, D)\mu_2(dx) = \int \Pi(x, \mu_2, D)\mu_1(dx).$$

GPo5) **semicontinuity** *in* μ*: if* $\mu_n \to \mu$ *in* $\mathcal{D}'(\mathbb{R}^m)$*, then*

$$\liminf_{n\to\infty} \Pi(x, \mu_n, D) \geq \Pi(x, \mu, D).$$

GPo6) **continuity** *in* μ *in* $\mathcal{D}'$*: if* $\mu_n \to \mu$*, then* $\Pi(\bullet, \mu_n, D) \to \Pi(\bullet, \mu, D)$ *in* $\mathcal{D}'(\mathbb{R}^m)$ *and in* $\mathcal{D}'(S_R)$*, where* S_R *is the sphere* $\{|x| = R\}$*.*

Proof. Let us prove GPo1). Let $N > 0$. Set $G_N(x, y) := \max(G(x, y), -N)$, a truncation of the function $G(x, y)$.

The functions G_N are continuous in $\mathbb{R}^m \times \mathbb{R}^m$ and $G_N(x, y) \downarrow G(x, y)$ for every (x, y) when $N \to \infty$. Set

$$\Pi_N(x, \mu, D) := -\int G_N(x, y, D)\mu(dy).$$

The functions Π_N are continuous and $\Pi_N(x, \bullet) \uparrow \Pi(x, \bullet)$ by the B. Levy theorem (Theorem 2.2.2.2). Then $\Pi_N(x, \bullet)$ is lower semicontinuous by the Second Criterion of semicontinuity (Theorem 2.1.2.9).

Let us prove GPo5). From Theorem 2.3.4.4 ($\mathcal{D}'$ and C^* convergences)

$$\lim_{n\to\infty} \Pi_N(x, \mu_n, D) = \Pi_N(x, \mu, D).$$

Further $\Pi(x, \mu_n, D) \geq \Pi_N(x, \mu_n, D)$, hence

$$\liminf_{n\to\infty} \Pi(x, \mu_n, D) \geq \Pi_N(x, \mu, D).$$

Passing to the limit while $N \to \infty$, we obtain GPo5).

The assertion GPo2) follows from the local summability of the function $|x|^{2-m}$ that can be checked directly.

Let us prove GPo3). For $\phi \in \mathcal{D}(D)$ we have

$$\begin{aligned}
\langle \Delta\Pi, \phi\rangle := \langle \Pi, \Delta\phi\rangle &= -\int \mu(dy) \int G(x, y, D)\Delta\phi(x)dx \\
&= -\int \langle \Delta_x G(\bullet, y, D), \phi\rangle \mu(dy) = -\theta_m \int \phi(y)\mu(dy) \\
&= -\theta_m \langle \mu, \phi\rangle,
\end{aligned}$$

since

$$\langle \Delta_x G(\bullet, y, D), \phi\rangle = \theta_m \phi(y)$$

by Theorem 2.4.1.3. The property GPo4) follows from the symmetry of $G(x, y, \bullet)$ (property (g2)).

Let us prove GPo6). Note that integral $\int |x|^{m-1}dx$ converges locally in $\mathbb{R}^m$ and in $\mathbb{R}^{m-1}$. From this one can obtain by some simple estimates that functions

$\Psi(y) := \int G(x,y,D)\psi(x)dx$ while $\psi \in \mathcal{D}(\mathbb{R}^m)$ and $\Theta(y) := \int_{S_R} G(x,y,D)\theta(x)ds_x$ while $\theta \in \mathcal{D}(S_R)$ are continuous on $y \in \mathbb{R}^m$.

Now we have

$$\langle \Pi(\bullet, \mu_n, D), \psi \rangle = \int \Psi(y)\mu_n(dy) \to \int \Psi(y)\mu(dy) = \langle \Pi(\bullet, \mu, D), \psi \rangle.$$

Thus the first assertion in GPo6) is proved. The second one can be proved in the same way. □

Set $\nu := \mu_1 - \mu_2$, and let $\Pi(x,\nu,D) := \Pi(x,\mu_1,D) - \Pi(x,\mu_2,D)$ be a potential of this charge. Consider the boundary problem of the form

$$\Delta u = \mu_1 - \mu_2, \text{ in } \mathcal{D}'(D), \qquad u|_{\partial D} = f, \tag{2.5.1.2}$$

where f is a continuous function.

Theorem 2.5.1.2 (Solution of Poisson Equation) *The solution of the boundary problem* (2.5.1.2) *is given by the formula*

$$u(x) = H(x,f) - \theta_m^{-1}\Pi(x,\nu,D),$$

where $H(x,f)$ *is the harmonic function from* (2.4.1.4).

Proof. Since $\Pi(x,\nu,D)|_{\partial D} = 0$, the function $u(x)$ satisfies the boundary condition. Using GPo3) we obtain

$$\Delta u = \Delta H - [\theta_m]^{-1}\Delta\Pi = \mu_1 - \mu_2.$$ □

A potential of the form

$$\Pi(x,\mu) := \int \frac{\mu(dy)}{|x-y|^{m-2}}$$

is called a *Newton* potential. It is the Green potential for $D = \mathbb{R}^m$. The potential

$$\Pi(z,\mu) = -\int \log|z-\zeta|\mu(d\zeta)$$

is called *logarithmic.*

2.5.2 Let $K \Subset D$. The quantity

$$\mathbf{cap}_G(K,D) := \sup \mu(K) \tag{2.5.2.1}$$

where the supremum is taken over all mass distributions μ for which the following conditions are satisfied:

$$\Pi(x,\mu,D) \le 1, \tag{2.5.2.2}$$

$$\operatorname{supp}\mu \subset K, \tag{2.5.2.3}$$

is called the *Green capacity* of the compact set K relative to the domain D.

Theorem 2.5.2.1 (Properties of $\mathbf{cap}_G$) *For* $\mathbf{cap}_G$ *the following properties hold:*

capG1) *monotonicity with respect to* K*:* $K_1 \subset K_2$ *implies* $\mathbf{cap}_G(K_1, D) \leq \mathbf{cap}_G(K_2, D)$.

capG2) *monotonicity with respect to* D*:* $K \Subset D_1 \subset D_2$ *implies* $\mathbf{cap}_G(K, D_1) \geq \mathbf{cap}_G(K, D_2)$

capG3) *subadditivity with respect to* K*:*

$$\mathbf{cap}_G(K_1 \cup K_2, D) \leq \mathbf{cap}_G(K_1, D) + \mathbf{cap}_G(K_2, D).$$

Proof. The set of all mass distributions that satisfy (2.5.2.2) for $K = K_1$ is not less than the analogous set for $K = K_2$. Thus capG1) holds.

By the Green function property (g3) (see § 2.4.1) $-G(x,y,D_1) \leq -G(x,y,D_2)$. Thus the set of all μ that satisfy (2.5.2.2) for $D = D_1$ is wider than for $D = D_2$. Hence capG2) holds.

Let $\operatorname{supp} \mu \subset K_1 \cup K_2$ and let μ_1, μ_2 be the restrictions of μ to K_1, K_2 respectively.

If μ satisfies (2.5.2.2) for $K := K_1 \cup K_2$ then μ_1, μ_2 satisfy (2.5.2.2) for $K := K_1, K_2$ respectively.

From the inequality

$$\mu(K_1 \cup K_2) \leq \mu(K_1) + \mu(K_2)$$

we obtain that

$$\mu(K_1 \cup K_2) \leq \mathbf{cap}_G(K_1, D) + \mathbf{cap}_G(K_2, D)$$

for any μ with $\operatorname{supp} \mu \subset K_1 \cup K_2$. Thus capG3) holds.

The equivalent definition of the Green capacity is given by

Theorem 2.5.2.2 (Dual Property) *The following holds:*

$$\mathbf{cap}_G(K, D) = [\inf_{\mu} \sup_{x \in D} \Pi(x, \mu, D)]^{-1} \tag{2.5.2.4}$$

where the infimum is taken over all mass distributions μ *such that* $\mu(K) = 1$.

For the proof see, e.g., [La, Ch. 2, § 4 it. 18]. For $D = \mathbb{R}^m$, $m \geq 3$, the Green capacity is called *Wiener* capacity ($\mathbf{cap}_m(K)$). It has the following properties in addition to those of the Green capacity:

capW1) invariance with respect to translations and rotations, i.e.,

$$\mathbf{cap}_m(V(K + x_0)) = \mathbf{cap}_m(K),$$

where VK and $K + x_0$ are the rotation and the translation of K respectively.

The presence of these properties brings the notion of capacity closer to the notion of measure. Thus it is natural to extend the capacity to the Borel algebra of sets.

The Wiener capacity of an open set is defined as

$$\mathbf{cap}_m(D) := \sup_K \mathbf{cap}_m(K),$$

where the supremum is taken over all compact $K \Subset D$.

The *outer* and *inner* capacity of any set E can be defined by the equalities

$$\overline{\mathbf{cap}}_m(E) := \inf_{D \supset E} \mathbf{cap}_m(D); \ \underline{\mathbf{cap}}_m(E) := \sup_{K \subset E} \mathbf{cap}_m(K).$$

A set E is called *capacible* if $\overline{\mathbf{cap}}_m(E) = \underline{\mathbf{cap}}_m(E)$.

Theorem 2.5.2.3 (Choquet) *Every set E belonging to the Borel ring is capacible.*

For the proof see, e.g., [La, Ch2, Thm. 2.8].

Sets which have "small size" are sets of zero capacity. We emphasize the following properties of these sets:

capZ1) If $\mathbf{cap}_m(E^j) = 0$, $j = 1, 2, \ldots$. then $\mathbf{cap}_m(\cup_1^\infty E^j) = 0$;

capZ2) Having the property of zero capacity does not depend on the type of capacity: Green, Wiener or logarithmic capacity that we define below.

Example 2.5.2.1 Using Theorem 2.5.2.2 we obtain that any point has zero capacity, because for every mass distribution concentrated in the point the potential is equal to infinity. The same holds for any set of zero $m-2$ Hausdorff measure (see 2.5.4).

Example 2.5.2.2 Any $(m-1)$-hyperplane or smooth hypersurface has positive capacity, because the potential with masses uniformly distributed over the surface is bounded.

The Wiener 2-capacity can be defined naturally only for sets with diameter less then 1, because the logarithmic potential is positive only when this condition holds.

Instead, one can use the *logarithmic* capacity which is defined by the formulae

$$\mathbf{cap}_l(K) := \exp[-\mathbf{cap}_2(K)] \tag{2.5.2.5}$$

for $K \subset \{|z| < 1\}$ and

$$\mathbf{cap}_l(K) := t^{-1}\mathbf{cap}_l(tK)$$

for any other bounded K, where t is chosen in such a way that $tK \subset \{|z| < 1\}$.

One can check that this definition is correct, i.e., it does not depend on t.

2.5.3

Theorem 2.5.3.1 (Balayage; sweeping) *Let D be a domain such that $\partial D \Subset \mathbb{R}^m$, and* $\operatorname{supp}\mu \Subset D$. *Then there exists a mass distribution μ_b such that for $m \geq 3$, or for $m = 2$ and for D which is a bounded domain, the following holds:*

bal1) $\Pi(x, \mu_b) < \Pi(x, \mu)$ *for* $x \in D$;

bal2) $\Pi(x, \mu_b) = \Pi(x, \mu)$ *for* $x \notin \overline{D}$;

bal3) $\operatorname{supp} \mu_b \subset \partial D$;

bal4) $\mu_b(\partial D) = \mu(D)$.

If $m = 2$ and the domain is unbounded, a potential of the form

$$\hat{\Pi}(z, \mu) := - \int \log |1 - z/\zeta| \mu(d\zeta)$$

satisfies all the properties.

Proof. We will prove this theorem when ∂D is smooth enough. For $y \in D, x \in \mathbb{R}^m \backslash \overline{D}$ the function $|x - y|^{2-m}$ is a harmonic function of y on D.

Since $|x - y|^{2-m} \to 0$ as $y \to \infty$ we can apply the Poisson formula (2.4.1.4) even if D is unbounded. Thus

$$|x - y|^{2-m} = \int_{\partial D} |x - y'|^{2-m} \frac{\partial G}{\partial n_{y'}}(y, y') ds_{y'} \tag{2.5.3.1}$$

where G is the Green function of D. From this we have

$$\int_D |x - y|^{2-m} \mu(dy) = \int_{\partial D} |x - y'|^{2-m} ds_{y'} \left(\int_D \frac{\partial G}{\partial n_{y'}}(y, y') \mu(dy) \right).$$

The inner integral is nonnegative, because $\frac{\partial G}{\partial n} > 0$ for $y' \in \partial D$. Let us denote

$$\mu_b(dy') := \left(\int_D \frac{\partial G}{\partial n}(y, y') \mu(dy) \right) ds_{y'}.$$

Then we obtain the properties bal2) and bal3).

The potential $\Pi(x, \mu_b)$ is harmonic in D. Thus the function

$$u(x) := \Pi(x, \mu_b) - \Pi(x, \mu)$$

is a *subharmonic* function (see Theorem 2.6.4.1). Every subharmonic function satisfies the *maximum principle* (see Theorem 2.6.1.2), i.e.,

$$u(x) < \sup_{y \in \partial D} u(y) = 0.$$

Thus the property bal1) is fulfilled. To prove bal4) we can write the identity

$$\int_{\partial G} \mu_b(dy') = \int_G \mu(dy) \int_{\partial G} \frac{\partial G}{\partial n_{y'}}(y, y') dy'.$$

The inner integral is equal to 1 identically, because the function $\equiv 1, \ y \in G$ is harmonic. Thus bal4) is true.

Consider now the special case when $m = 2$, and D is an unbounded domain. Since $\log |1 - z/\zeta| \to 0$ when $\zeta \to \infty$, we obtain an equality like (2.5.3.1). Repeating the previous reasoning we obtain the last assertion for D with a smooth boundary.

Exercise 2.5.3.1 Check this in detail. □

For the general case see [La, Ch. 4, § 1]; [Ca, Ch. 3, Thm. 4].

Pay attention that the swept potential $\Pi(x, \mu_b)$ is also a solution of the Dirichlet problem in the domain D and the boundary function $f(x) = \Pi(x, \mu)$ in the following sense:

Theorem 2.5.3.2 (Wiener) *The equality* bal2) *holds in the points* $x \in \partial D$ *which can be reached by the top of a cone placed outside* D. *For* $m = 2$ *it can fail only for isolated points.*

For the proof see [He], [La, Ch. 4, § 1, Thm. 4.3.].

The points of ∂D where the equality bal2) does not hold are called *irregular.*

Theorem 2.5.3.3 (Kellogg's Lemma) *The set of all the irregular points of* ∂D *has zero capacity.*

For the proof see, e.g., [He], [La, Ch. 4, § 2, it. 10].

One can often compute the capacity using the following

Theorem 2.5.3.4 (Equilibrium distribution) *For any compact* K *with* $\mathbf{cap}_m(K) > 0$ *there exists a mass distribution* λ_K *such that the following holds:*

eq1) $\Pi(x, \lambda) = 1, \ x \in \overline{D} \backslash E, \ \mathbf{cap}_m(E) = 0;$

eq2) $\operatorname{supp} \lambda_K \subset \partial K;$

eq3) $\lambda_K(\partial K) = \mathbf{cap}_m(K).$

For the proof see [He], [La, Ch. 2, § 1, it. 3, Thm. 2.3].

Let us note that the set E in the previous theorem is a set of irregular points.

The mass distribution λ_K is called *equilibrium distribution*, and the corresponding potential is called *equilibrium potential.*

2.5.4 Let $h(x)$, $x \geq 0$ be a positive continuous, monotonically increasing function which satisfies the condition $h(0) = 0$. Let $\{K_j^\epsilon\}$ be a family of balls such that their diameters $d_j := d(K_j^\epsilon)$ are no bigger then ϵ. Let us denote

$$m_h(E, \epsilon) := \inf \sum h\left(\frac{1}{2} d(K_j^\epsilon)\right),$$

where the infimum is taken over all coverings of the set E by the families $\{K_j^\epsilon\}$.

The quantity

$$m_h(E) := \lim_{\epsilon \to 0} m_h(E, \epsilon)$$

is called *h-Hausdorff measure* [Ca, Ch. II].

Theorem 2.5.4.1 (Properties of m_h) *The following properties hold:*

h1) *monotonicity:*

$$E_1 \subset E_2 \Longrightarrow m_h(E_1) \leq m_h(E_2);$$

h2) *countable additivity:*

$$m_h(\cup E_j) = \sum m_h(E_j); \ E_j \cap E_i = \varnothing, \ \textit{for } i \neq j; \ E_j \in \sigma(\mathbb{R}^m).$$

We will quote two conditions (necessary and sufficient) that connect the h-measure to the capacity (see, [La, Ch. 3, § 4, it. 9, 10].

Theorem 2.5.4.2 *Let* $\mathbf{cap}E = 0$. *Then* $m_h(E) = 0$ *for all* h *such that*

$$\int_0 \frac{h(r)}{r^{m-1}} dr < \infty.$$

Theorem 2.5.4.3 *Let* $h(r) = r^{m-2}$ *for* $m \geq 3$ *and* $h(r) = (\log 1/r)^{-1}$ *for* $m = 2$. *If the* h*-measure of a set* E *is finite, then* $\mathbf{cap}_m(E) = 0$.

Side by side with the Hausdorff measure the *Carleson measure* (see, [Ca, Ch. II], is often considered. It is defined by

$$m_h^C(E) := \inf \sum h(0.5d_j),$$

where the infimum is taken over all coverings of the set E with balls of radii $0.5d_j$. The inequality $m_h^C(E) \leq m_h(E)$ obviously holds. Let $\beta - \text{mes}_C\, E$ be the Carleson measure for $h = r^\beta$. The following assertion connects the $\beta - \text{mes}_C$ to capacity.

Theorem 2.5.4.4 *The following inequalities hold:*

$$\beta - \text{mes}_C\, E \leq N(m)(\mathbf{cap}_m(E))^{\beta/m-2}, \ \textit{for}\ m \geq 3,\ \beta > m-2;$$
$$\beta - \text{mes}_C(E) \leq 18\mathbf{cap}_l(E), \ \textit{for}\ m = 2,\ \beta > 0,$$

where N *depends only on the dimension of the space.*

For the proof see [La, Ch. III,§ 4, it. 10, Cor. 2].

2.5.5 Now we will formulate an analog of the Luzin theorem for potentials.

Theorem 2.5.5.1 *Let* $\operatorname{supp}\mu = K$ *and let the potential* $\Pi(x,\mu)$ *be bounded on* K. *Then for any* $\delta > 0$ *there exists a compact set* $K' \subset K$ *such that* $\mu(K \backslash K') < \delta$ *and the potential* $\Pi(x,\mu')$ *of the measure* $\mu' := \mu\mid_K$ *(the restriction of* μ *to* K*) is continuous.*

For the proof see, e.g., [La, Ch. 3, § 2, it. 3, Thm. 3.6].

Let us prove the following assertion:

Theorem 2.5.5.2 *Let* $\mathbf{cap}K > 0$. *Then for arbitrarily small* $\epsilon > 0$ *there exists a measure* μ *such that* $\operatorname{supp}\mu \subset K$, *the potential* $\Pi(x,\mu)$ *is continuous and* $\mu(K) > \mathbf{cap}(K) - \epsilon$.

Proof. Consider the equilibrium distribution λ_K on K. Its potential is bounded by Theorem 2.5.3.4. By Theorem 2.5.5.1 we can find a mass distribution μ such that $\Pi(x,\mu)$ is continuous, $\operatorname{supp}\mu \subset K$ and $\mu(K) > \lambda_K(K) - \epsilon = \mathbf{cap}(K) - \epsilon$. □

2.6 Subharmonic functions

2.6.1 Let $u(x),\ x \in D \subset \mathbb{R}^m$ be a measurable function bounded from above which can be $-\infty$ on a set of no more than zero measure.

Let us denote as

$$\mathcal{M}(x, r, u) := \frac{1}{\sigma_m r^{m-1}} \int_{S_{x,r}} u(y) ds_y \tag{2.6.1.1}$$

the *mean value* of $u(x)$ on the sphere $S_{x,r} := \{y : |y - x| = r\}$.

The function $\mathcal{M}(x, r, u)$ is defined if $S_{x,r} \subset D$, but it can be $-\infty$ a priori.

A function $u(x)$ is called *subharmonic* if it is upper semicontinuous, $\not\equiv -\infty$, and for any $x \in D$ there exists $\epsilon = \epsilon(x)$ such that the inequality

$$u(x) \leq \mathcal{M}(x, r, u) \tag{2.6.1.2}$$

holds for all $r < \epsilon$.

The class of functions subharmonic in D will be denoted as $SH(D)$.

Example 2.6.1.1 The function

$$u(x) := -|x|^{2-m}, x \in \mathbb{R}^m$$

belongs to $SH(\mathbb{R}^m)$ for $m \geq 3$, and the function

$$u(z) := \log |z|,\ z \in \mathbb{R}^2$$

is subharmonic in $\mathbb{R}^2$.

Example 2.6.1.2 Let $f(z)$ be a holomorphic function in a plane domain D. Then $\log |f(z)| \in SH(D)$.

Example 2.6.1.3 Let $f = f(z_1, z_2, \dots, z_n)$ be a holomorphic function of $z = (z_1, \dots, z_n)$. Then $u(x_1, y_1, \dots, x_n, y_n) := \log |f(x_1 + iy_1, \dots, x_n + iy_n)|$ is subharmonic in every pair (x_j, y_j), and, as we can see later, in all the variables.

Example 2.6.1.4 Every harmonic function is subharmonic, as follows from Theorem 2.4.1.6. (Mean Value).

Theorem 2.6.1.1 (Elementary Properties) *The following holds:*

sh1) *if $u \in SH(D)$, then $Cu \in SH(D)$ for any constant $C \geq 0$;*

sh2) *if $u_1, u_2 \in SH(D)$, then $u_1 + u_2$, $\max[u_1, u_2] \in SH(D)$;*

sh3) *suppose $u_n \in SH(D)$, $n = 1, 2, \dots,$ and the sequence converges to u monotonically decreasing or uniformly on every compact set in D. Then $u \in SH(D)$;*

sh4) *suppose* $u(x, y) \in SH(D_1)$ *for all* $y \in D_2$, *and be upper semicontinuous in* $D_1 \times D_2$. *Let* μ *be a measure in* D_2 *such that* $\mu(D_2) < \infty$. *Then the function* $u(x) := \int u(x, y)\mu(dy)$ *is subharmonic in* D_1.

sh5) *let* $V \in SO(m)$ *be an orthogonal transformation of the space* $\mathbb{R}^m$ *and* $u \in SH(\mathbb{R}^m)$. *Then* $u(V\bullet) \in SH(\mathbb{R}^m)$.

All the assertions follow directly from the definition of subharmonic functions, properties of semicontinuous functions and properties of the Lebesgue integral. For a detailed proof see, e.g., [HK, Ch. 2].

Theorem 2.6.1.2 (Maximum Principle) *Let* $u \in SH(D)$, $G \subset \mathbb{R}^m$ *and* $u(x) \not\equiv$ const. *Then the inequality*

$$u(x) < \sup_{x' \in \partial D} \limsup_{y \to x', y \in D} u(y), \ x \in D$$

holds.

I.e., the maximum is not attained inside the domain.

The assertion follows from (2.6.1.2) and the upper semicontinuity of $u(x)$. For details see [HK, Ch. 2].

Let $K \Subset D$ be a compact set with nonempty interior $\overset{\circ}{K}$, and let f_n be a decreasing sequence of functions continuous in K that tends to $u \in SH(D)$. Such a sequence exists by Theorem 2.1.2.9. (The second criterion of semicontinuity).

Consider a sequence $\{H(x, u_n)\}$ of functions which are harmonic in $\overset{\circ}{K}$ and $H\mid_{\partial K} = f_n$. The sequence converges monotonically to a function $H(x)$ harmonic in $\overset{\circ}{K}$ by Theorem 2.3.4.3. (Connection between convergences), Theorem 2.4.1.8. (Uniform and $\mathcal{D}'$-convergences) and Theorem 2.6.1.2. The limit depends only on u as one can see, i.e., it does not depend on the sequence f_n. This harmonic function $H(x) := H(x, u, K)$ is called *the least harmonic majorant* of u in K.

This name is justified because of the following

Theorem 2.6.1.3. (Least Harmonic Majorant) *Let* $u \in SH(D)$. *Then for any* $K \Subset D$, $u(x) \leq H(x, u, K)$, $x \in K$. *If* $h(x)$ *is harmonic in* $\overset{\circ}{K}$ *and satisfies the condition* $h(x) \geq u(x)$, $x \in \overset{\circ}{K}$, *then* $H(x, u, K) \leq h(x)$, $x \in \overset{\circ}{K}$.

For the proof see [HK, Ch. 3].

2.6.2 Let us study properties of the mean values of subharmonic functions. Let $\mathcal{M}(x, r.u)$ be defined by (2.6.1.1) and $\mathcal{N}(x, r, u)$ by

$$\mathcal{N}(x, r, u) := \frac{1}{\omega_m r^m} \int_{K_{x,r}} u(y)dy,$$

where ω_m is the volume of the ball $K_{0,1}$.

Theorem 2.6.2.1 (Properties of Mean Values) *The following holds:*

me1) $\mathcal{M}(x,r,u)$ *and* $\mathcal{N}(x,r,u)$ *non-decreases in* r *monotonically;*

me2) $u(x) \leq \mathcal{N}(x,\bullet) \leq \mathcal{M}(x,\bullet)$*;*

me3) $\lim_{r\to 0} \mathcal{M}(x,r,u) = \lim_{r\to 0} \mathcal{N}(x,r,u) = u(x)$*.*

Proof. For simplicity let us prove me1) for $m = 2$. We have

$$\mathcal{M}(z_0,|z|,u) = \frac{1}{2\pi}\int_0^{2\pi} u(z_0 + ze^{i\phi})d\phi.$$

Since $u(z,\phi) := u(z_0 + ze^{i\phi})$ is a family of subharmonic functions that satisfies the condition sh4) of Theorem 2.6.1.1, $\mathcal{M}(z_0,|z|,u)$ is subharmonic in z on any $K_{0,r}$. By Maximum Principle (Theorem 2.6.1.2) we have

$$\mathcal{M}(z_0,r_1,u) = \max_{S_{0,r_1}} \mathcal{M}(z_0,|z|,u) \leq \max_{S_{0,r_2}} \mathcal{M}(z_0,|z|,u) = \mathcal{M}(z_0,r_2,u)$$

for $r_1 < r_2$.

Monotonicity of $\mathcal{N}(x,r,u)$ follows from the equality

$$\mathcal{N}(x,r,u) = m\int_0^1 s^{m-1}\mathcal{M}(x,rs.u)ds \tag{2.6.2.1}$$

and monotonicity of $\mathcal{M}(x,r,u)$.

The property me2) follows now from the definition of a subharmonic function and (2.6.2.1).

Let us prove me3). Let $M(u,x,r)$ be defined by (2.1.1.1). We have

$$\mathcal{M}(x,r,u) \leq M(u,x,r) \quad \text{and} \quad M(u,x,r) \to u(x)$$

because of upper semicontinuity of $u(x)$. Thus me2) implies me3). □

It is clear from me2) that a subharmonic function is locally summable. From me3) we have the corollary

Theorem 2.6.2.2 (Uniqueness of subharmonic function) *If* $u, v \in SH(D)$ *and* $u = v$ *almost everywhere, then* $u \equiv v$.

Let $\alpha(t)$ be defined by the equality (2.3.1.1), $\alpha_\epsilon(x)$ by (2.3.1.3). For a Borel set E let

$$E^\epsilon := \{x : \exists y \in E : |x-y| < \epsilon\}.$$

This is the ϵ-extension of E; this is, of course, an open set. For an open set D we set

$$D^{-\epsilon} := \bigcup_{E^\epsilon \subset D} E^\epsilon.$$

This is the maximal set such that its ϵ-extension is a subset of D. One can see that $D^{-\epsilon}$ is not empty for small ϵ and $D^{-\epsilon} \uparrow D$ when $\epsilon \downarrow 0$. Therefore for any $D_1 \Subset D$ there exists ϵ such that $D_1 \Subset D^{-\epsilon}$.

For $u \in SH(D)$ set

$$u_\epsilon(x) := \int u(x+y)\alpha_\epsilon(y)dy \tag{2.6.2.2}$$

which is defined in $D^{-\epsilon}$.

Theorem 2.6.2.3 (Smooth Approximation) *The following holds:*

ap1) u_ϵ *is an infinitely differentiable subharmonic function in any open set* $D_1 \subset D^{-\epsilon}$*;*

ap2) $u_\epsilon \downarrow u(x)$ *while* $\epsilon \downarrow 0$ *for all* $x \in D$.

Proof. The property ap1) follows from sh4) (Theorem 2.6.1.1) and the following equality that one can obtain from (2.6.2.2):

$$u_\epsilon(x) = \int u(y)\alpha_\epsilon(x-y)dy. \tag{2.6.2.3}$$

Exercise 2.6.2.1 Prove this.

Let us prove ap2). From (2.6.2.2) we obtain

$$u_\epsilon(x) = \int_0^1 \alpha(s)s^{m-1}\mathcal{M}(x,\epsilon s,u)ds. \tag{2.6.2.4}$$

It follows from the property me1) (Theorem 2.6.2.1) that $u_{\epsilon_1} \le u_{\epsilon_2}$ while $\epsilon_1 < \epsilon_2$. Now we pass to the limit in (2.6.2.4). Using me3) we have $\mathcal{M}(x,\epsilon s,u) \downarrow u(x)$. We can pass to the limit under the integral because of Theorem 2.2.2.2. Thus

$$\lim_{\epsilon\downarrow 0} u_\epsilon(x) = \int_0^1 \alpha(s)s^{m-1}u(x)ds = u(x). \qquad \square$$

Theorem 2.6.2.4 (Symmetry of u_ϵ) *If $u(x)$ depends only on $|x|$ then u_ϵ depends only on $|x|$.*

Proof. Let $V \in SO(m)$ be a rotation of $\mathbb{R}^m$. Then

$$u_\epsilon(Vx) = \int u(y)\alpha_\epsilon(Vx-y)dy.$$

Set $y = Vy'$ and change the variables. We obtain

$$u_\epsilon(Vx) = \int u(Vy')\alpha_\epsilon(V(x-y'))dy.$$

Since $\alpha_\epsilon = \alpha_\epsilon(|x|)$ and $u = u(|x|)$, $\alpha_\epsilon(Vy) = \alpha_\epsilon(y)$ and $u(Vy) = u(y)$. Thus $u_\epsilon(Vx) = u_\epsilon(x)$ for any V and thus $u_\epsilon(x) = u_\epsilon(|x|)$. $\square$

2.6.3 Since a subharmonic function is locally summable and defined uniquely by its values almost everywhere,every $u \in SH(D)$ corresponds to a (unique) distribution

$$\langle u, \phi \rangle := \int u(x)\phi(x)dx, \ \phi \in \mathcal{D}'.$$

Theorem 2.6.3.1 (Necessary Differential Condition for Subharmonicity) *If $u \in SH(D)$, then Δu is a positive distribution in $\mathcal{D}'(D)$.*

Proof. Suppose to begin that $u(x)$ has second continuous derivatives. By using (2.4.1.5) and (2.4.1.6) we can represent $u(x)$ in the form

$$u(x) = \mathcal{M}(x, r, u) + \int_{K_{x,r}} G(x, y, K_{x,r})\Delta u(y)dy, \tag{2.6.3.1}$$

where G is negative for all r. Suppose $\Delta u(x) < 0$. Then it is negative in $K_{x,r}$ for some r. Thus the integral in (2.6.3.1) is positive and we obtain that $u(x) - \mathcal{M}(x, r, u) > 0$. This contradicts the subharmonicity of $u(x)$.

Now suppose $u(x)$ is an arbitrarily subharmonic function. Then $\Delta u_\epsilon(x) \geq 0$ for every $x \in D$ when ϵ is small enough.For each x there is a neighborhood D_x such that every u_ϵ defines a distribution from $\mathcal{D}'(D_x)$. Hence $\Delta u_\epsilon(x)$ defines a positive distribution from $\mathcal{D}'(D_x)$. Passing to the limit in u_ϵ when $\epsilon \downarrow 0$ we obtain in $\mathcal{D}'(D_x)$ a distribution that is defined by function $u(x)$. Since the Laplace operator is continuous in any $\mathcal{D}'$ (Theorem 2.3.4.2), $\Delta u > 0$ in $\mathcal{D}'(D_x)$. From Theorem 2.3.5.1 we obtain that Δu is a positive distribution in $\mathcal{D}'(D)$. □

The distribution Δu can be realized as a measure by Theorem 2.3.2.2. The measure $(\theta_m)^{-1}\Delta u$ is called the *Riesz* measure of the subharmonic function u.

Theorem 2.6.3.2 (Subharmonicity and Convexity) *Let $u(|x|)$ be subharmonic in x on $K_{0,R}$. Then $u(r)$ is convex with respect to $-r^{2-m}$ for $m \geq 3$ and with respect to $\log r$ for $m = 2$.*

Proof. By Theorem 2.6.2.4, $u_\epsilon(x)$ depends on $|x|$ only, i.e., $u_\epsilon(x) = u_\epsilon(|x|)$, and the function $u_\epsilon(r)$ is smooth. Passing to the spherical coordinates we obtain

$$\Delta u_\epsilon = \frac{1}{r^{m-1}} \frac{\partial}{\partial r} r^{m-1} \frac{\partial}{\partial r} u_\epsilon(r) \geq 0.$$

By changing variables, $r = e^v$ for $m = 2$ or $r = (-v)^{\frac{1}{2-m}}$for $m \geq 3$, we obtain $[u_\epsilon(r(v)]'' \geq 0$, i.e., $u_\epsilon(r(v))$ is convex in v.

Passing to the limit on $\epsilon \downarrow 0$ we obtain that $u(r(v))$ is convex too, as a monotonic limit of convex functions. □

2.6.4 Now we will consider the connection between subharmonicity and potentials.

Theorem 2.6.4.1 (Subharmonicity of $-\Pi$) $-\Pi(x, \mu, D) \in SH(D)$

It is because of GPo1) and GPo3) (Theorem 2.5.1.1).

The following theorem is inverse to Theorem 2.6.3.1.

Theorem 2.6.4.2 (Sufficient Differential Condition of Subharmonicity) *Let $\Delta u \in \mathcal{D}'(D)$ be a positive distribution. Then there exists $u_1 \in SH(D)$ that realizes u.*

Proof. Set $\mu := \theta_m^{-1}\Delta u$. Let $\Omega_1 \Subset \Omega \Subset D$ and $\Pi(x, \mu_\Omega)$ be the Newtonian (or logarithmic) potential of $\mu\mid_\Omega$. By GPo5) (Theorem 2.5.1.1) the sum $H := u{+}\Pi$ is a harmonic distribution in $\mathcal{D}'(\Omega_1)$. Hence there exists a "natural" harmonic function H_1 that realizes H (Theorem 2.4.1.1). Thus the function $u_1 := H_1 - \Pi \in SH(\Omega_1)$ and realizes u in $\mathcal{D}'(\Omega)$. Since Ω and Ω_1 can be chosen such that a neighborhood of any $x \in D$ belongs to Ω_1, the assertion holds for D. □

By the way, we showed in this theorem that every subharmonic function can be represented inside its domain of subharmonicity as a difference of a harmonic function and a Newton potential. Thus all the smooth properties of a subharmonic function depend on the smooth properties of the potential only because any harmonic function is infinitely differentiable.

The following representation determines the harmonic function completely.

Theorem 2.6.4.3 (F. Riesz representation) *Let $u \in SH(D)$ and let K be a compact Lipschitz subdomain of D. Then*

$$u(x) = H(x, u, K) - \Pi(x, \mu_u, K)$$

where μ_u is the Riesz measure of u and $H(x, u, K)$ the least subharmonic majorant.

Proof. We can prove as above that the function $H(x) := u(x) + \Pi(x, \mu_u, K)$ is harmonic in $\overset{\circ}{K}$. Since $H(x) \geq u(x)$ we have $H(x) \geq H(x, u, K)$. So we need the reverse inequality.

Let us write the same equality for u_ϵ that is smooth.

$$u_\epsilon := H(x, u_\epsilon) - \Pi(x, \mu_{u_\epsilon}, K).$$

Passing to the limit as $\epsilon \downarrow 0$ we obtain

$$u(x) = H(x, u, K) - \lim_{\epsilon \downarrow 0} \Pi(x, \mu_{u_\epsilon}, K),$$

and the potentials converge because other summands converge. By Gpo5)

$$\lim_{\epsilon \downarrow 0} \Pi(x, \mu_{u_\epsilon}, K) \geq \Pi(x, \mu_u, K).$$

Hence $H(x) \leq H(x, u, K)$. □

2.6.5 In this item we will consider subharmonic functions in the ball $K_R := K_{0,R}$ which are *harmonic* in some neighborhood of the origin and write $u \in SH(R)$.

Set

$$M(r,u) := \max\{u(x) : |x| = r\},$$
$$\mu(r,u) := \mu_u(K_r),$$
$$\mathcal{M}(r,u) := \mathcal{M}(0,r,u),$$
$$N(r,u) := A(m)\int_0^r \frac{\mu(t,u)}{t^{m-1}}dt, \text{ where } A(m) = \max(1, m-2). \tag{2.6.5.1}$$

Theorem 2.6.5.1 (Jensen-Privalov) *For $u \in SH(R)$,*

$$\mathcal{M}(r,u) - u(0) = N(r,u), \text{ for } 0 < r < R. \tag{2.6.5.2}$$

Proof. By Theorem 2.6.4.3 we have

$$u(x) = \frac{1}{\sigma_m r}\int_{|y|=r} u(y)\frac{r^2 - |x|^2}{|x-y|^m}ds_y + \int_{K_r} G(x,y,K_r)\mu(dy).$$

For $x = 0$ we obtain

$$u(0) = \begin{cases} -\int_0^r \left(\frac{1}{t^{m-2}} - \frac{1}{r^{m-2}}\right)\mu(dt,u) + \mathcal{M}(r,u), & \text{for } m \geq 3; \\ -\int_0^r \log\frac{r}{t}\mu(dt,u) + \mathcal{M}(r,u), & \text{for } m = 2. \end{cases}$$

Integrating by parts gives

$$u(0) - \mathcal{M}(r,u) = \begin{cases} -\mu(t,u)\left(\frac{1}{t^{m-2}} - \frac{1}{r^{m-2}}\right)\Big|_0^r + (m-2)\int_0^r \frac{\mu(t,u)}{t^{m-1}}dt, & \text{for } m \geq 3; \\ -\mu(t,u)\log\frac{r}{t}\Big|_0^r + \int_0^r \frac{\mu(t,u)}{t}dt, & \text{for } m = 2. \end{cases} \tag{2.6.5.3}$$

We have $\mu(t,u) = 0$ for small t because of harmonicity of $u(x)$. Thus (2.6.5.3) implies (2.6.5.2). □

Theorem 2.6.5.2 (Convexity of $M(r,u)$ and $\mathcal{M}(r,u)$) *These functions increase monotonically and are convex with respect to $\log r$ for $m = 2$ and $-r^{2-m}$ for $m \geq 3$.*

Proof. Consider the case $m = 2$. Set $M(z) := \max_\phi u(ze^{i\phi})$. One can see that $M(r) = M(r,u)$.

Let u be a continuous subharmonic function. Then $M(z)$ is subharmonic (Theorem 2.6.1.1, sh5) and continuous because the family $\{u_\phi(z) := u(ze^{i\phi})\}$ is uniformly continuous. The function $M(z)$ depends only on $|z|$. Thus it is convex with respect to $\log r$ by Theorem 2.6.3.2.

Let $u(z)$ be an arbitrarily subharmonic function and $u_\epsilon \downarrow u$ while $\epsilon \downarrow 0$. Then $M(r,u_\epsilon) \downarrow M(r,u)$ by Proposition 2.1.2.7 and is convex with respect to $\log r$ by sh3), Theorem 2.6.1.1.

If $m \geq 3$ you should consider the function $M(x) := \max_{|y|=|x|} u(V_y x)$ where V_y is a rotation of $\mathbb{R}^m$ transferring x into y.

The convexity of $\mathcal{M}(r,u)$ is proved analogously.

Exercise 2.6.5.1 Prove it.

The monotonicity of $M(r,u)$ follows from the Maximum Principle (Theorem 2.6.1.2). The monotonicity of $\mathcal{M}(r,u)$ was proved in Theorem 2.6.2.1. □

The following classical assertion is a direct corollary of Theorem 2.6.5.2.

Theorem 2.6.5.3 (Three Circles Theorem of Hadamard) *Let $f(z)$ be a holomorphic function in the disc K_R and let $M_f(r)$ be its maximum on the circle $\{|z| = r\}$. Then*

$$M_f(r) \leq \left([M_f(r_1)]^{\log \frac{r_2}{r}} [M_f(r_2)]^{\log \frac{r}{r_1}}\right)^{\frac{1}{\log \frac{r_2}{r_1}}}$$

for $0 < r_1 \leq r \leq r_2 < R$.

For the proof you should write down the condition of convexity with respect to $\log r$ of the function $\log M_f(r)$ which is the maximum of the subharmonic function $\log |f(z)|$.

Exercise 2.6.5.2 Do this.

For details see [PS, Part I, Sec. III, Ch. 6, Problem 304].

2.7 Sequences of subharmonic functions

2.7.1 We will formulate the following analogue for the Montel theorem of normal families of holomorphic functions.

The family

$$\{u_\alpha,\ \alpha \in A\} \subset SH(D) \tag{2.7.1.1}$$

is called *precompact* in $\mathcal{D}'(D)$ if, for any sequence $\{\alpha_n,\ n = 1, 2, \ldots\} \subset A$, there exists a subsequence $\alpha_{n_j},\ j = 1, 2, \ldots$ and a function $u \in SH(D)$ such that $u_{\alpha_{n_j}} \to u$ in $\mathcal{D}'(D)$.

Example 2.7.1.1 $u_\alpha := \log |z - \alpha|,\ |\alpha| < 1$ form a precompact family.

Example 2.7.1.2 $u_\alpha := \log |f_\alpha|$ where $\{f_\alpha\}$ is a family of holomorphic functions bounded in a domain D form a precompact family.

A criterion of precompactness is given by

Theorem 2.7.1.1 (Precompactness in $\mathcal{D}'$) *A family* (2.7.1.1) *is precompact iff the following conditions hold:*

compl) *for any compact set $K \subset D$ a constant $C(K)$ exists such that*

$$u_\alpha(x) \leq C(K) \tag{2.7.1.2}$$

for all $\alpha \in A$ and $x \in K$;

comp2) *there exists a compact set* $K_1 \Subset D$ *such that*

$$\inf_{\alpha\in A} \max\{u_\alpha(x) : x \in K_1\} > -\infty. \tag{2.7.1.3}$$

For the proof see [Hö, Thm. 4.1.9].

Theorem 2.7.1.2 *Let* $u_n \to u$ *in* $\mathcal{D}'(K_R)$. *Then* $u_n \to u$ *in* $\mathcal{D}'(S_r)$ *for any* $r < R$.

Proof. We have $\mu_n \to \mu$. Let us choose R_1 such that $r < R_1 < R$. Then

$$u_n(x) = H(x, u_n, K_{R_1}) - \Pi(x, \mu_n, K_{R_1})$$

by the F. Riesz theorem (Theorem 2.6.4.3).

Now, we have $\Pi(x, \mu_n, K_{R_1}) \to \Pi(x, \mu, K_{R_1})$ in $\mathcal{D}'(R_1)$ by GPo6), Theorem 2.5.1.1. Thus $H(x, u_n, K_{R_1}) \to H(x, u, K_{R_1})$ in $\mathcal{D}'(R_1)$.

By Theorem 2.4.1.8, $H(x, u_n, K_{R_1}) \to H(x, u, K_{R_1})$ uniformly on any compact set in K_{R_1}, in particular, on S_r. Hence $H(x, u_n, K_{R_1}) \to H(x, u, K_{R_1})$ in $\mathcal{D}'(S_r)$. Also $\Pi(x, \mu_n, K_{R_1}) \to \Pi(x, \mu, K_{R_1})$ in $\mathcal{D}'(S_r)$ by GPo6), Theorem 2.5.1.1. Hence, $u_n \to u$ in $\mathcal{D}'(S_r)$. □

We say that a sequence f_n of locally summable functions *converges in* L_{loc} to a locally summable function f if for any $x \in D$ there exists a neighborhood $V \ni x$ such that $\int_V |f_n - f| dx \to 0$.

Theorem 2.7.1.3 (Compactness in L_{loc}) *Under conditions of Theorem* 2.7.1.1 *the family* (2.7.1.1) *is precompact in* L_{loc}.

For the proof see [Hö, Thm. 4.1.9].

Theorem 2.7.1.4 *Let* $u_n \to u$ *in* $\mathcal{D}'(K_R)$. *Then* $u_n^+ \to u^+$ *in* $\mathcal{D}'(K_R)$.

This is because $u_n^+(x) \le M,\ x \in K$, for all compact sets $K \Subset K_R$.

2.7.2 The following theorem shows that a subharmonic function is much more "flexible" than a harmonic or analytic function.

Theorem 2.7.2.1 *Let* $D \Subset \mathbb{R}^m$ *be a Lipschitz domain and let* $u \in SH(D)$ *satisfy the condition* $u(x) < C$ *for* $x \in D$. *Then for any closed domain* $D_1 \Subset D$ *there exists a function* $\tilde{u}(x) := \tilde{u}(x, D_1)$ *such that:*

ext1) $u(x) = \tilde{u}(x)$ *for* $x \in D_1$;
ext2) $\tilde{u}(x) = C$ *for* $x \in \partial D$;
ext3) $\tilde{u} \in SH(D)$ *and is harmonic in* $D \setminus \overline{D}_1$;
ext4) $u(x) \le \tilde{u}(x)$ *for* $x \in D$.

The function $\tilde{u}$ *is defined uniquely.*

Proof. We can suppose without loss of generality that $C = 0$, because we can consider the function $u - C$.

Let $u(x)$ be continuous in $\overline{D}_1$. Consider a harmonic function $H(x)$ which is zero on ∂D and $u(x)$ on ∂D_1. We have $H(x) \geq u(x)$ for $x \in D \setminus D_1$ because of Theorem 2.6.1.3. Set

$$\tilde{u}(x) = \begin{cases} H(x), & x \in D \setminus D_1; \\ u(x), & x \in D_1. \end{cases}$$

The function $\tilde{u}(x)$ is subharmonic in D. For $x \notin \partial D_1$ it is obvious, and for $x \in \partial D_1$ it follows from

$$u(x) = \tilde{u}(x) \leq \mathcal{M}(x, r, u) \leq \mathcal{M}(x, r, \tilde{u})$$

for r small enough.

It is easy to check that all the assertions of the theorem are fulfilled for the function $\tilde{u}$.

Exercise 2.7.1.1 Check this.

Let $u(x)$ be an arbitrarily subharmonic function. Consider the family u_ϵ of smooth subharmonic functions that converges to $u(x)$ decreasing monotonically in a neighborhood of $\overline{D}_1$. The sequence $\widetilde{(u_\epsilon)}$ converges monotonically to a subharmonic function that has all the properties ext1)–ext4). □

Theorem 2.7.2.2 (Continuity of $\tilde{\bullet}$) *Let $u_n \to u$ in $\mathcal{D}'(D)$ and $u_n(x) < 0$ in D. Then for any $K \Subset D$ with a smooth boundary ∂K $\widetilde{u_n}(\bullet, K) \to \tilde{u}(\bullet, K)$ in $\mathcal{D}'(D)$.*

For proving, we need the following auxiliary statement:

Theorem 2.7.2.3 *Let $u_n \to u$ in $\mathcal{D}'(D)$. Then for any smooth surface $S \Subset D$ and any function $g(x)$ continuous in a neighborhood of S the assertion*

$$\int_S u_n(x) g(x) ds_x \to \int_S u(x) g(x) ds_x \tag{2.7.2.1}$$

holds.

Proof. Since $u_n \to u$ in $\mathcal{D}'(D)$ also the Riesz measures of the functions converge. Hence $\mu_n(K) \leq C(K)$ for some $K \Supset S$. Thus, for the sequence of potentials $\Pi(x, \mu_n)$, we have

$$\int_S \Pi(x, \mu_n) g(x) ds_x = \int \mu_n(dy) \int_S \frac{g(x) ds_x}{|x - y|^{m-2}}.$$

The inner integral is a continuous function of y as can be seen by simple estimates. Thus the assertion (2.7.2.1) holds for potentials. Now, one can represent u_n in the form

$$u_n(x) = H_n(x) - \Pi(x, \mu_n)$$

in K. The sequence H_n convergences in $\mathcal{D}'$ and, hence, uniformly on S. Thus (2.7.2.1) holds for every u_n. □

Proof of Theorem 2.7.2.2. Let $\phi \in \mathcal{D}(D)$ and $\operatorname{supp}\phi \subset \overset{\circ}{K}$. Then

$$\langle \widetilde{u_n}, \phi\rangle = \langle u_n, \phi\rangle \to \langle u, \phi\rangle = \langle \tilde{u}, \phi\rangle.$$

Let $x \in D\backslash K$. Then

$$\widetilde{u}_n(x) = \int_{\partial K} \frac{\partial G}{\partial n_y}(x,y)u_n(y)ds_y.$$

By Theorem 2.7.2.3, $\widetilde{u_n}(x) \to \tilde{u}(x)$ for $x \in D\backslash K$. The sequence $\widetilde{u}_n$ is precompact in $\mathcal{D}'(D)$. Thus every limit u_0 of the $\widetilde{u}_n$ coincides with $\tilde{u}(x)$ in $\overset{\circ}{K}$ and in $D\backslash K$. Hence, $u_0 \equiv \tilde{u}$ in $\mathcal{D}'(D)$. □

2.7.3 The property sh2), Theorem 2.6.1.1, shows that the maximum of any finite number of subharmonic functions is a subharmonic function too. However, it is not so if the number is not finite.

Example 2.7.3.1 Set $u_n(z) = \frac{1}{n}\log|z|$, $n = 1, 2\ldots$. The functions $u_n \in SH(K_1)$. Taking the supremum in n we obtain

$$u(z) =: \sup_n u_n(z) = \begin{cases} 0, & \text{for } z \neq 0; \\ -\infty & \text{for } z = 0. \end{cases}$$

The function is not semicontinuous, thus it is not subharmonic. However, it differs from a subharmonic function on a set of zero capacity. The following theorem shows that this holds in general.

Theorem 2.7.3.1 (H. Cartan) *Let a family $\{u_\alpha \in SH(D),\ \alpha \in A\}$ be bounded from above and $u(x) := \sup_{\alpha\in A} u_\alpha(x)$. Then $u^* \in SH(D)$ and the set $E := \{x : u^*(x) > u(x)\}$ is a zero capacity set.*

For proving this theorem we need an auxiliary assertion

Theorem 2.7.3.2 *Let $\Pi(x, \mu_n, D)$ be a monotonically decreasing sequence of Green potentials and $\operatorname{supp}\mu_n \subset K \Subset D$. Then there exists a measure μ such that the inequality*

$$\lim_{n\to\infty} \Pi(x, \mu_n, D) \geq \Pi(x, \mu, D)$$

holds for all $x \in D$ with equality outside some set of zero capacity.

Proof. The sequence $\Pi(x, \mu_n, D)$ converges monotonically and thus in $\mathcal{D}'$ (Theorem 2.3.4.3). Then $\mu_n \to \mu$ in $\mathcal{D}'$ (Theorem 2.2.4.2.) and thus in C^*- topology (Theorem 2.3.4.4). By GPo5) (Theorem 2.5.1.1) we have

$$\lim_{n\to\infty} \Pi(x, \mu_n, D) \geq \Pi(x, \mu, D).$$

Suppose that the strict inequality holds on some set E of a positive capacity. By Theorem 2.5.2.3 one can find a compact set $K \subset E$ such that $\mathbf{cap}(K) > 0$. Then there exists a measure ν concentrated on E such that its potential $\Pi(x, \nu, D)$ is continuous (Theorem 2.5.5.2). Thus we have

$$\int \Pi(x,\mu,D)\nu(dx) < \int \lim_{n\to\infty} \Pi(x,\mu_n,D)\nu(dx) = \lim_{n\to\infty} \int \Pi(x,\mu_n,D)\nu(dx),$$

$$\lim_{n\to\infty} \int \Pi(x,\nu,D)\mu_n(dx) = \int \Pi(x,\nu,D)\mu(dx) = \int \Pi(x,\mu,D)\nu(dx).$$

The equalities use Theorem 2.2.2.2 (B. Levy), reciprocity law (GPo4), Theorem 2.5.1.1, C^*-convergence of μ_n and once more the reciprocity law, respectively. So we have a contradiction. □

Proof of Theorem 2.7.3.1. Suppose that $u_n(x) \uparrow u(x)$. We can suppose also that $u_n < 0$. For any domain $G \Subset D$ the sequence $\tilde{u}_n(x) \to u(x)$ for $x \in G$ (see Theorem 2.7.2.1), because $u_n(x) = \tilde{u}_n(x)$ for $x \in G$. Since $\tilde{u}_n = \Pi(x, \tilde{\mu}_n, D)$ for $x \in D$, $\tilde{u}(x) = \Pi(x, \tilde{\mu}, D) = u(x)$ for $x \in G$ and coincides with $\lim_{n\to\infty} u_n(x)$ outside some set E_G of zero capacity. Consider a sequence of domains G_n that exhaust D. Then $u(x) = \lim_{n\to\infty} u_n(x)$ outside the set $E := \cup_{n=1}^{\infty} E_{G_n}$ which has zero capacity by capZ1) (see item 2.5.2).

Now let $\{u_n,\ n = 1, 2 \dots\}$ be a general countable set that satisfies the conditions of the theorem. Then the sequence $v_n := \max\{u_k : k = 1, 2 \dots, n\} \in SH(D)$ and $v_n \uparrow u$. Applying the previous reasoning we obtain the assertion of the theorem also in this case.

Let $\{u_\alpha,\ \alpha \in A\}$ be an arbitrary set satisfying the condition of the theorem. By Theorem 2.1.3.2 (Choquet's Lemma) one can find a countable set $A_0 \subset A$ such that

$$(\sup_{A_0} u_\alpha)^* = (\sup_{A} u_\alpha)^*.$$

Since $\sup_{A_0} u_\alpha \le \sup_A u_\alpha$, we have

$$E := \{x : (\sup_{A} u_\alpha)^* > \sup_{A} u_\alpha\} \subset E_0 := \{x : (\sup_{A_0} u_\alpha)^* > \sup_{A_0} u_\alpha\}.$$

Thus **cap** $(E) \le$**cap** $(E_0) = 0$. □

Corollary of Theorem 2.7.3.1 is

Theorem 2.7.3.3 (H. Cartan +) *Let* $\{u_t,\ t \in (0;\infty)\} \subset SH(D)$ *be a bounded from above family, and* $v := \limsup_{t\to\infty} u_t$. *Then* $v^* \in SH(D)$ *and the set* $E := \{x : v^*(x) > v(x)\}$ *has zero capacity.*

Proof. Set $u_n := \sup_{t\ge n} u_t$, $E_n := \{x : (u_n)^* > u_n\}$, $E := \cup E_n$. Since **cap**$(E_n) = 0$, **cap**$E = 0$ too.

Let $x \notin E$. Then

$$v(x) = \lim_{n\to\infty} \sup_{t\ge n} u_t(x) = \lim_{n\to\infty} (u_n)^*(x).$$

The function

$$v^* := \lim_{n\to\infty} (u_n)^*(x)$$

is the upper semicontinuous regularization of $v(x)$ for all $x \in D$. □

In spite of Example 2.7.3.1 we have

Theorem 2.7.3.4 (Sigurdsson's Lemma) [Si] *Let $S \subset SH(D)$ be compact in $\mathcal{D}'$. Then*

$$v(x) := \sup\{u(x) : u \in S\}$$

is upper semicontinuous

and, hence, subharmonic.

Proof. Note that

$$u_\epsilon(x) = \langle u, \alpha(x - \bullet)\rangle$$

(see (2.6.2.3), (2.3.2.1)); and it is continuous in (u, x) with respect to the product topology on $(SH(D) \cap \mathcal{D}') \times \mathbb{R}^m$ (Theorem 2.3.4.6).

Let $x_0 \in D, a \in \mathbb{R}$ and assume that $v(x_0) < a$. We have to prove that there exists a neighborhood X of x_0 such that

$$v(x) < a, \ x \in X. \tag{2.7.3.1}$$

We choose $\delta > 0$ such that $v(x_0) < a-\delta$. If $u^0 \in SH(D)$ and ϵ is chosen sufficiently small, then

$$u^0(x_0) \le u^0_\epsilon(x_0) < a - \delta$$

by Theorem 2.6.2.3 (Smooth Approximation).

Since $u_\epsilon(x)$ is continuous, there exists an open neighborhood U_0 of u^0 in $SH(D)$ and an open neighborhood X_0 of x_0 such that

$$u_\epsilon(x) < a - \delta, \ u \in U_0, \ x \in X_0.$$

The property ap2) (Theorem 2.6.2.3) implies

$$u(x) < a - \delta, \ u \in U_0, \ x \in X_0. \tag{2.7.3.2}$$

Since u^0 is arbitrary and S is compact, there exists a finite covering $U_1, U_2, \dots, U_n$ of S and open neighborhoods $X_1, X_2, \dots, X_n$ of x_0 such that (2.7.3.2) holds for all $(u, x) : u \in U_j, \ x \in X_j, \ j = 1, \dots, n$. Set $X := \cap_j X_j$. Then (2.7.3.1) holds. □

2.7.4 Now we are going to connect $\mathcal{D}'$-convergence to convergence outside a zero capacity set, the so-called *quasi-everywhere* convergence.

Theorem 2.7.4.1 ($\mathcal{D}'$ and Quasi-everywhere Convergence) *Let $u_n, \ u \in SH(D)$ and $u_n \to u$ in $\mathcal{D}'(D)$. Then $u(x) = \limsup_{n\to\infty} u_n(x)$ quasi-everywhere and $u(x) = (\limsup_{n\to\infty} u_n(x))^*$ everywhere in D.*

For the proof we need the following assertion in the spirit Theorem 2.7.3.2.

Theorem 2.7.4.2 *Let* $\mu_n \to \mu$ *in* $\mathcal{D}'(D)$ *and* $\operatorname{supp}\mu_n \subset K \Subset D$. *Then*

$$\liminf_{n\to\infty} \Pi(x, \mu_n, D) \geq \Pi(\mu, D)$$

with equality quasi-everywhere.

Proof. The inequality was in GPo5), Theorem 2.5.1.1.

Suppose the set

$$E := \{x : \liminf_{n\to\infty} \Pi(x, \mu_n, D) > \Pi(x, \mu, D)$$

has a positive capacity. By Theorem 2.5.2.3 one can find a compact set $K \subset E$ such that $\mathbf{cap}(K) > 0$. By Theorem 2.5.5.2 one can find a measure ν concentrated on K with continuous potential. As in the proof of Theorem 2.7.3.2 we have

$$\int \Pi(x, \mu, D)\nu(dx) < \int \liminf_{n\to\infty} \Pi(x, \mu_n, D)\nu(dx) \leq \liminf_{n\to\infty} \int \Pi(x, \mu_n, D)\nu(dx)$$

$$= \liminf_{n\to\infty} \int \Pi(x, \nu, D)\mu_n(dx) = \int \Pi(x, \nu, D)\mu(dx) = \int \Pi(x, \mu, D)\nu(dx).$$

The second inequality uses Theorem 2.2.2.3 (Fatou's Lemma). The equalities use the reciprocity law (GPo4), Theorem 2.5.1.1, C^*-convergence of μ_n and once more the reciprocity law, respectively. So we have a contradiction. □

Proof of Theorem 2.7.4.1. Let $D_1 \Subset D$. Then the sequence u_n is bounded in D_1 by Theorem 2.7.1.1. We can assume that $u_n(x) < 0$ for $x \in D_1$.

For any domain $G \Subset D_1$ the sequence $\tilde{u}_n(x, G) \to u(x)$ in $\mathcal{D}'(D_1)$ by Theorem 2.7.2.2. We also have the equality $\tilde{u}_n = -\Pi(x, \tilde{\mu}_n, D_1)$. Thus $\tilde{\mu}_n \to \tilde{\mu}$ in $\mathcal{D}'(D_1)$. By Theorem 2.7.4.2, $\liminf_{n\to\infty} \Pi(x, \tilde{\mu}_n, D_1) = \Pi(x, \tilde{\mu}, D_1)$ quasi-everywhere in D_1. Hence

$$\limsup_{n\to\infty} u_n = u \tag{2.7.4.1}$$

quasi-everywhere in G because $u_n(x) = \tilde{u}_n(x)$ for $x \in G$.

Consider a sequence of domains G_n that exhaust D. Then (2.7.4.1) holds outside a set E_n of zero capacity and (2.7.4.1) holds in D outside the set $E := \cup_{n=1}^{\infty} E_n$ which has zero capacity by capZ1) (see item 2.5.2), i.e., quasi-everywhere. □

2.7.5 Now we connect the convergence of subharmonic functions in $\mathcal{D}'$ to the convergence relative to the Carleson measure (see 2.5.4).

We say that a sequence of functions u_n *converges* to a function u *relative* to the α-Carleson measure if the sets $E_n := \{x : |u_n(x) - u(x)| > \epsilon\}$ possess the property

$$\alpha - \operatorname{mes}_C E_n \to 0. \tag{2.7.5.1}$$

Theorem 2.7.5.1 ($\mathcal{D}'$ and α-mes$_C$ Convergences) *Let $u_n, u \in SH(D)$ and $u_n \to u$ in $\mathcal{D}'(D)$. Then for an every $\alpha > 0$ and every domain $G \Subset D$ $u_n \to u$ relative to the $(\alpha + m - 2)$-Carleson measure.*

For proving this theorem we need some auxiliary definitions and assertions.

Let μ be a measure in $\mathbb{R}^m$. We will call a point $x \in \mathbb{R}^m$ $(\alpha, \alpha', \epsilon)$-*normal* with respect to the measure μ, $(\alpha < \alpha')$ if the inequality

$$\mu_x(t) := \mu(K_{x,t}) < \epsilon^{-\alpha'} t^{\alpha+m-2}$$

holds for all $t < \epsilon$.

Theorem 2.7.5.2 *In any $(\alpha, \alpha', \epsilon)$-normal point the following inequality holds:*

$$-\int_{K_{z,\epsilon}} [\log|z-\zeta| - \log\epsilon]d\mu_\zeta \le C\epsilon^{\alpha-\alpha'}, \ \textit{for } m = 2;$$

$$\int_{K_{x,t}} [|x-y|^{2-m} - \epsilon^{2-m}]d\mu_y \le C\epsilon^{\alpha-\alpha'}, \ \textit{for } m \ge 3;$$

while $C = C(\alpha, m)$ depends on α and m only.

Proof. Let us consider the case $m = 2$. We have

$$\int_{K_{z,\epsilon}} \log\frac{\epsilon}{|z-\zeta|} d\mu_\zeta = \int_0^\epsilon \log\frac{\epsilon}{t} d\mu_z(t).$$

Integrating by parts we obtain

$$\int_{K_{z,\epsilon}} \log\frac{\epsilon}{|z-\zeta|} d\mu_\zeta = \log\frac{\epsilon}{t}\mu_z(t)\ |_0^\epsilon + \int_0^\epsilon \frac{\mu_z(t)}{t}dt$$
$$\le \epsilon^{-\alpha'} \int_0^\epsilon t^{\alpha-1}dt = \frac{1}{\alpha}\epsilon^{\alpha-\alpha'}.$$

Let us consider the case $m \ge 3$. We have

$$\int_{K_{x,t}} [|x-y|^{2-m} - \epsilon^{2-m}]d\mu_y = \int_0^\epsilon (t^{2-m} - \epsilon^{2-m})d\mu_x(t)$$
$$= (t^{2-m} - \epsilon^{2-m})\mu_x(t)\ |_0^\epsilon + (m-2)\int_0^\epsilon \frac{\mu_x(t)}{t^{m-2}}dt$$
$$\le \frac{m-2}{\epsilon^{\alpha'}} \int_0^\epsilon t^{\alpha-1}dt = \frac{m-2}{\alpha}\epsilon^{\alpha-\alpha'}.$$

□

Theorem 2.7.5.3 (Ahlfors-Landkof Lemma) *Let a set $E \subset \mathbb{R}^m$ be covered by balls with bounded radii such that every point is a center of a ball. Then there exists an at most countable subcovering of the same set with maximal multiplicity $cr = cr(m)$.*

I.e., every point of E is covered no more than cr times. Let us note that $cr(2) = 6$.

For the proof see [La, Ch. III, § 4, Lem. 3.2].

Theorem 2.7.5.4 *Let $K \Subset D$. The set $E := E(\alpha, \alpha', \epsilon, \mu)$ of points that belong to K and are not $(\alpha, \alpha', \epsilon)$-normal with respect to μ satisfies the condition*

$$(\alpha + m - 2) - \operatorname{mes}_C E \leq cr(m)\epsilon^{\alpha'} \mu(K^\epsilon) \tag{2.7.5.2}$$

where K^ϵ is the 2ϵ-extension of K.

Proof. Let $x \in E$. Then there exists t_x such that

$$\mu_x(t_x) \geq t_x^{\alpha+m-2}\epsilon^{-\alpha'}.$$

Thus every point of E is covered by a ball K_{x,t_x}. By the Ahlfors-Landkof lemma (Theorem 2.7.5.3) one can find a no more than $cr(m)$-multiple subcovering $\{K_{x_j,t_{x_j}}\}$. Then we have

$$\sum_j t_{x_j}^{\alpha+m-2} \leq cr(m)\epsilon^{\alpha'} \mu(K^\epsilon).$$

By definition of the Carleson measure we obtain (2.7.5.2). □

Theorem 2.7.5.5 *Let $\mu_n \to \mu$ in $\mathcal{D}'(\mathbb{R}^m)$ and* $\operatorname{supp} \mu_n \subset K \Subset \mathbb{R}^m$*. Then for every $\alpha > 0$ and $G \Subset \mathbb{R}^m$, $\Pi(x, \mu_n) \to \Pi(x, \mu)$ relative to the $(\alpha + m - 2)$-Carleson measure.*

Proof. Let $m = 2$. Set

$$\log_\epsilon |z - \zeta| = \begin{cases} \log |z - \zeta|, & \text{for } |z - \zeta| > \epsilon, \\ \log \epsilon, & \text{for } |z - \zeta| \leq \epsilon. \end{cases}$$

This function is continuous for $(z, \zeta) \in K \times K$.

Set $\nu_n := \mu_n - \mu$. Then we have

$$-\int \log |z - \zeta| \mu_n(d\zeta) + \int \log |z - \zeta| \mu(d\zeta) = -\int \log |z - \zeta| \nu_n(d\zeta)$$

$$= -\int \log_\epsilon |z - \zeta| \nu_n(d\zeta) - \int_{K_{z,\epsilon}} [\log |z - \zeta| - \log \epsilon] \nu_n(d\zeta).$$

The function $\log_\epsilon |z-\zeta|$ is continuous in ζ uniformly over $z \in K$. Thus the sequence

$$\Pi_\epsilon(z) := \int \log_\epsilon |z - \zeta| \nu_n(d\zeta)$$

converges uniformly to zero on K. Suppose now that

$$z \notin E(\alpha, \alpha', \epsilon, \mu) \cup E(\alpha, \alpha', \epsilon, \mu_n),$$

i.e., it is an $(\alpha, \alpha', \epsilon)$- normal point for μ and μ_n. By Theorem 2.7.5.2 we have

$$\int_{K_{z,\epsilon}} [\log|z-\zeta| - \log\epsilon]\nu_n(d\zeta) < 2C\epsilon^{\alpha-\alpha'}.$$

Thus for sufficiently large $n > n_0(\epsilon)$,

$$|\Pi(z,\mu_n) - \Pi(z,\mu)| = \left|\int \log|z-\zeta|\mu_n(d\zeta) - \int \log|z-\zeta|\mu(d\zeta)\right| < \delta = \delta(\epsilon)$$

while $z \notin E(\alpha, \alpha', \epsilon, \mu) \cup E(\alpha, \alpha', \epsilon, \mu_n) := E_n(\epsilon)$.

By Theorem 2.7.5.3 the Carleson measure of $E_n(\epsilon)$ satisfies the inequality

$$\alpha - \mathrm{mes}_C\, E_n(\epsilon) \le cr(m)\epsilon^{\alpha'}[\mu(K) + \mu_n(K)] \le C\epsilon^{\alpha'} := \gamma(\epsilon)$$

where $C = C(K)$ does not depend on n because $\mu_n(K)$ are bounded uniformly.

Hence, for any $\epsilon > 0$ the set

$$E'_n(\epsilon) := \{z : |\Pi(z,\mu_n) - \Pi(z,\mu)| > \delta(\epsilon)\}$$

satisfies the condition

$$\alpha - \mathrm{mes}_C\, E'_n(\epsilon) \le \gamma(\epsilon) \tag{2.7.5.3}$$

while $n > n_0 = n_0(\epsilon)$.

Let us show that $\Pi(z,\mu_n) \to \Pi(z,\mu)$ relative to $\alpha - \mathrm{mes}_C$ on K. Let γ_0, δ_0 be arbitrarily small. One can find ϵ such that $\delta(\epsilon) < \delta_0, \gamma(\epsilon) < \gamma_0$. One can find $n_0 = n_0(\epsilon)$ such that (2.7.5.3) is fulfilled. Now the set

$$E_{n,\delta_0} := \{z : |\Pi(z,\mu_n) - \Pi(z,\mu)| > \delta_0\}$$

is contained in $E'_n(\epsilon)$. Thus $\alpha - \mathrm{mes}_C\, E_{n,\delta_0} < \gamma_0$ and this implies the convergence relative to $\alpha - \mathrm{mes}_C$. An analogous reasoning works for $m \ge 3$. □

Proof of Theorem 2.7.5.1. Let $u_n \to u$ in $\mathcal{D}'$. One can assume that u_n, u are potentials on any compact set (Theorem 2.7.2.2). Hence, by Theorem 2.7.5.5 it converges relative $(\alpha + m - 2) - \mathrm{mes}_C$. □

2.8 Scale of growth. Growth characteristics of subharmonic functions

2.8.1 Let A be a class of nondecreasing functions $a(r)$, $r \in (0, \infty)$ such that $a(r) \ge 0$ and $a(r) \to \infty$ when $r \to \infty$. The quantity

$$\rho[a] := \limsup_{r\to\infty} \frac{\log a(r)}{\log r} \tag{2.8.1.1}$$

is called the *order* of $a(r)$.

Suppose $\rho := \rho[a] < \infty$. The number

$$\sigma[a] := \limsup_{r\to\infty} \frac{a(r)}{r^{\rho}} \tag{2.8.1.2}$$

is called the *type number.*

If $\sigma[a] = 0$, we say $a(r)$ has *minimal type.* If $0 < \sigma[a] < \infty$, $a(r)$ has *normal type.* If $\sigma[a] = \infty$, it has *maximal type.*

Example 2.8.1.1 Set $a(r) := \sigma_0 r^{\rho_0}$. Then $\rho[a] = \rho_0$, $\sigma[a] = \sigma_0$.

Example 2.8.1.2 Set $a(r) := (\log r)^{-1} r^{\rho_0}$. Then $\rho[a] = \rho_0$, $\sigma[a] = 0$.

Example 2.8.1.3 Set $a(r) := (\log r) r^{\rho_0}$. Then $\rho[a] = \rho_0$, $\sigma[a] = \infty$.

Theorem 2.8.1.1 (Convergence Exponent) *The following equality holds:*

$$\rho[a] = \inf\left\{\lambda : \int^{\infty} \frac{a(r)dr}{r^{\lambda+1}} < \infty\right\}. \tag{2.8.1.3}$$

If the integral converges for $\lambda = \rho[a]$, $a(r)$ has minimal type.

Exercise 2.8.1.1 Prove this.

For the proof see, e.g., [HK, § 4.2].

Example 2.8.1.4 Let r_j, $j = 1, 2, \dots$ be a nondecreasing sequence of positive numbers. Let us concentrate the unit mass in every point r_j and define a mass distribution

$$n(E) := \{\text{the number points of the sequence } \{r_j\} \text{ in } E\},\ E \subset \mathbb{R}.$$

Then

$$\int_0^{\infty} \frac{dn}{r^{\lambda}} = \sum_1^{\infty} \frac{1}{r_j^{\lambda}}. \tag{2.8.1.4}$$

The infimum of λ for which the series in (2.8.1.4) converges is usually called the *convergence exponent* for the sequence $\{r_j\}$ [PS, Part I, Sec. 1, Ch. III, § 2]. Integrating by parts one can transform the integral in (2.8.1.4) to an integral of the form (2.8.1.3) where $a(r) = n((-\infty, r))$. Theorem 2.8.1.1 shows that the convergence exponent coincides with the order of this $a(r)$.

A function $\rho(r)$ is called a *proximate order* with respect to order ρ if

po1) $\rho(r) \geq 0$,

po2) $\lim_{r\to\infty} \rho(r) = \rho$,

po3) $\rho(r)$ has a continuous derivative on $(0, \infty)$,

po4) $\lim_{r\to\infty} r \log r \rho'(r) = 0$.

Two proximate orders $\rho_1(r)$ and $\rho_2(r)$ are called *equivalent*, if

$$\rho_1(r) - \rho_2(r) = o\left(\frac{1}{\log r}\right). \tag{2.8.1.5}$$

For $a \in A$ set

$$\sigma[a, \rho(r)] := \limsup_{r\to\infty} \frac{a(r)}{r^{\rho(r)}}. \tag{2.8.1.6}$$

It is called a *type number with respect to a proximate order* $\rho(r)$. It is clear that this type number is the same for equivalent proximate orders.

Theorem 2.8.1.2 (Proper Proximate Order) *Let $a \in A$ and $\rho[a] = \rho < \infty$. Then there exists a proximate order $\rho(r)$ such that*

$$0 < \sigma[a, \rho(r)] < \infty. \tag{2.8.1.7}$$

For the proof see [Le, Ch. 1, Sec. 12, Thm. 16].

If a proximate order satisfies the condition (2.8.1.7), we will call it the *proper* proximate order of $a(r)$ (p.p.o.). The function $r^{\rho(r)}$ inherits a lot of useful properties of the power function r^ρ.

Theorem 2.8.1.3 (Properties of P.O) *The following holds:*

ppo1) *the function $V(r) := r^{\rho(r)}$ increases monotonically for sufficiently large values of r.*

ppo2) *for $q < \rho + 1$,*

$$\int_1^r t^{\rho(t)-q}dt \sim \frac{r^{\rho(r)+1-q}}{\rho+1-q}$$

and for $q > \rho + 1$,

$$\int_r^\infty t^{\rho(t)-q}dt \sim \frac{r^{\rho(r)+1-q}}{q-\rho-1}$$

as $r \to \infty$.

ppo3) *the function $L(r) := r^{\rho(r)-\rho}$ satisfies the condition*

$$\forall\delta > 0,\ L(kr)/L(r) \to 1$$

when $r \to \infty$ uniformly for $k \in [\frac{1}{\delta}, \delta]$.

Exercise 2.8.1.2 Prove these properties.

For the proof see, e.g., [Le, Ch. 2, Sec. 12]. The following assertion allows us to replace any p.o. with a smooth one.

Theorem 2.8.1.4 (Smooth P.O) *Let $\rho(r)$ be an arbitrary p.o. There exists an infinitely differentiable equivalent p.o. $\rho_1(r)$ such that*

$$r^k \log r \rho_1^{(k)}(r) \to 0,\ k = 1, 2, \ldots \tag{2.8.1.8}$$

when $r \to \infty$.

Proof. Let α_ϵ be defined by (2.3.1.3). Set $\epsilon := 0.5$, $po(x) := \rho(e^x)$ and

$$po_1(x) := po(n) + [po(n+1) - po(n)] \int_n^x \alpha_{0.5}(t+0.5)dt$$

for $x \in [n, n+1)$. The function $po_1(x)$ is continuous and infinitely differentiable due to properties of α_ϵ and $po_1(n) = po(n)$ for $n = 1, 2, \ldots$. By property po3) of p.o. we have

$$(n+1)|po(n+1) - po(n)| \le \frac{n+1}{n} \max_{y\in[n,n+1]} |y \cdot po'(y)| \to 0$$

as $n \to \infty$. Thus

$$\max_{y\in[n,n+1]} |y \cdot po_1^{(k)}(y)| \le \text{const} \cdot (n+1)|po(n+1) - po(n)| \to 0$$

as $n \to \infty$.

So $\rho_1(r) := po_1(\log r)$ is a p.o. that satisfies (2.8.1.8). Let us show that it is equivalent to $\rho(r)$. Indeed

$$\begin{aligned}|po(x) - po_1(x)| &= \left| \int_n^x [po(y) - po_1(y)]' y \frac{dy}{y} \right| \\ &\le \max_{y\in[n,n+1]} [|y \cdot po'(y)| + |y \cdot po_1'(y)|] \log \frac{n+1}{n} = o\left(\frac{1}{x}\right),\end{aligned}$$

when $x \in [n, n+1]$ and $n \to \infty$. □

We will further need (in 2.9.3) the following assertion.

Theorem 2.8.1.5. (A.A. Gol′dberg) *Let $\rho(r) \to \rho$ be a p.o., and let $f(t)$ be a function that is locally summable on $(0, \infty)$ and such that*

$$\lim_{t\to 0} t^{\rho+\delta} f(t) = \lim_{t\to\infty} t^{\rho+1+\gamma} f(t) = 0 \tag{2.8.1.9}$$

for some $0 < \delta, \gamma < 1$.

Then

$$\begin{aligned}\lim_{r\to\infty} r^{-\rho(r)} \int_{cr^{-1}}^x (rt)^{\rho(rt)} f(t)dt &= \int_0^x t^\rho f(t)dt, \\ \lim_{r\to\infty} r^{-\rho(r)} \int_x^\infty (rt)^{\rho(rt)} f(t)dt &= \int_x^\infty t^\rho f(t)dt\end{aligned} \tag{2.8.1.10}$$

for any $c > 0$ and any $x \in (0, \infty)$.

Proof. Set

$$I(r) := \int_{cr^{-1}}^{\infty} \frac{(rt)^{\rho(rt)}}{r^{\rho(r)}} f(t)dt.$$

It will be enough to prove that

$$\lim_{r\to\infty} I(r) = \int_0^{\infty} t^{\rho} f(t)dt \tag{2.8.1.11}$$

because both functions

$$f_0(t,x) := \begin{cases} f(t), & \text{for } t \in (0,x), \\ 0 & \text{for } t \in [x,\infty) \end{cases}$$

and $f_\infty(t,x) := f(t) - f_0(t,x)$ also satisfy the condition of the theorem.

Let us represent the integral as the following sum:

$$I(r) := \int_{cr^{-1}}^{\infty} \frac{(rt)^{\rho(rt)}}{r^{\rho(r)}} f(t)dt = I_1(r,\epsilon) + I_2(r,\epsilon) + I_3(r,\epsilon), \tag{2.8.1.12}$$

where

$$I_1(r,\epsilon) := \int_{cr^{-1}}^{\epsilon} \frac{(rt)^{\rho(rt)}}{r^{\rho(r)}} f(t)dt,$$

$$I_2(r,\epsilon) := \int_{\epsilon}^{\epsilon^{-1}} \frac{(rt)^{\rho(rt)}}{r^{\rho(r)}} f(t)dt,$$

$$I_3(r,\epsilon) := \int_{\epsilon^{-1}}^{\infty} \frac{(rt)^{\rho(rt)}}{r^{\rho(r)}} f(t)dt.$$

We can represent $I_2(r,\epsilon)$ in the form

$$I_2(r,\epsilon) = \int_{\epsilon}^{\epsilon^{-1}} \frac{L(rt)}{L(r)} t^{\rho} f(t)dt.$$

By ppo3) (Theorem 2.8.1.3),

$$\lim_{r\to\infty} I_2(r,\epsilon) = \int_{\epsilon}^{\epsilon^{-1}} t^{\rho} f(t)dt. \tag{2.8.1.13}$$

Let us estimate the "tails". From (2.8.1.9) we have

$$|f(t)| \le Ct^{-\rho-\delta}$$

for $0 < t \le \epsilon$ where C does not depend on ϵ and

$$|f(t)| \le Ct^{-\rho-1-\gamma}$$

for $t \geq \epsilon^{-1}$. We have

$$|I_1(r,\epsilon)| \leq C \int_{cr^{-1}}^{\epsilon} \frac{(rt)^{\rho(rt)}}{r^{\rho(r)}} t^{-\rho-\delta} dt := CJ_1(r,\epsilon)$$

and

$$\limsup_{r\to\infty} |I_1(r,\epsilon)| \leq C \lim_{r\to\infty} J_1(r,\epsilon). \tag{2.8.1.14}$$

Let us calculate the last limit.We perform the change $x = tr$:

$$J_1(r,\epsilon) = r^{-\rho(r)+\rho+\delta-1} \int_c^{\epsilon r} t^{-\rho(x)-(\rho+\delta)} dx.$$

Now we use ppo2) for $q = \rho + \delta$ and ppo3):

$$\begin{aligned}\lim_{r\to\infty} J_1(r,\epsilon) &= \frac{1}{1-\delta} \lim_{r\to\infty} \frac{(\epsilon r)^{\rho(\epsilon r)-(\rho+\delta)+1}}{r^{\rho(r)-(\rho+\delta)+1}} \\ &= \frac{\epsilon^{1-\delta}}{1-\delta} \lim_{r\to\infty} \frac{L(\epsilon r)}{L(r)} = \frac{\epsilon^{1-\delta}}{1-\delta}.\end{aligned}$$

Substituting in (2.8.1.14) we obtain

$$\limsup_{r\to\infty} |I_1(r,\epsilon)| \leq C \frac{\epsilon^{1-\delta}}{1-\delta}. \tag{2.8.1.15}$$

Analogously one can obtain

$$\limsup_{r\to\infty} |I_3(r,\epsilon)| \leq C \frac{\epsilon^{\gamma}}{\gamma}. \tag{2.8.1.16}$$

Using (2.8.1.13), (2.8.1.15) and (2.8.1.16) one can pass to the limit in (2.8.1.12) as $r \to \infty$, then let $\epsilon \to 0$ and obtain (2.8.1.11). □

2.8.2 Let

$$u(x) := u_1(x) - u_2(x) \tag{2.8.2.1}$$

where $u_1, u_2 \in SH(\mathbb{R}^m)$, $u_1(0) > -\infty$, $u_2(0) = 0$ and $\mu_1 := \mu_{u_1}$, $\mu_2 := \mu_{u_2}$ are concentrated on disjoint sets.

Let $m = 2$, $u_j(z) := \log|f_j(z)|$, $j = 1,2$ where $f_j(z)$, $j = 1,2$ are entire functions. Then the function $u(z) = \log|f(z)|$, where $f(z) := f_1(z)/f_2(z)$, is meromorphic. The condition for masses means that f_1 and f_2 have no common zeros, $u_2(0) = 0$ corresponds to $f_2(0) = 1$ and $u_1(0) > -\infty$ means $f_1(0) \neq 0$.

The class of such functions is denoted as $\delta SH(\mathbb{R}^m)$. In spite of the standardization conditions the representation (2.8.2.1) is not unique. However for any pair of representations $u_1 - u_2$ and $u_1' - u_2'$,

$$u_j(x) - u_j'(x) = H_j(x),\ j = 1,2 \tag{2.8.2.2}$$

where H_j are harmonic and $H_2(0) = 0$.

Really, from the equality $u_1 - u_2 = u_1' - u_2'$ we obtain $\mu_1 - \mu_2 = \mu_1' - \mu_2'$. Using Theorem 2.2.1.2 (Jordan decomposition) we obtain $\mu_1 = \mu_1'$, $\mu_2 = \mu_2'$. Thus (2.8.2.2) holds. Obviously $H_2(0) = 0$.

Set

$$T(r,u) := \frac{1}{\sigma_m}\int_{|y|=1} \max(u_1,u_2)(ry)dy \tag{2.8.2.3}$$

where σ_m is the surface square of the unit sphere. It is called the *Nevanlinna characteristic* of $u \in \delta SH(\mathbb{R}^m)$.

The Nevanlinna characteristic does not depend on the representation (2.8.2.1). Indeed,

$$\begin{aligned}\int_{|y|=1} \max(u_1,u_2)(ry)dy &= \int_{|y|=1} [(u_1-u_2)^+(ry) - u_2(ry)]dy\\ &= \int_{|y|=1} [(u_1'-u_2')^+(ry) - u_2'(ry) + H_2(rx)]dy\\ &= \int_{|y|=1} [\max(u_1',u_2')(ry) + H_2(rx)]dy\\ &= \int_{|y|=1} \max(u_1',u_2')(ry)dy + H_2(0)\\ &= \int_{|y|=1} \max(u_1',u_2')(ry)dy.\end{aligned}$$

Note also that the class $\delta SH(\mathbb{R}^m)$ is linear.

Actually, let $u \in \delta SH(\mathbb{R}^m)$. Then $\lambda u \in \delta SH(\mathbb{R}^m)$ for $\lambda > 0$. The function $-u \in \delta SH(\mathbb{R}^m)$, since

$$-u(x) = [u_2(x) - u_1(0)] - [u_1(x) - u_1(0)].$$

Let us show that $u_1 + u_2 \in \delta SH(\mathbb{R}^m)$ if $u, v \in \delta SH(\mathbb{R}^m)$.

Set $\nu := \nu_u + \nu_v$, where ν_u, ν_v are the corresponding charges. By Theorem 2.2.1.2 (Jordan decomposition) $\nu = \nu^+ - \nu^-$, where ν^+, ν^- are measures concentrated on disjoint sets.

Let u_1 be a subharmonic function in $\mathbb{R}^m$ the mass distribution of which coincides with ν^+.[2] Then $u_2 := u_1 - (u+v)$ is a subharmonic function with the mass distribution ν^-. Hence $u(x) + v(x) = [u_1(x) - u_2(0)] - [u_2(x) - u_2(0)]$.

Theorem 2.8.2.1 (Properties $T(r,u)$) *The following holds:*

t1) *$T(r,u)$ increases monotonically and is convex with respect to $-r^{m-2}$ for $m =$ 2 and with respect to $\log r$ for $m = 2$.*

[2]We will give the construction of such a function for the case of finite order (item 2.9.2), but it is possible actually always, see, for example, [HK, Thm. 4.1]

t2) *For* $u \in SH(\mathbb{R}^m)$, *(i.e.,* $u_2 \equiv 0$*)*

$$T(r,u) = \frac{1}{\sigma_m}\int_{|y|=1} u^+(ry)dy.$$

t3) $T(r,u) = T(r,-u) - u_1(0)$.

t4) $T(r,u+u') \le T(r,u) + T(r,u')$, $T(r,\lambda u) = \lambda T(r,u)$ *for* $\lambda > 0$.

Proof. Since $v(x) := \max(u_1,u_2)(x)$ is subharmonic, t1) follows from Theorem 2.6.5.2 (Convexity of $M(r,u)$ and $\mathcal{M}(r,u)$).

The property t2) is obvious, t3) follows from the equality $-u(x) = u_2(x) - [u_1(x) - u_1(0)] - u_1(0)$.

The properties t4) follow from the properties of maximum and t3). □

Set $\rho_T[u] := \rho[a]$ (see, (2.8.1.1)) where $a(r) := T(r,u)$. It is called the *order of* $u(x)$ *with respect to* $T(r)$.

Theorem 2.8.2.2 (ρ_T-property) *For* $u_1, u_2 \in \delta SH(\mathbb{R}^m)$ *the following inequality holds:*

$$\rho_T[u_1+u_2] \le \max(\rho_T[u_1], \rho_T[u_2]). \tag{2.8.2.4}$$

Equality in (2.8.2.4) *is attained if* $\rho_T[u_1] \ne \rho_T[u_2]$.

Proof. Set $u := u_1 + u_2$. From t3) and t4)

$$T(r,u) \le T(r,u_1) + T(r,u_2) + O(1) \le 2\max[T(r,u_1), T(r,u_2)] + O(1).$$

From the definition of ρ_T we obtain (2.8.2.4).

Suppose, for example, $\rho_T[u_1] > \rho_T[u_2]$. Let us show that $\rho_T[u] = \rho_T[u_1]$. From the equality $u_1 = u + (-u_2)$ we obtain $\rho_T[u_1] \le \max(\rho_T[u], \rho_T[u_2]$ If $\rho_T[u] < \rho_T[u_1]$, then from the previous inequality we would have the contradiction $\rho_T[u_1] < \rho_T[u_1]$. □

Let us define $\sigma_T[u]$ by (2.8.1.2) while $\rho := \rho_T[u]$. Set also $\sigma_T[u,\rho(r)] := \sigma[a,\rho(r)]$ (see (2.8.1.6)), where $a(r) := T(r,u)$.

The characteristics $\rho_T[u]$, $\sigma_T[u]$, $\sigma_T[u,\rho(r)]$ are defined for $u \in \delta SH(\mathbb{R}^m)$. For the class of subharmonic functions we have the inclusion $SH(\mathbb{R}^m) \subset \delta SH(\mathbb{R}^m)$ and, of course, all these characteristics can be applied to a subharmonic function. However, for the class $SH(\mathbb{R}^m)$ the standard characteristic of growth is $M(r,u)$ that we can not apply to a δ-subharmonic function $u \in \delta SH(\mathbb{R}^m)$. Thus for $u \in SH(\mathbb{R}^m)$ we define new *characteristics* $\rho_M[u]$, $\sigma_M[u]$, $\sigma_M[u,\rho(r)]$ in the same way by replacing $T(r,u)$ for $M(r,u)$. The following theorem shows that there is not a big difference between characteristics with respect to T and M for $u \in SH(\mathbb{R}^m)$.

Theorem 2.8.2.3 (T and M-characteristics) *Let $u \in SH(\mathbb{R}^m)$ and $\rho(r)(\to \rho)$ any p.o. Then*

ρMT1) $\rho_T[u]$ *and* $\rho_M[u]$ *are finite simultaneously and* $\rho_T[u] = \rho_M[u] := \rho[u]$

ρMT2) *there exists* $A := A(\rho, m)$ *such that*

$$A\sigma_M[u, \rho(r)] \le \sigma_T[u, \rho(r)] \le \sigma_M[u, \rho(r)].$$

In particular, the last property means that the types with respect to $T(r)$ and $M(r)$ for the same p.o. are minimal, normal or maximal at the same time.

Proof. From t2), Theorem 2.8.2.1 we have $T(r, u) \le M(r, u)$ for $u \in SH(\mathbb{R}^m)$. Thus $\rho_T[u] \le \rho_M[u]$, proving the second part of ρMT2).

Let $H(x)$ be the least harmonic majorant of $u(x)$ in the ball K_{2R}. By the Poisson formula (Theorem 2.4.1.5) and Theorem 2.6.1.3,

$$\begin{aligned} M(R, u) \le M(R, H) &= \max_{|x|=R} \frac{1}{\sigma_m 2R} \int_{|y|=2R} u(y) \frac{(4R^2 - |x|^2)}{|x - y|^m} ds_y \\ &\le \frac{2^{m-2}}{\sigma_m} \int_{|y|=1} |u(2Ry)| ds_y \qquad (2.8.2.5) \\ &= 2^{m-2}[T(2R, u) + T(2R, -u)] = 2^{m-2}[2T(2R, u) - u(0)]. \end{aligned}$$

From here one can obtain $\rho_T[u] \ge \rho_M[u]$. The left side of ρMT2) with $A(\rho, m) := 2^{-\rho-m+2}$ follows from the properties of p.o. □

Exercise 2.8.2.1 Prove the first inequality from ρMT2).

2.8.3 Let μ be a mass distribution (measure) in $\mathbb{R}^m$ ($\mu \in \mathcal{M}(\mathbb{R}^m)$). The characteristic

$$\rho[\mu] := \rho[a] - m + 2$$

for $a(r) := \mu(K_r)$ (see (2.8.1.1)) is called the *convergence exponent* of μ, and

$$\bar{\Delta}[\mu] := \sigma[a]$$

for the same a (see (2.8.1.2)) is called the *upper density* of μ.

The least integer number p for which the integral

$$\int^{\infty} \frac{\mu(t)}{t^{p+m}} dt \qquad (2.8.3.1)$$

converges is called the *genus* of μ and is denoted $p[\mu]$.

Theorem 2.8.3.1 (Convergence Exponent and Genus) *The following holds:*

ceg1) $p[\mu] \le \rho[\mu] \le p[\mu] + 1$,

ceg2) *for* $\rho[\mu] = p[\mu] + 1$, $\bar{\Delta}[\mu] = 0$.

Proof. From Theorem 2.8.1.1 (Convergence Exponent) we have $\rho[\mu]+1+m-2 \leq p[\mu]+m$. Thus $\rho[\mu] \leq p[\mu]+1$. The same theorem implies $\rho[\mu]+m-2+1 \geq p[\mu]+m-1$. Thus $p[\mu] \leq \rho[\mu]$, and ceg1) is proved.

Let $\rho(\mu) = p[\mu]+1$. Then the integral (2.8.3.1) converges for $p[\mu] = \rho[\mu]-1$. We use the inequality

$$\int_r^\infty \frac{\mu(t)}{t^{\rho[\mu]+m-1}} dt \geq \mu(r) \int_r^\infty \frac{dt}{t^{\rho[\mu]+m-1}} dt = \frac{\mu(r)}{r^{\rho[\mu]+m-2}} (\rho[\mu]+m-2)^{-1}.$$

Since the left side of the inequality tends to zero we obtain

$$\bar{\Delta}[\mu] = \lim_{r\to\infty} \frac{\mu(r)}{r^{\rho[\mu]+m-2}} = 0. \qquad \square$$

Set

$$\bar{\Delta}[\mu, \rho(r)] := \sigma[a, \rho(r)+m-2], \tag{2.8.3.2}$$

where $a(r) := \mu(r)$ (see (2.8.1.6)). It is clear that $\rho(r)+m-2$ is also a p.o. Set as in (2.6.5.1),

$$N(r,\mu) := A(m) \int_0^r \frac{\mu(t)}{t^{m-1}} dt,$$

where $A(m) = \max(1, m-2)$. Set also

$$\rho_N[\mu] := \rho[a], \; \bar{\Delta}_N[\mu, \rho(r)] := \sigma[a, \rho(r)],$$

where $a(r) := N(r,\mu)$. This is the *N-order of* μ and the *N-type of* μ with respect to p.o. $\rho(r)$.

Theorem 2.8.3.2 (N-order and Convergence Exponent) *The following holds:*

Nce1) $\rho_N[\mu]$ *and* $\rho[\mu]$ *are finite simultaneously and* $\rho_N[\mu] = \rho[\mu]$,

Nce2) *for* $\rho > 0$ *there exists such* $A_j := A_j(\rho, m)$, $j = 1, 2$, *that*

$$A_1\bar{\Delta}[\mu, \rho(r)] \leq \bar{\Delta}_N[\mu, \rho(r)] \leq A_2\bar{\Delta}[\mu, \rho(r)].$$

Proof. We have the inequality

$$N(2r,\mu) \geq A(m) \int_r^{2r} \frac{\mu(t)}{t^{m-1}} dt \geq A(m)\mu(r) \int_r^{2r} \frac{dt}{t^{m-1}} \geq A(m)B(m) \frac{\mu(r)}{(2r)^{m-2}},$$

where $B(m) := 1 - 2^{2-m}$ for $m \geq 3$ and $B(2) := \log 2$.

From here one can obtain the inequality $\rho[\mu] \geq \rho_N[\mu]$ and the left side of Nce2) for $A_1(\rho, m) := A(m)B(m)2^{-\rho}$. For proving the opposite inequalities we use the l'Hôspital Rule (slightly improved):

$$\limsup_{r\to\infty} \frac{N(r,\mu)}{r^{\rho(r)}} \leq \limsup_{r\to\infty} \frac{N'(r,\mu)}{(r^{\rho(r)})'} = \limsup_{r\to\infty} \frac{\mu(r)r^{2-m}}{r^{\rho(r)}[\rho(r) + r\log r\rho'(r)]} = \frac{1}{\rho}\bar{\Delta}[\mu].$$

Thus $\rho_N[\mu] \leq \rho[\mu]$ and the right side of Nce2) holds. $\square$

We shall denote as $\delta\mathcal{M}(\mathbb{R}^m)$ the set of charges (signed measures) of the form $\nu := \mu_1 - \mu_2$ where $\mu_1, \mu_2 \in \mathcal{M}(\mathbb{R}^m)$. Let us remember that $|\nu| \in \mathcal{M}(\mathbb{R}^m)$ is the full variation of ν (see 2.2.1).

Theorem 2.8.3.3 (Jensen) *Let* $u := u_1 - u_2 \in \delta SH(\mathbb{R}^m)$ *and* $\nu := \mu_1 - \mu_2$ *be a corresponding charge. Then*

J1) $\rho[|\nu|] \le \max(\rho[\mu_1], \rho[\mu_2]) \le \rho[u]$,

J2) $\bar{\Delta}[|\nu|, \rho(r)] \le \bar{\Delta}[\mu_1, \rho(r)] + \bar{\Delta}[\mu_2, \rho(r)] \le A\sigma_T[u, \rho(r)]$ *for some* $A := A(\rho, m)$.

Proof. We can suppose without loss of generality that $u(0) = 0$ because the function $u(x) - u(0)$ has the same order and the same number type if $\rho > 0$. We apply the Jensen-Privalov formula (Theorem 2.6.5.1) to the functions u_1, u_2 and obtain

$$N(r, \mu_j) \le \mathcal{M}(r, u_j) \le T(r, u).$$

Thus $N(r, |\nu|) \le N(r, \mu_1) + N(r, \mu_2) \le 2T(r, u)$. From here one can obtain J1) and J2) for $\rho_N[|\nu|]$ and $\bar{\Delta}_N[|\nu|, \rho(r)]$. However, we can delete the subscript N because of Theorem 2.8.3.2. □

2.9 The representation theorem of subharmonic functions in $\mathbb{R}^m$

2.9.1 Set

$$H(z, \cos\gamma, m) := \begin{cases} -\frac{1}{2}\log(z^2 - 2z\cos\gamma + 1), & \text{for } m = 2, \\ (z^2 - 2z\cos\gamma + 1)^{-\frac{m-2}{2}}, & \text{for } m \ge 3. \end{cases} \tag{2.9.1.1}$$

The function $H(z, \cos\gamma, m)$ is holomorphic on z in the disk $\{|z| < 1\}$. It can be represented there in the form

$$H(z, \cos\gamma, m) = \sum_{k=0}^{\infty} C_k^{\frac{m-2}{2}}(\cos\gamma) z^k \tag{2.9.1.2}$$

where every coefficient $C_k^{\beta}(\bullet)$, $k = 0, 1, \ldots$ is a polynomial of degree k.

Such polynomials are called the *Gegenbauer* polynomials. Note that $C_k^{\frac{1}{2}}(\bullet)$ are the Legendre polynomials and

$$C_k^0(\lambda) = \begin{cases} 0, & \text{for } k = 0, \\ \frac{1}{k}\cos(k \arccos\lambda), & \text{for } k \ge 1, \end{cases}$$

i.e., they are proportional to the *Chebyshev* polynomials.

Thus for $m = 2$ we have the equality

$$-\frac{1}{2}\log(z^2 - 2z\cos\gamma + 1) = \sum_{k=1}^{\infty} \frac{\cos k\gamma}{k} z^k$$

that can be checked directly.

Let $x \in \mathbb{R}^m$. Set $x^0 := x/|x|$. Then the scalar product (x^0, y^0) is equal to $\cos\gamma$ where γ is the angle between x and y.

Let $\mathcal{E}_m(x)$ be defined by (2.4.1.1). For $m \geq 3$ the function $\mathcal{E}_m(x-y)$ is the Green function for $\mathbb{R}^m$. One can see that it is represented in the form

$$G(x, y, \mathbb{R}^m) := \mathcal{E}_m(x - y) = -|y|^{2-m} H(|x|/|y|, \cos\gamma, m)$$

where $\cos\gamma = (x^0, y^0)$.

For $m = 2$ the function $-H(|x|/|y|, \cos\gamma, 2)$ plays the same role. Thus we will denote it as $G(x, y, \mathbb{R}^2)$.

Theorem 2.9.1.1 (Expansion of $G(x, y, \mathbb{R}^m)$) *The following holds:*

$$G(x, y, \mathbb{R}^m) = -\sum_{k=0}^{\infty} C_k^{\frac{m-2}{2}}(\cos\gamma) \frac{|x|^k}{|y|^{k+m-2}}, \tag{2.9.1.3}$$

for $|x| < |y|$, and the functions

$$D_k(x, y) := C_k^{\frac{m-2}{2}}(\cos\gamma) \frac{|x|^k}{|y|^{k+m-2}} \tag{2.9.1.4}$$

are homogeneous harmonic functions in x and harmonic in y for $y \neq 0$.

Proof. The expansion (2.9.1.3) follows from (2.9.1.2). The function $G(zx, y, \mathbb{R}^m)$ is harmonic for $|x| < |y|$ and, hence, for any real $0 \leq z < 1$. Hence, for any $\psi \in \mathcal{D}(K_r)$ while $r := 0.5|y|$ the function $g(z) := \langle G(z\bullet, y, \mathbb{R}^m), \Delta\psi \rangle = 0$ for $z \in (0, 1)$. The function g is holomorphic for all complex $z \in \{|z| < 1\}$ because $G(zx, y, \mathbb{R}^m)$ is holomorphic. Thus $g(z) \equiv 0$, i.e., all its coefficients are zero.

From the expansion (2.9.1.3) we can see that the coefficients of $G(zx, y, \mathbb{R}^m)$ are $D_k(x, y)$. Hence, $\langle D_k(\bullet, y), \Delta\psi \rangle = 0$ for every $\psi \in \mathcal{D}(K_r)$. Thus $D_k(\bullet, y)$ is a harmonic distribution. By Theorem 2.4.1.1 it is an ordinary harmonic function for $|x| < 0.5|y|$.

$C_k^{\frac{m-2}{2}}(\cos\gamma)$ is a polynomial of degree k with respect to (x^0, y^0). Thus $D_k(x, y)$ is a homogeneous polynomial of x and is harmonic for all x.

Let us prove the harmonicity in y.

By Theorem 2.4.1.10 the function $D_k(y^*, x^0)|y|^{2-m}$ ($*$ stands for inversion) is harmonic in y. We have

$$D_k(y^*, x^0)|y|^{2-m} = |y|^{2-m} D_k(y/|y|^2, x^0) = D_k(x^0, y). \qquad \square$$

Set

$$H(z,\cos\gamma,m,p) = H(z,\cos\gamma,m) - \sum_{k=0}^{p} C_k^{\frac{m-2}{2}}(\cos\gamma)z^k. \tag{2.9.1.5}$$

Theorem 2.9.1.2 *The following holds:*

$$|H(z,\cos\gamma,m,p)| \leq A_1(m,p)|z|^{p+1} \tag{2.9.1.6}$$

for $|z| \leq 1/2$, *and*

$$|H(z,\cos\gamma,m,p)| \leq A_2(m,p)|z|^p \tag{2.9.1.7}$$

for $|z| \geq 2, \ -\pi < \arg z \leq \pi$.

The factor $|z|^p$ *should be replaced by* $\log|z|$ *if* $m = 2, \ p = 0$.

Proof. Consider the function $\phi(z) := H(z,\cos\gamma,m,p)z^{-p-1}$. It is holomorphic in the disk $\{|z| \leq 1/2\}$. We apply the maximum principle and obtain (2.9.1.6) where

$$A_1(m,p) = 2^{p+1} \max_{|z|=1/2} |\phi(z)|.$$

For proving (2.9.1.7) we consider the function $\psi(z) := H(z,\cos\gamma,m,p)z^{-p}$ that is holomorphic in the domain $D := \{z : |z| \geq 2, \ -\pi < \arg z \leq \pi\}$ and continuous in its closure. Applying the maximum principle we obtain (2.9.1.7) where

$$A_2(m,p) = 2^p \max_{z\in\partial D} |\psi(z)|.$$ □

Set

$$G_p(x,y,m) := -|y|^{2-m} H(|x|/|y|,\cos\gamma,m,p)$$

where $\cos\gamma = (x^0,y^0)$.

Note the equality

$$G_p(x,y,m) = G(x,y,\mathbb{R}^m) + \sum_{k=0}^{p} D_k(x,y).$$

Exercise 2.9.1.1 Check this using (2.9.1.3), (2.9.1.4) and (2.9.1.5).

It looks like a Green function for $\mathbb{R}^m$ but it tends more quickly to zero at infinity and generally speaking it is not negative.

For $m = 2$ it can be represented in the form

$$G_p(z,\zeta,2) = \log|E(z/\zeta,p)|$$

where $E(z/\zeta,p)$ is the primary Weierstrass factor:

$$E(z/\zeta,p) := \left(1 - \frac{z}{\zeta}\right)\exp\left[\left(\frac{z}{\zeta}\right) + \frac{1}{2}\left(\frac{z}{\zeta}\right)^2 + \cdots + \frac{1}{p}\left(\frac{z}{\zeta}\right)^p\right].$$

We will call it the *primary kernel* analogously to the primary factor.

Theorem 2.9.1.3 (Estimate of Primary Kernel) *The following holds:*

$$|G_p(x,y,m)| \leq A(m,p)\frac{|x|^{p+1}}{|y|^{p+m-1}} \tag{2.9.1.8}$$

for $|x| < 2|y|$,

$$|G_p(x,y,m)| \leq A(m,p)\frac{|x|^{p}}{|y|^{p+m-2}} \tag{2.9.1.9}$$

for $|y| < 2|x|$, *and*

$$G_p(x,y,m) \leq A(m,p)\min\left(\frac{|x|^{p+1}}{|y|^{p+m-1}}, \frac{|x|^{p}}{|y|^{p+m-2}}\right) \tag{2.9.1.10}$$

for all $x, y \in \mathbb{R}^m$, *where* $A(m,p)$ *does not depend on* x, y.

For $m = 2,\ p = 0$ *we have* $G_p(z, \zeta, 2) \leq A(2,0)\log(1 + \frac{|z|}{|\zeta|})$.

Proof. The inequality (2.9.1.8) follows directly from (2.9.1.6) and (2.9.1.9) follows from (2.9.1.7). By the condition $2 \leq |x|/|y|$ (2.9.1.10) follows from (2.9.1.9).

Suppose $1/2 \leq |x|/|y| \leq 2$. Since all the summands in (2.9.1.5) are bounded from below, for $1/2 \leq z \leq 2$ we have

$$G_p(x,y,m) \leq A_1(m,p)|y|^{2-m} \leq A(m,p)\min\left(\frac{|x|^{p+1}}{|y|^{p+m-1}}, \frac{|x|^{p}}{|y|^{p+m-2}}\right)$$

also under these conditions.

The case $m = 2,\ p = 0$ is obvious. □

2.9.2 Let $\mu \in \mathcal{M}(\mathbb{R}^m)$. We suppose below that its support does not contain the origin.

We will say that the integral $\int_{\mathbb{R}^m} f(x,y)\mu(dy)$ converges uniformly on $x \in D$ if

$$\sup_{x\in D}\left|\int_{|y|>R} f(x,y)\mu(dy)\right| \to 0$$

when $R \to \infty$.

Hence, the integral is permitted to be equal to infinity for some finite x.

Let μ have genus p (see, 2.8.3). Set

$$\Pi(x,\mu,p) := \int_{\mathbb{R}^m} G_p(x,y,m)\mu(dy). \tag{2.9.2.1}$$

It is called the *canonical potential.*

In particular, let $m = 2$ and $\mu := n$ be a *zero distribution*, i.e., it has unit masses concentrated on a discrete point set $\{z_j : j = 1, 2, \dots\}$. Then

$$\Pi(z, n, p) = \log \left| \prod_{j=1}^{\infty} E\left(\frac{z}{z_j}, p\right) \right|$$

where

$$\prod_{j=1}^{\infty} E\left(\frac{z}{z_j}, p\right)$$

is the *canonical Weierstrass product.*

Theorem 2.9.2.1 (Brelot-Weierstrass) *The canonical potential* (2.9.2.1) *converges uniformly on any bounded domain. It is a subharmonic function with* μ *as its Riesz measure.*

Proof. Let $|x| < R_0$ and $|y| > R$. From the estimate of the primary kernel (Theorem 2.9.1.3) we have

$$\begin{aligned}\left| \int_{|y|>R} G_p(x, y, m)\mu(dy) \right| &\leq A(m, p)|x|^{p+1} \int_{|y|>R} |y|^{-p-m+1}\mu(dy) \\ &= A(m, p)|x|^{p+1} \int_R^{\infty} t^{-p-m+1}\mu(dt).\end{aligned}$$

Integrating by part we obtain

$$\int_R^{\infty} t^{-p-m+1}\mu(dt) = \frac{\mu(R)}{R^{p+m-1}} + (p + m - 1)\int_R^{\infty} \frac{\mu(t)}{t^{p+m}} dt.$$

The last integral converges since the genus of μ is p. Hence, both summands tend to zero when $R \to \infty$. Thus

$$\sup_{|x|<R_0} \left| \int_{|y|>R} G_p(x, y, m)\mu(dy) \right| \to 0$$

while R_0 is fixed and $R \to \infty$, i.e., the canonical potential converges uniformly on any bounded domain.

Let us represent the canonical potential for $R > R_0$ in the form

$$\begin{aligned}\Pi(x, \mu, p) = \int_{|y|<R} G(x, y, \mathbb{R}^m)\mu(dy) &+ \int_{|y|<R} \sum_{k=0}^{p} D_k(x, y)\mu(dy) \\ &+ \int_{|y|>R} G_p(x, y, m)\mu(dy).\end{aligned}$$

The first summand is a potential, hence a subharmonic function and its Riesz measure coincide with μ. The other summands are harmonic for $|x| < R_0$. □

The following proposition estimates the growth of the canonical potential in terms of its masses.

Theorem 2.9.2.2 (Estimation of Canonical Potential) *The following inequality holds:*

$$M(r, \Pi(\bullet, \mu, p)) \leq A\left[\int_0^\infty \frac{\mu(r\tau)}{r^{m-2}} \frac{\min(1, \tau^{-1})}{\tau^{p+m-1}} d\tau + \frac{\mu(r)}{r^{m-1}}\right] \tag{2.9.2.2}$$

where $A := A(m,p)$ does not depend on r and μ.

Proof. From (2.9.1.10),

$$\Pi(x, \mu, p) \leq A(m,p) \int_{\mathbb{R}^m} \min\left(\frac{|x|^{p+1}}{|y|^{p+m-1}}, \frac{|x|^p}{|y|^{p+m-2}}\right) \mu(dy).$$

Set $r := |x|,\ t := |y|$. Then we have

$$M(r, \Pi(\bullet, \mu, p)) \leq A \int_0^\infty \min\left(\frac{r^{p+1}}{t^{p+m-1}}, \frac{r^p}{t^{p+m-2}}\right) \mu(dt). \tag{2.9.2.3}$$

The integral on the right side of (2.9.2.3) can be represented in the form

$$\int_0^r \frac{r^p}{t^{p+m-2}} \mu(dt) + \int_r^\infty \frac{r^{p+1}}{t^{p+m-1}} \mu(dt).$$

Integrating every integral by parts we obtain

$$\begin{aligned}(p+m-2) \int_0^r \frac{r^p}{t^{p+m-1}} \mu(t)dt + (p+m-1) \int_r^\infty \frac{r^{p+1}}{t^{p+m}} \mu(t)dt + \frac{\mu(r)}{r^{m-1}} \\ \leq (p+m-1) \int_0^\infty \min\left(1, \frac{r}{t}\right) \frac{r^p}{t^{p+m-1}} \mu(t)dt + \frac{\mu(r)}{r^{m-1}}.\end{aligned}$$

After the change $t = r\tau$ we obtain (2.9.2.2) where the new $A(m,p)$ is equal to $A(m,p)(p+m-1)$. □

Theorem 2.9.2.3 (Brelot-Borel) *The order of the canonical potential is equal to the convergence exponent of its mass distribution, i.e.,*

$$\rho[\Pi(\bullet, \mu, p)] = \rho[\mu],$$

if the genus of μ is equal to p.

Proof. First assume $\rho[\mu] < p+1$. Let us choose λ such that $\rho[\mu] < \lambda < p+1$.

For some constant C that does not depend on t we have $\mu(t) \leq Ct^{\lambda+m-2}$.

Actually, $\mu(t)/t^{\lambda+m-2} \to 0$, because $\lambda > \rho[\mu]$. Since $\mu(t) = 0$ for small t, this function is bounded and we can take its lower bound as C.

Now we have

$$f(r,\tau) := \frac{\mu(r\tau)}{r^{\lambda+m-2}} \frac{\min(1,\tau^{-1})}{\tau^{p+m-1}} \leq C\tau^{\lambda-p-1}\min(1,1/\tau) \tag{2.9.2.4}$$

for all $\tau \in (0,\infty)$.

We also have

$$\lim_{r\to\infty} f(r,\tau) = 0 \tag{2.9.2.5}$$

because of $\lambda > \rho[\mu]$.

Let us divide (2.9.2.2) by r^λ and pass to the upper limit. By Fatou's lemma (Theorem 2.2.2.3)

$$\limsup_{r\to\infty} \frac{M(r,\Pi(\bullet,\mu,p))}{r^\lambda} \leq A(m,p)\left[\int_0^\infty \limsup_{r\to\infty} f(r,\tau)d\tau + \limsup_{r\to\infty}\frac{\mu(r)}{r^{\lambda+m-1}}\right] = 0. \tag{2.9.2.6}$$

Hence,

$$\lambda \geq \rho[\Pi(\bullet,\mu,p)]. \tag{2.9.2.7}$$

Since this holds for any $\lambda > \rho[\mu]$, we have $\rho[\mu] \geq \rho[\Pi(\bullet,\mu,p)]$ under the assumption $\lambda < p[\mu]+1$.

Let $\rho[\mu] = p[\mu]+1$. By Theorem 2.8.3.1, $\bar{\Delta}[\mu] = 0$. Hence, $\mu(t)t^{-p-m+1} \leq C$ and

$$f(r,\tau) := \frac{\mu(r\tau)}{(r\tau)^{p+m-1}}\min(1,\tau^{-1}) \leq C\min(1,1/\tau).$$

The function $\min(1,1/\tau)$ is not summable on $(0,\infty)$. Therefore we will act in a slightly different way. From Theorem 2.9.2.2 we have

$$\limsup_{r\to\infty}\frac{M(r,\Pi(\bullet,\mu,p))}{r^{p+1}} \leq A(m,p)\left[\int_0^1 \limsup_{r\to\infty} f(r,\tau)d\tau\right]$$
$$+ A(m,p)\left[\limsup_{r\to\infty}\int_r^\infty \frac{\mu(t)}{t^{p+m}}dt + + \limsup_{r\to\infty}\frac{\mu(r)}{r^{p+m}}\right].$$

The first integral is equal to zero because $\bar{\Delta}[\mu] = 0$. The second addend vanishes since the integral converges. Thus we have $p+1 = \rho[\mu] \geq \rho[\Pi(\bullet,\mu,p)]$.

The reverse inequality holds for any subharmonic function in $\mathbb{R}^m$ by the Jensen theorem (Theorem 2.8.3.3). □

2.9.3 Let us denote as $\delta SH(\rho)$ the class of functions $u \in \delta SH(\mathbb{R}^m)$ for which $\rho_T[u] \leq \rho$.

Theorem 2.9.3.1 (Brelot-Hadamard) *Let $u = u_1 - u_2 \in \delta SH(\rho)$, and let p_1, p_2 be the genuses of the mass distributions $\mu_j := \mu_{u_j}$, $j = 1,2$. Suppose* $\operatorname{supp}[\mu_1-\mu_2] \cap \{0\} = \varnothing$.

Then the following equality holds:

$$u(x) = \Pi(x,\mu_1,p_1) - \Pi(x,\mu_2,p_2) + \Phi_q(x)$$

where $\Phi_q(x)$ is a harmonic polynomial of degree $q \leq \rho$.

Proof. The function $v(x) := u(x) - \Pi(x, \mu_1, p_1) + \Pi(x, \mu_2, p_2)$ is harmonic by the Brelot-Weierstrass theorem (Theorem 2.9.2.1). We also have the inequality

$$\rho_T[v] \le \max(\rho_T[u], \rho_T[\Pi(\bullet, \mu_1, p_1)], \rho_T[\Pi(\bullet, \mu_2, p_2)]) \tag{2.9.3.1}$$

by Theorem 2.8.2.2 (ρ_T-properties). The property ρMT1) (Theorem 2.8.2.3) implies

$$\rho_T[\Pi(\bullet, \mu_j, p_j)] = \rho_M[\Pi(\bullet, \mu_j, p_j)] := \rho[\Pi(\bullet, \mu_j, p_j)], \ j = 1, 2.$$

The Brelot-Borel theorem (Theorem 2.9.2.3) implies

$$\rho[\Pi(\bullet, \mu_j, p_j)] = \rho[\mu_j], \ j = 1, 2.$$

The Jensen theorem (Theorem 2.8.3.3) implies

$$\max(\rho[\mu_1], \rho[\mu_2]) \le \rho_T[u].$$

From (2.9.3.1) we have

$$\rho_T[v] \le \rho_T[u] \le \rho.$$

Since v is subharmonic, $\rho_T[v] = \rho_M[v] := \rho[v]$ by Theorem 2.8.2.3, and $\rho[v] \le \rho$. Therefore

$$\lim_{r \to \infty} \frac{M(r, v)}{r^{\rho+\epsilon}} = 0$$

for arbitrarily small $\epsilon > 0$.

By the Liouville theorem (Theorem 2.4.2.3) $v(x)$ is a harmonic polynomial of degree $q \le \rho + \epsilon$, and thus $v(x) = \Phi_q(x)$ for $q \le \rho$. □

For a non-integer ρ the Brelot-Hadamard theorem allows us to connect the growth of functions and masses more tightly than in the Jensen theorem.

Theorem 2.9.3.2 (Sharpening of Jensen) *Let $\rho > 0$ and be non-integer, $u = u_1 - u_2 \in \delta SH(\mathbb{R}^m)$ with $\rho_T[u] = \rho$, and let $\nu_u = \mu_1 - \mu_2$ the corresponding charge. Then*

pJ1) $\rho[\nu_u] = \max(\rho[\mu_1], \rho[\mu_2]) = \rho$,

pJ2) $A_1\sigma_T[u, \rho(r)] \le \bar{\Delta}[\nu_u, \rho(r)] \le \bar{\Delta}[\mu_1, \rho(r)] + \bar{\Delta}[\mu_2, \rho(r)] \le A_2\sigma_T[u, \rho(r)]$, *where $A_j = A_j(m, \rho)$ and $\rho(r)$ is an arbitrarily proximate order such that $\rho(r) \to \rho$ when $r \to \infty$.*

For proving this theorem we need

Theorem 2.9.3.3 *Let $\Pi(x, \mu, p)$ be a canonical potential with non-integer $\rho[\mu] := [\rho]$, and let $\rho(r)(\to \rho)$ be a proximate order. Then*

$$\sigma[\Pi(\bullet, \mu, p), \rho(r)] \le A(m, \rho, p)\bar{\Delta}[\mu, \rho(r)]. \tag{2.9.3.2}$$

Proof. We can suppose without loss of generality that $\bar{\Delta}[\mu,\rho(r)]<\infty$. By this condition and since $\mu(t)=0$, $0<t<c$ for some $c>0$, we have the inequality

$$\mu(t)t^{-\rho(t)-m+2}\le C$$

for all $t\in(0,\infty)$ and some $C>0$ that does not depend on t. Set

$$I(r):=\int_{c/r}^{\infty}\frac{\mu(rt)}{r^{\rho(r)+m-2}}\frac{\min(1,1/t)}{t^{p+m-1}}dt.$$

By Theorem 2.9.2.2 we have

$$\sigma[\Pi(\bullet,\mu,p),\rho(r)]=\limsup_{r\to\infty}\frac{M(r,\Pi(\bullet,\mu,p)}{r^{\rho(r)}}\le A(m,p)\limsup_{r\to\infty}I(r). \tag{2.9.3.3}$$

Let us choose r_ϵ such that

$$\sup_{r>r_\epsilon}\frac{\mu(r\epsilon)}{(r\epsilon)^{\rho(r\epsilon)+m-2}}\le\bar{\Delta}[\mu,\rho(r)]+\epsilon.$$

For such r we have

$$\begin{aligned}I(r)&=\int_{c/r}^{\infty}\frac{\mu(rt)}{(rt)^{\rho(rt)+m-2}}\frac{(rt)^{\rho(rt)}}{r^{\rho(r)}}\frac{\min(1,1/t)}{t^{p+1}}dt\\&\le\sup_{c/r\le t\le\epsilon}\frac{\mu(rt)}{(rt)^{\rho(rt)+m-2}}\int_{c/r}^{\epsilon}\frac{(rt)^{\rho(rt)}}{r^{\rho(r)}}\frac{\min(1,1/t)}{t^{p+1}}dt\\&\quad+\sup_{\epsilon\le t\le 1/\epsilon}\cdots\int_{\epsilon}^{1/\epsilon}\cdots dt+\sup_{1/\epsilon\le t\le\infty}\cdots\int_{1/\epsilon}^{\infty}\dots dt\\&\le C\int_{c/r}^{\epsilon}\frac{(rt)^{\rho(rt)}}{r^{\rho(r)}}\frac{\min(1,1/t)}{t^{p+1}}dt+(\bar{\Delta}[\mu,\rho(r)]+\epsilon)\int_{\epsilon}^{1/\epsilon}\cdots dt+C\int_{1/\epsilon}^{\infty}\cdots dt.\end{aligned}$$

The function

$$f(t):=\frac{\min(1,1/t)}{t^{p+1}}$$

satisfies the conditions of Gol′dberg's theorem (Theorem 2.8.1.5) with $p+1-\rho<\delta<1$ and $0<\gamma<p+1-\rho$. Passing to the limit we have

$$\begin{aligned}\limsup_{r\to\infty}I(r)\le C\int_0^{\epsilon}t^{\rho-p}dt&+(\bar{\Delta}[\mu,\rho(r)]+\epsilon)\\&\times\int_{\epsilon}^{1/\epsilon}t^{\rho-p-1}\min(1,1/t)dt+C\int_{1/\epsilon}^{\infty}t^{\rho-p-2}dt.\end{aligned}$$

Passing to the limit as $\epsilon\to 0$ we obtain with the help of (2.9.3.3)

$$\sigma[\Pi(\bullet,\mu,p),\rho(r)]\le A(m,p)\bar{\Delta}[\mu,\rho(r)]\int_0^{\infty}t^{\rho-p-1}\min(1,1/t)dt.$$

□

Proof of Theorem 2.9.3.2. The inequality $\rho[\nu_u] \le \rho$ and the last inequality in pJ2) follow from the Jensen theorem (Theorem 2.8.3.3). Let us prove the reverse inequality and the left side.

Since ρ is non-integer, $q < \rho$ in the Brelot-Hadamard theorem (Theorem 2.9.3.1). Hence $M(r, \Phi_q) = o(r^\rho)$ and

$$T(r,u) \le T(r, \Pi(\bullet, \mu_1, p)) + T(r, \Pi(\bullet, \mu_2, p)) + o(r^\rho).$$

Thus

$$\rho_T[u] \le \max(\rho[\Pi(\bullet, \mu_1, p)], \rho[\Pi(\bullet, \mu_2, p]),$$
$$\sigma_T[u, \rho(r)] \le \max(\sigma_T[\Pi(\bullet, \mu_1, p), \rho(r)], \sigma_T[\Pi(\bullet, \mu_2, p], \rho(r)]).$$

From Theorem 2.9.3.3 we obtain

$$\begin{aligned}\rho_T[u] &\le \max(\rho[\mu_1], \rho[\mu_2]);\\ \sigma_T[u, \rho(r)] &\le A(m, \rho, p) \max(\bar{\Delta}[\mu_1, \rho(r)], \bar{\Delta}[\mu_2, \rho(r)]\\ &= A(m, \rho, p)\bar{\Delta}[|\nu|, \rho(r)].\end{aligned}$$

We can set $A_1 := A^{-1}(m, \rho, p)$ and obtain the left side of pJ2). □

2.9.4 Let $u \in \delta SH(\mathbb{R}^m)$ and $\rho := \rho_T[u]$ be an integer number. We can always represent the function u in the form

$$u(x) = \Pi(x, \nu, \rho) + \Phi_\rho(x) \tag{2.9.4.1}$$

where $\Phi_\rho(x)$ is a harmonic polynomial of degree at most ρ. Actually, such a representation can be obtained from Theorem 2.9.3.1 by addition and subtraction of terms of the form

$$\Phi_{k_j}(x) := \int_{\mathbb{R}^m} D_{k_j}(x, y)\mu_j(dy), \ j = 1, 2$$

where $p_j < k_j \le \rho$. All $\Phi_{k_j}(x)$ of such a kind are harmonic polynomials of degree at most ρ. Set

$$\Pi^R_<(x, \nu, \rho - 1) := \int_{|y|<R} G_{\rho-1}(x, y, m)\nu(dy), \tag{2.9.4.2}$$

$$\Pi^R_>(x, \nu, \rho) := \int_{|y|\ge R} G_\rho(x, y, m)\nu(dy), \tag{2.9.4.3}$$

$$\delta_R(x, \nu, \rho) := \int_{|y|<R} D_\rho(x, y)\nu(dy). \tag{2.9.4.4}$$

In particular, for $m = 2$,

$$\delta_R(z, \nu, \rho) := \frac{1}{\rho} \int_{|\zeta|<R} \Re \left(\frac{z}{\zeta}\right)^\rho \nu(d\zeta). \tag{2.9.4.4a}$$

Let $Y_\rho(x)$ be the homogeneous polynomial of degree ρ from the polynomial Φ_ρ in (2.9.4.1). Set also

$$\begin{aligned} \delta_R(x,u,\rho) &:= \delta_R(x,\nu,\rho) + Y_\rho(x), \\ M(r,\delta) &:= \max_{|y|=1} |\delta_r(ry,u,\rho)|, \\ \bar{\Delta}_\delta[u,\rho] &:= \limsup_{r\to\infty} M(r,\delta) r^{-\rho(r)}. \end{aligned} \tag{2.9.4.5}$$

The functions $\delta_R(x,\nu,\rho)$ are homogeneous polynomials that are determined completely by their values on the unit sphere. Thus, by the Harnack theorem (Theorem 2.4.1.7) we have

Theorem 2.9.4.1 $\bar{\Delta}_\delta[u,\rho(r)] < \infty$ *if and only if the family* $\delta_R(x,u,\rho)R^{\rho-\rho(R)}$, $R > 0$ *is precompact in* $\mathcal{D}'(\mathbb{R}^m)$.

Let ρ be an integer number and $\rho(r) \to \rho$ be a p.o. Set

$$\Omega[u,\rho(r)] := \max(\bar{\Delta}_\delta[u,\rho(r)], \bar{\Delta}[|\nu_u|,\rho(r)].$$

Theorem 2.9.4.2 (Brelot-Lindelöf) *The following holds:*

$$A_1\Omega[u,\rho(r)] \le \sigma_T[u,\rho(r)] \le A_2\Omega[u,\rho(r)],$$

where $A_j := A_j(m,\rho)$.

For proving this theorem we will first study the function $\Pi^R_<$ and $\Pi^R_>$. Set

$$\begin{aligned} T(r,\lambda,>) &:= T(r,\Pi^{\lambda r}_>(\bullet,\nu,\rho)), \\ T(r,\lambda,<) &:= T(r,\Pi^{\lambda r}_<(\bullet,\nu,\rho-1)). \end{aligned}$$

Theorem 2.9.4.3 (Estimate of $T(\bullet,>,T(\bullet,<)$) *The following holds:*

$$T(r,\lambda,>) \le A\left(\int_\lambda^\infty \frac{|\nu|(rt)}{r^{m-2}}\frac{\min(1,t^{-1})}{t^{\rho+m-1}}dt + \frac{|\nu|(r)}{r^{m-1}}\right), \tag{2.9.4.6}$$

$$\begin{aligned} T(r,\lambda,<) \le{}& A\left(\int_0^\lambda \frac{|\nu|(rt)}{r^{m-2}}\frac{\min(1,t^{-1})}{t^{\rho+m-2}}dt\right) \\ &+ A\left(\frac{|\nu|(r\lambda)}{r^{m-2}}\int_\lambda^\infty \frac{\min(1,t^{-1})}{t^{\rho+m-2}}dt + \frac{|\nu|(r)}{r^{m-1}}\right), \end{aligned} \tag{2.9.4.7}$$

where $A := A(m,\rho)$.

Proof. Let $\nu = \mu_1 - \mu_2$. Then $|\nu| = \mu_1 + \mu_2$. We have

$$\Pi^R_<(x,\nu,\rho-1) = \Pi^R_<(x,\mu_1,\rho-1) - \Pi^R_<(x,\mu_2,\rho-1). \tag{2.9.4.8}$$

Since $\Pi^R_<(0,\mu_2,\rho-1) = 0$, we have (see t3),t4), Theorem 2.8.2.1)

$$T(r,\Pi^R_<(\bullet,\nu,\rho-1)) \le T(r,\Pi^R_<(\bullet,\mu_1,\rho-1)) + T(r,\Pi^R_<(\bullet,\mu_2,\rho-1)). \tag{2.9.4.9}$$

Set

$$\Pi_1 := \Pi^R_<(\bullet, \mu_1, \rho - 1), \ \Pi_2 := \Pi^R_<(\bullet, \mu_2, \rho - 1).$$

Let us estimate, for example, $T(r, \Pi_1)$. The masses of the canonical potential Π_1 are concentrated in K_R. Applying Theorem 2.9.2.2 (Estimation of Canonical Potential) for $p = \rho - 1$ we obtain

$$\begin{aligned} T(r,\Pi_1) &\le M(r,\Pi_1) \\ &\le A\int_0^{\frac{R}{r}} \frac{\mu_1(rt)}{r^{m-2}}\frac{\min(1,t^{-1})}{t^{\rho+m-2}}dt + A\frac{\mu_1(R)}{r^{m-2}}\int_{\frac{R}{r}}^{\infty}\frac{\min(1,t^{-1})}{t^{\rho+m-2}}dt + \frac{\mu_1(r)}{r^{m-1}}. \end{aligned}$$

Set $R := r\lambda$. Then we obtain the inequality (2.9.4.7) for $\nu := \mu_1$. Analogously one can do the same for $\nu := \mu_2$. The inequality (2.9.4.9) allows us to pass to the limit in (2.9.4.7) in the general case.

Set $\Pi_1 := \Pi^R_>(\bullet, \mu_1, \rho)$. Applying (2.9.2.2) for $p = \rho$ we obtain

$$T(r,\Pi_1) \le M(r,\Pi_1) \le \int_{\frac{R}{r}}^{\infty} \frac{\mu_1(rt)}{r^{m-2}}\frac{\min(1,t^{-1})}{t^{\rho+m-1}}dt + \frac{\mu_1(r)}{r^{m-1}}.$$

In the same way we obtain (2.9.4.6). □

Set

$$\sigma[\Pi_>, \rho(r)] := \limsup_{r\to\infty} \frac{T(r, \Pi^r_>(\bullet,\nu,\rho))}{r^{\rho(r)}},$$
$$\sigma[\Pi_<, \rho(r)] := \limsup_{r\to\infty} \frac{T(r, \Pi^r_<(\bullet,\nu,\rho))}{r^{\rho(r)}}.$$

Theorem 2.9.4.4 *Let $\nu := \mu_1 - \mu_2 \in \delta\mathcal{M}(\rho)$ and ρ an integer number. Then for any p.o. $\rho(r) \to \rho$,*

$$\max(\sigma[\Pi_>, \rho(r)], \sigma[\Pi_<, \rho(r)]) \le A\bar{\Delta}[|\nu|, \rho(r)]$$

where $A := A(m, \rho)$.

Proof. From (2.9.4.6) we have

$$T(r, \Pi^r_>(\bullet,\nu,\rho)) = T(r, 1, >) \le A\int_1^{\infty} \frac{|\nu|(rt)}{r^{m-2}}\frac{1}{t^{\rho+m}}dt + \frac{|\nu|(r)}{r^{m-1}}.$$

Now we repeat the reasoning of Theorem 2.9.3.3 for $\mu := |\nu|$ and $p := \rho$. We will obtain

$$\sigma[\Pi_>, \rho(r)] \le A\bar{\Delta}[|\nu|, \rho(r)]\int_1^{\infty} t^{-2}dt.$$

For the other case we have from (2.9.4.7),

$$T(r,\Pi^r_<(\bullet,\nu,\rho-1)) = T(r,1,<)$$
$$\leq A\int_0^1 \frac{|\nu|(rt)}{r^{m-2}}\frac{1}{t^{\rho+m-1}}dt + A\frac{|\nu|(r)}{r^{m-2}}\left(\int_1^\infty t^{-\rho-m+1}dt + r^{-1}\right).$$

We divide this inequality by $r^{\rho(r)}$ and pass to the upper limit while $r\to\infty$.

The first summand of the right side gives

$$A\bar{\Delta}[|\nu|,\rho(r)]\int_0^1 dt$$

by the reasoning of Theorem 2.9.3.3.

The second one can be computed directly, yielding

$$A\bar{\Delta}[|\nu|,\rho(r)]\int_1^\infty t^{-\rho-m+1}dt.$$

Combining all these inequalities we obtain the assertion of the theorem. □

Proof of Theorem 2.9.4.2. Let us represent $u(x)$ in the form

$$u(ry) = \Pi^r_<(ry,\nu_u,\rho-1) + \Pi^r_>(ry,\nu_u,\rho) + \delta_r(ry,u,\rho) + o(r^{\rho-1}) \qquad (2.9.4.10)$$

where $|y| = 1$.

Then we have

$$T(r,u) \leq T(r,\Pi^r_<(\bullet,\nu_u,\rho-1)) + T(r,\Pi^r_>(\bullet,\nu_u,\rho)) + M(r,\delta) + o(r^{\rho-1}).$$

Let us divide this by $r^{\rho(r)}$ and pass to the upper limit. By Theorem 2.9.4.4 we obtain

$$\sigma_T[u,\rho(r)] \leq A\max(\bar{\Delta}[|\nu|,\rho(r)],\bar{\Delta}_\delta[u,\rho(r)]) = A_2\Omega[u,\rho(r)]$$

where $A_2 = A(m,\rho)$. Let us write (2.9.4.11) in the form

$$\delta_r(ry,u,\rho) = u(ry) - \Pi^r_<(ry,\nu_u,\rho-1) - \Pi^r_>(ry,\nu_u,\rho) + o(r^{\rho-1}).$$

We obtain

$$T(r,\delta_r(\bullet,u,\rho)) \leq T(r,u) + T(r,\Pi^r_<(\bullet,\nu_u,\rho-1)) + T(r,\Pi^r_>(\bullet,\nu_u,\rho) + o(r^{\rho-1}).$$

Since $\delta_R(\bullet,u,\rho)$ is harmonic and homogeneous, we have by (2.8.2.5)

$$M(r,\delta_R) \leq 2^{m-1}T(2r,\delta_R) = 2^{m-1+\rho}T(r,\delta_R).$$

Therefore we obtain the inequality

$$\bar{\Delta}_\delta[u,\rho(r)] \leq \sigma_T[u,\rho(r)] + 2A\bar{\Delta}[|\nu|,\rho(r)].$$

By the Jensen theorem (Theorem 2.8.3.3) we have

$$\Omega[u,\rho(r)] \leq A_1^{-1}\sigma_T[u,\rho(r)]$$

for some $A_1 = A_1(m,\rho)$. □

Chapter 3

Asymptotic Behavior of Subharmonic Functions of Finite Order

3.1 Limit sets

3.1.1 Let $\{V_t : t \in (0,\infty)\}$ be a family of rotations of $\mathbb{R}^m$ that form a one-parametric group, i.e.,

$$V_{t_1}V_{t_2} = V_{t_1t_2}, \; V_1 = I, \tag{3.1.1.0}$$

where I is the identity map.

The family of linear transformations

$$P_t := tV_t \tag{3.1.1.1}$$

is also a one-parametric group.

In particular, for $m = 2$ the general form of the rotations is

$$V_t z = z\exp(i\gamma \log t),$$

where γ is real.

The orbit $\{P_t z : t \in (0,\infty)\}$ of every point $z \neq 0$ is a logarithmic spiral if $\gamma \neq 0$ and a ray if $\gamma = 0$.

For $m \geq 3$ and $V_t \equiv I$, $t \in (0,\infty)$ the orbit of every point $x \neq 0$ is a ray from the origin. For other $V_\bullet$ it is a spiral connecting the origin to infinity.

It is clear that only one orbit $\{P_t x : t \in (0,\infty)\}$ passes through every $x \neq 0$. The behavior of every point $y(t) := P_t x$ is completely determined by a system of differential equations with constant coefficients:

$$\frac{d}{dt}y = (I + V')y, \quad V' := \frac{d}{dt}V_t \mid_{t=1}$$

with the initial condition of $y(1) = x$.

3.1.2 Let $u \in SH(\rho)$ and $\sigma_M[u, \rho(r)] < \infty$ for some p.o. $\rho(r) \to \rho$. We will write $u \in SH(\mathbb{R}^m, \rho, \rho(r))$ or shorter, $u \in SH(\rho(r))$.

For $u \in SH(\rho(r))$ set

$$u_t(x) := u(P_t x)t^{-\rho(t)}. \tag{3.1.2.1}$$

We will denote this transformation as $(\bullet)_t$.

Theorem 3.1.2.1 (Existence of Limit Set) *The following holds:*

els1) $u_t \in SH(\rho(r))$ *for any* $t \in (0, \infty)$,

els2) *the family* $\{u_t\}$ *is precompact at infinity.*

I.e., for any sequence $t_k \to \infty$ there exists a subsequence $t_{k_j} \to \infty$ and a function $v \in SH(\mathbb{R}^m)$ such that $u_{t_{k_j}} \to v$ in $\mathcal{D}'(\mathbb{R}^m)$ (see Section 2.7.1).

Proof. The functions u_t are subharmonic by sh1) and sh5), Theorem 2.6.1.1. (Elementary Properties), and

$$M(r, u_t) = M(rt, u)t^{-\rho(t)}.$$

Now we have

$$\sigma_M[u_t, \rho(r)] = t^{-\rho(t)} \limsup_{r\to\infty} \frac{M(rt, u)}{(rt)^{\rho(rt)}} \cdot \lim_{r\to\infty} \frac{(rt)^{\rho(rt)}}{r^{\rho(r)}} = \sigma_M[u, \rho(r)]t^{\rho-\rho(t)},$$

because

$$\lim_{r\to\infty} \frac{(rt)^{\rho(rt)}}{r^{\rho(r)}} = t^{\rho} \lim_{r\to\infty} \frac{L(rt)}{L(r)} = t^{\rho} \tag{3.1.2.2}$$

(see, ppo3), Theorem 2.8.1.3 (Properties of P.O)). Therefore els1) is proved.

Let us check the conditions of Theorem 2.7.1.1 (Compactness in $\mathcal{D}'$). We have

$$\limsup_{t\to\infty} M(r, u_t) = \limsup_{t\to\infty} \frac{M(rt, u)}{(rt)^{\rho(rt)}} \cdot \lim_{t\to\infty} \frac{(rt)^{\rho(rt)}}{t^{\rho(t)}} = \sigma_M[u, \rho(r)]r^{\rho}. \tag{3.1.2.3}$$

Thus, the family is bounded from above on every compact set and

$$\lim_{t\to\infty} u_t(0) = \lim_{t\to\infty} u(0)t^{-\rho(t)} = 0.$$

Therefore $u_t(0)$ are bounded from below for large t. □

We will call the set of all functions v from Theorem 3.1.2.1 the *limit set* of the function $u(x)$ with respect to $V_\bullet$ and denote it by $\mathbf{Fr}[u, \rho(r), V_\bullet, \mathbb{R}^m]$ or shortly $\mathbf{Fr}[u]$.

The limit set does not depend on values of the subharmonic function on a bounded set, hence, it is a characteristic of asymptotic behavior.

Set

$$U[\rho, \sigma] := \{v \in SH(\mathbb{R}^m) : M(r, v) \leq \sigma r^\rho,\ r \in [0, \infty);\ v(0) = 0\},$$

$$U[\rho] := \bigcup_{\sigma > 0} U[\rho, \sigma] \tag{3.1.2.4}$$

and

$$v_{[t]}(x) := t^{-\rho} v(P_t x),\ t \in (0, \infty). \tag{3.1.2.4a}$$

Let us emphasize that the transformation $(\bullet)_{[t]}$ coincides with $(\bullet)_t$ from (3.1.2.1) for $\rho(r) \equiv \rho$ and satisfies the condition

$$(\bullet)_{[t\tau]} = ((\bullet)_{[t]})_{[\tau]} \tag{3.1.2.4b}$$

Theorem 3.1.2.2 (Properties of Fr) *The following holds:*

fr1) $\mathbf{Fr}[u]$ *is a connected compact set;*

fr2) $\mathbf{Fr}[u] \subset U[\rho, \sigma]$, *for* $\sigma \geq \sigma_M[u]$;

fr3) $(\mathbf{Fr}[u])_{[t]} = \mathbf{Fr}[u],\ t \in (0, \infty)$. *I.e.,* $v \in \mathbf{Fr}[u]$ *implies* $v_{[t]} \in \mathbf{Fr}[u]$;

fr4) *if* $\rho_1(r)$ *and* $\rho(r)$ *are equivalent* (*see* (2.8.1.5)), *then*

$$\mathbf{Fr}[u, \rho_1(r), \bullet] = \mathbf{Fr}[u, \rho(r), \bullet].$$

We need the following assertion.

Theorem 3.1.2.3 (Continuity u_t) *The functions*

$$u_t, v_{[t]} : (0, \infty) \times \mathcal{D}'(\mathbb{R}^m) \mapsto \mathcal{D}'(\mathbb{R}^m)$$

are continuous in the natural topology.

Proof. For any $\psi \in \mathcal{D}(\mathbb{R}^m)$ consider

$$\langle u_t, \psi \rangle := \int u_t(x)\psi(x)dx = \int u(y)\psi(y/t)t^{m-\rho(t)}dy := \langle u, \psi(\bullet, t)\rangle,$$

where $\psi(y, t) := \psi(y/t)t^{m-\rho(t)}$.

The function $\psi(\bullet, t)$ is continuous in t in $\mathcal{D}(\mathbb{R}^m)$. By Theorem 2.3.4.6 (Continuity $\langle \bullet, \bullet \rangle$) $\langle u, \psi(\bullet, t)\rangle$ is continuous in (u, t). □

Proof of Theorem 3.1.2.2. Let us denote as clos$\{\bullet\}$ the closure in $\mathcal{D}'$-topology.

The set $F_N := \operatorname{clos}\{u_t : t \geq N\} \supset \mathbf{Fr}[u]$ is compact in $\mathcal{D}'$-topology. Indeed, let $t_j \to t$ and $t < \infty$; then $u_{t_j} \to u_t$ because of Theorem 3.1.2.3. If $t_j \to \infty$ and $u_{t_j} \to v$, then $v \in \mathbf{Fr}[u]$ by its definition, hence, $v \in F_N$. Since $\mathbf{Fr}[u] = \cap_{N=1}^{\infty} F_N$, it is compact.

Let us prove the connectedness. Suppose $\mathbf{Fr}[u]$ is not connected. Then it can be written as a union of two disjoint nonempty closed sets F^1 and F^2. Let V^1, V^2 be disjoint open neighborhoods of F^1, F^2 respectively in $\mathcal{D}'(\mathbb{R}^m)$. Since F^1, F^2 are nonempty there exist sequences $\{s_j\}, \{t_j\}$ such that $s_j < t_j$, $s_j \to \infty$, $u_{s_j} \in V^1$, $u_{t_j} \in V^2$. Since the mapping $u_t : (0, \infty) \mapsto \mathcal{D}'(\mathbb{R}^m)$ is continuous, by Theorem 3.1.2.3 its image is connected. Thus there exists a sequence $\{p_j\}$ with $s_j < p_j < t_j$ such that $u_{p_j} \notin V^1 \cup V^2$. This sequence has a subsequence that converges to a function $v \in \mathbf{Fr}[u]$ and $v \notin F^1 \cup F^2$. This is a contradiction. Hence, $\mathbf{Fr}[u]$ is connected and fr1) is proved.

Set

$$\psi(r) := \limsup_{r \to \infty} M(r, u_t).$$

This function is convex with respect to $-r^{2-m}$ for $m \geq 3$ and with respect to $\log r$ for $m = 2$ and hence continuous.

Indeed, $M(|x|, u_t)$ are subharmonic (see Theorem 2.6.5.2 (Convexity $M(\bullet, u)$ and $\mathcal{M}(r, u)$). By Theorem 2.7.3.3 (H.Cartan +) the function $\psi^*(|x|)$ is subharmonic and $\psi(|x|) = \psi^*(|x|)$ quasi-everywhere. However, if $\psi(|x|) < \psi^*(|x|)$ at some point, the same inequality holds on a sphere which has a positive capacity (see Example 2.5.2.2). Hence, $\psi(|x|) = \psi^*(|x|)$ everywhere, and $\psi(|x|)$ is subharmonic. Thus $\psi(r)$ is convex with respect to $-r^{2-m}$ for $m \geq 3$ and with respect to $\log r$ for $m = 2$ by Theorem 2.6.3.2 (Subharmonicity and Convexity).

One can also see that for $u \in SH(\mathbb{R}^m)$,

$$M(r, u_\epsilon) \leq M(r + \epsilon, u),$$

where $(\bullet)_\epsilon$ is defined by (2.6.2.3).

Let $v \in \mathbf{Fr}[u]$ and $u_{t_j} \to v$ in $\mathcal{D}'(\mathbb{R}^m)$. By property reg3), Theorem 2.3.4.5 $(u_{t_j})_\epsilon \to v_\epsilon$ uniformly on any compact set. Thus

$$\begin{aligned} v_\epsilon(x) = \lim_{j \to \infty} (u_{t_j})_\epsilon &\leq \limsup_{t \to \infty)} M(|x|, (u_t)_\epsilon) \\ &\leq \limsup_{t \to \infty)} M(|x| + \epsilon, u_t) = \psi(|x| + \epsilon). \end{aligned} \tag{3.1.2.5}$$

If $\epsilon \downarrow 0$, then $v_\epsilon \downarrow v$ by Theorem 2.6.2.3 and $\psi(r + \epsilon) \to \psi(r)$ because of continuity. Passing to the limit in (3.1.2.5) and using (3.1.2.3) we obtain

$$v(x) \leq \sigma_M[u, \rho(r)]|x|^\rho. \tag{3.1.2.6}$$

Since $u(0) \le u_\epsilon(0)$ we have $u(0)t^{-\rho(t)} \le (u_t)_\epsilon(0)$. Let us pass to the limit as $t := t_j \to \infty$. We obtain $v_\epsilon(0) \ge 0$. Passing to the limit as $\epsilon \downarrow 0$ we have

$$v(0) \ge 0. \tag{3.1.2.7}$$

The inequalities (3.1.2.6) and (3.1.2.7) imply fr2).

One can check the equality

$$(u_t)_{[\tau]} = u_{t\tau} \cdot \frac{(t\tau)^{\rho(t\tau)}}{t^{\rho(t)}\tau^{\rho}}. \tag{3.1.2.8}$$

By using properties of p.o. we have

$$\lim_{t\to\infty} \frac{(t\tau)^{\rho(t\tau)}}{t^{\rho(t)}\tau^{\rho}} = 1$$

(compare (3.1.2.2)).

Let $v \in \mathbf{Fr}[u]$ and $u_{t_j} \to v$. Set $t := t_j$, $\tau := t$ in (3.1.2.8) and pass to the limit. Then

$$v_{[t]} = \mathcal{D}' - \lim_{j\to\infty} u_{t_j t}.$$

Thus $v_{[t]} \in \mathbf{Fr}[u]$. The property f3) is proved.

Let us prove f4).We have

$$\frac{u(P_t x)}{t^{\rho_1(t)}} = \frac{u(P_t x)}{t^{\rho(t)}} \times e^{(\rho_1(t)-\rho(t))\log t} = \frac{u(P_t x)}{t^{\rho(t)}} \times (1 + o(1))$$

as $t \to \infty$ because of (2.8.1.5).

This implies f4).

Exercise 3.1.2.1 Check this in detail. □

We can consider the limit sets as a mapping $u \mapsto \mathbf{Fr}[u]$. The following theorem describes some properties of this mapping.

Set

$$U[\rho] := \bigcup_{\sigma>0} U[\rho,\sigma] \tag{3.1.2.9}$$

where $U[\rho,\sigma]$ is defined by (3.1.2.4).

Let X, Y be subsets of a cone (i.e., a subset of a linear space that is closed with respect to sum and multiplication by a positive number). The set $U[\rho]$ is such a cone. Set

$$X + Y := \{z = x + y : x \in X,\ y \in Y\};\ \lambda X := \{z = \lambda x : x \in X\}. \tag{3.1.2.10}$$

Theorem 3.1.2.4 (Properties of $u \mapsto \mathbf{Fr}[u]$) *The following holds:*

fru1) $\mathbf{Fr}[u_1 + u_2] \subset \mathbf{Fr}[u_1] + \mathbf{Fr}[u_2]$,

fru2) $\mathbf{Fr}[\lambda u] = \lambda \mathbf{Fr}[u]$.

Proof. Let $v \in \mathbf{Fr}[u_1 + u_2]$. Then there exists $t_j \to \infty$ such that $(u_1 + u_2)_{t_j} \to v$ in $\mathcal{D}'$. We can find a subsequence t_{j_k} such that $(u_1)_{t_{j_k}} \to v_1$ and $(u_2)_{t_{j_k}} \to v_2$. Then $v = v_1 + v_2$. The property fru1) has been proved.

The property fru2) is proved analogically. □

3.1.3 We will write $\mu \in \mathcal{M}(\mathbb{R}^m, \rho(r))$ or shortly, $\mu \in \mathcal{M}(\rho(r))$ if $\mu \in \mathcal{M}(\mathbb{R}^m)$ (see 2.8.3) and $\bar{\Delta}[\mu, \rho(r)] < \infty$ (see 2.8.3.2).

Let us define a distribution μ_t for $\mu \in \mathcal{M}(\rho(r))$ by

$$\langle \mu_t, \phi \rangle := t^{-\rho(t)-m+2} \int \phi(P_t^{-1}x)\mu(dx) \tag{3.1.3.1}$$

for $\phi \in \mathcal{D}(\mathbb{R}^m)$.

It is positive. Hence, it defines uniquely a measure μ_t.

Theorem 3.1.3.1 (Explicit form of μ_t) *For any $E \in \sigma(\mathbb{R}^m)$ the following holds:*

$$\mu_t(E) = t^{-\rho(t)-m+2}\mu(P_t E). \tag{3.1.3.2}$$

Proof. It is enough to prove the assertion for some dense ring (see Theorem 2.2.3.5), for example, for all compact sets.

Let χ_K be a characteristic function of a compact set K and let $\phi_\epsilon \downarrow \chi_K$ be a monotonically converging sequence of functions that belong to $\mathcal{D}(\mathbb{R}^m)$ (see Theorems 2.1.2.1, 2.1.2.9 and 2.3.4.4). Then

$$\int \phi_\epsilon(x)\mu_t(dx) = t^{-\rho(t)-m+2} \int \phi_\epsilon(P_t^{-1}x)\mu(dx).$$

Since $\phi_\epsilon(P_t^{-1}x) \downarrow \chi_{P_t K}(x)$,

$$\mu_t(K) = \int \chi_K(x)\mu_t(dx) = t^{-\rho(t)-m+2} \int \chi_{P_t K}(x)\mu(dx) = t^{-\rho(t)-m+2}\mu(P_t K). \quad \square$$

Theorem 3.1.3.2 (Existence of μ-Limit Set) *The following holds:*

mels1) *$\mu_t \in \mathcal{M}(\rho(r))$ for any $t \in (0, \infty)$;*

mels2) *the family $\{\mu_t\}$ is precompact in infinity.*

I.e., for any sequence $t_k \to \infty$ there exists a subsequence $t_{k_j} \to \infty$ and a measure $\nu \in \mathcal{M}(\mathbb{R}^m)$ such that $\mu_{t_{k_j}} \to \nu$ in $\mathcal{D}'(\mathbb{R}^m)$ (see Section 2.7.1).

Proof. We have

$$\mu_t(r) = \mu(rt)t^{-\rho(t)-m+2}.$$

Thus

$$\limsup_{r\to\infty} \frac{\mu_t(r)}{r^{\rho(r)+m-2}} = \limsup_{r\to\infty} \frac{\mu(rt)}{(rt)^{\rho(rt)+m-2}} \frac{(rt)^{\rho(rt)}}{t^{\rho(t)+m-2}r^{\rho(r)}}$$
$$= t^{\rho-\rho(t)-(m-2)}\bar{\Delta}[\mu, \rho(r)]. \tag{3.1.3.3}$$

Therefore mels1) holds.

We also have

$$\limsup_{t\to\infty} \mu_t(r) = \bar{\Delta}[\mu, \rho(r)]r^{\rho+m-2}.$$

Thus μ_t satisfies the assumption of the Helly theorem (Theorem 2.2.3.2). Using also Theorem 2.3.4.4 we obtain mels2). □

We will call the set of all measures ν from Theorem 3.1.2.1 the *limit set* of the mass distribution μ with respect to $V_\bullet$ and denote it $\mathbf{Fr}[\mu, \rho(r), V_\bullet, \mathbb{R}^m]$ or shortly, $\mathbf{Fr}[\mu]$.

Set

$$\mathcal{M}[\rho, \Delta] := \{\nu : \nu(r) \le \Delta r^{\rho+m-2},\ \forall r > 0\}. \tag{3.1.3.4}$$

$$\mathcal{M}[\rho] := \bigcup_{\Delta>0} \mathcal{M}[\rho, \Delta],$$

and

$$\nu_{[t]}(E) := t^{-\rho-m+2}\nu(P_t E) \tag{3.1.3.5}$$

for $E \in \sigma(\mathbb{R}^m)$.

Theorem 3.1.3.3 (Properties of $\mathbf{Fr}[\mu]$) *The following holds:*

frm1) $\mathbf{Fr}[\mu]$ *is connected and compact;*

frm2) $\mathbf{Fr}[\mu] \subset \mathcal{M}[\Delta, \rho]$, *for* $\Delta \ge \bar{\Delta}[\mu, \rho(r)]$;

frm3) $(\mathbf{Fr}[\mu])_{[t]} = \mathbf{Fr}[\mu],\ t \in (0, \infty)$.

Proof. We will only prove frm2) because frm1) and frm3) are proved word by word as in Theorem 3.1.2.2.

Suppose $t_n \to \infty$ and $\mu_{t_n} \to \nu \in \mathbf{Fr}[\mu]$. Let us choose $r' > r$ such that the open ball $K_{r'}$ is squarable with respect to ν. It is possible because of Theorem 2.2.3.3, sqr2). By Theorems 2.2.3.7 and 2.3.4.4, $\mu_{t_n}(r') \to \nu(r')$. Thus (compare with (3.1.2.3))

$$\nu(r') = \lim_{t_n\to\infty} \mu_{t_n}(r') \le \limsup_{t\to\infty} \mu_t(r') = \bar{\Delta}[\mu, \rho(r)](r')^{\rho+m-2}.$$

Choosing $r' \downarrow r$ we obtain

$$\lim_{r'\to r} \nu(r') = \nu(r)$$

because (2.2.3.3). Thus frm2) holds. □

The following assertion is a "copy" of Theorem 3.1.2.4.

Theorem 3.1.3.4 (Properties of $\mu \mapsto \mathbf{Fr}[\mu]$) *The following holds:*

frmu1) $\mathbf{Fr}[\mu_1 + \mu_2] \subset \mathbf{Fr}[\mu_1] + \mathbf{Fr}[\mu_2]$,

frmu2) $\mathbf{Fr}[\lambda\mu] = \lambda\mathbf{Fr}[\mu]$.

The proof is also a "copy" and we omit it.

Exercise 3.1.3.1 Prove Theorem 3.1.3.4

3.1.4 We are going to study the class $U[\rho]$ and obtain for it "non-asymptotic" analogies of Theorem 2.8.3.3 (Jensen), 2.9.2.3 (Brelot-Borel), 2.9.3.1 (Brelot-Hadamard)

Theorem 3.1.4.1 (*Jensen) *Let $v \in U[\rho]$. Then its Riesz measure $\nu_v \in \mathcal{M}[\rho]$.*

Proof. As in Theorem 2.8.3.2 we have an inequality

$$\frac{\nu_v(r)}{r^{m-2}} \leq A(m)N(2r, \nu_v). \tag{3.1.4.1}$$

Since $v(0) = 0$ we have (Theorem 2.6.5.1. (Jensen-Privalov))

$$N(2r, \nu_v) = \mathcal{M}(2r, v) \leq M(2r, v) \leq 2^\rho \sigma r^\rho. \tag{3.1.4.2}$$

Substituting (3.1.4.2) in (3.1.4.1) we obtain $\nu_v \in \mathcal{M}[\rho, \Delta]$ for some Δ. Thus $\nu_v \in \mathcal{M}[\rho]$. □

Let ρ be non-integer and $\nu \in \mathcal{M}[\rho]$. Consider the canonical potential $\Pi(x,\nu,p)$ where $p := [\rho]$ (see (2.9.2.1)). Let us emphasize that the support of ν may contain the origin but $\nu(0) = 0$, i.e., there is no concentrated mass in the origin. Thus we must also check its convergence in the origin.

Theorem 3.1.4.2 (*Brelot-Borel) *Let ρ be non-integer and let $\nu \in \mathcal{M}[\rho]$. Then $\Pi(x, \nu, p)$ converges and belongs to $U[\rho]$.*

Proof. Using (2.9.1.9) we have

$$\left| \int_{|y|<2|x|} G_p(x, y, m)\nu(dy) \right| \leq A(m, p)|x|^p \int_0^{2|x|} \frac{\nu(dt)}{t^{p+m-2}}. \tag{3.1.4.3}$$

Let us estimate the integral in (3.1.4.3). Integrating by parts we obtain

$$I_<(x) := \int_0^{2|x|} \frac{\nu(dt)}{t^{p+m-2}} = \frac{\nu(t)}{t^{p+m-2}}\Big|_0^{2|x|} + (p + m - 2) \int_0^{2|x|} \frac{\nu(t)dt}{t^{p+m-1}}.$$

Since $\nu \in \mathcal{M}[\rho, \Delta]$ for some Δ,

$$I_<(x) \leq A(m, \rho, p)\Delta|x|^{\rho-p}.$$

Substituting this in (3.1.4.3) we obtain

$$\left| \int_{|y|<2|x|} G_p(x, y, m)\nu(dy) \right| \leq A(m, \rho, p)\Delta|x|^\rho. \tag{3.1.4.4}$$

Analogously, using (2.9.1.8) we obtain

$$\left| \int_{|x|<2|y|} G_p(x, y, m)d\nu(dy) \right| \leq A(m, \rho, p)\Delta|x|^\rho. \tag{3.1.4.5}$$

In particular, these estimates show that $\Pi(x,\nu,p)$ exists. Now using (2.9.1.10) we have also

$$\int_{\frac{|x|}{2}\le|y|\le 2|x|} G_p(x,y,m)\nu(dy) \le A(m,p)\int_{\frac{|x|}{2}}^{2|x|}\nu(dt)\min\left(\frac{|x|^{p+1}}{t^{p+m-1}},\frac{|x|^p}{t^{p+m-2}}\right).$$

The latter integral can also be easily estimated by $\Delta A(m,p,\rho)|x|^\rho$. Thus we have

$$\int_{\frac{|x|}{2}\le|y|\le 2|x|} G_p(x,y,m)\nu(dy) \le A(m,p,\rho)|x|^\rho.$$

Therefore by (3.1.4.5) and (3.1.4.4) we obtain $M(r,\Pi)\le\sigma r^\rho$ for some σ.

Since $G_p(0,y,m)=0$ for all $y\neq 0$ and the integral converges, $\Pi(0,\nu,p)=0$. □

We will need an assertion that looks like the Liouville theorem (Theorem 2.4.2.3).

Theorem 3.1.4.3 (*Liouville) *Let H be a harmonic function in $\mathbb{R}^m$ and $H\in U[\rho]$. Then $H\equiv 0$ if ρ is non-integer and H is a homogeneous polynomial of degree p if $\rho = p$ is integer.*

In particular, for $m=2$ we have $H(re^{i\phi}) = r^p\Re(ce^{ip\phi})$.

Proof. Like in the proof of the Liouville theorem we obtain the inequality (2.4.2.9) and

$$|c_k|\le AR^{-k}\max_{|x|=R}H(x)\le A\sigma R^{\rho-k}$$

for some $\sigma>0$.

If $k>\rho$, we will pass to the limit when $R\to\infty$ and obtain $c_k=0$. If $k<\rho$, we will do that when $R\to 0$ and obtain $c_k=0$. □

The following theorem can be considered as an analogy of the Brelot-Hadamard theorem (Theorem 2.9.3.1):

Theorem 3.1.4.4 (*Hadamard) *Let ρ be non-integer and $v\in U[\rho]$. Then*

$$v(x) = \Pi(x,\nu_v,p) \tag{3.1.4.6}$$

for $p=[\rho]$.

Proof. Consider the function $H(x) := v(x) - \Pi(x,\nu_v,p)$. It is harmonic.We also have by(2.8.2.5)

$$M(r,H)\le A(m)T(r,H)\le A(m)[T(r,v)+T(r,\Pi)]\le\sigma r^\rho$$

for some σ.

Hence, $H(x)\equiv 0$ by Theorem 3.1.4.3. □

Let us consider the case of integer ρ. Let $\nu \in \mathcal{M}[\rho]$ for an integer $\rho = p$. Set

$$\Pi_<(x,\nu,\rho) := \int_{|y|<1} G_{p-1}(x,y,m)\nu(dy), \tag{3.1.4.7}$$

$$\Pi_>(x,\nu,\rho) := \int_{|y|\geq 1} G_p(x,y,m)\nu(dy). \tag{3.1.4.8}$$

Both potentials converge and belong to $U[\rho]$.

Theorem 3.1.4.5 (Hadamard)** *Let ρ be integer and let $v \in U[\rho]$. Then*

$$v = H_\rho(x) + \Pi_<(x,\nu,\rho) + \Pi_>(x,\nu,\rho), \tag{3.1.4.9}$$

where H_ρ is a homogeneous harmonic polynomial of degree ρ.

The proof is exactly the same as in the *Hadamard theorem, but we use the second case of Theorem 3.1.4.3. We also note that the polynomial may be equal to zero identically.

Exercise 3.1.4.1 Check this in detail.

Let as check that ν from (3.1.4.9) has the following property that is analogous to Theorem 2.9.4.2.

Theorem 3.1.4.6 (*Lindelöf) *Let ρ be integer and let $v \in U[\rho]$. Then*

$$\lim_{\epsilon\to 0} \int_{\epsilon\leq|y|<1} D_\rho(x,y)\mu(dy) = H_\rho(x). \tag{3.1.4.10}$$

Proof. Consider the function

$$v_\epsilon^*(x) := v(x) + \begin{cases} \int\limits_{|y|<\epsilon} \frac{\nu(dy)}{|x-y|^{m-2}} \; for \; m>2; \\ -\int\limits_{|y|<\epsilon} \log|x-y|\nu(dy), \; for \; m=2. \end{cases} \tag{3.1.4.11}$$

It is subharmonic with supp $\nu \cap \{0\} = \varnothing$. We represent this function as in (2.9.4.10) in the form

$$v_\epsilon^*(x) = \Pi_<^1(x,\nu_\epsilon^*,\rho) + \Pi_>^1(x,\nu_\epsilon^*,\rho) + P_{\rho-1}^*(x,v_\epsilon^*) + \delta_1(x,v_\epsilon^*,\rho).$$

In this representation we can pass to the limit as $\epsilon \to 0$ in the left side and in all the summands except perhaps the last two from the right side.

Exercise 3.1.4.2 Check this, using that all the integrals converge for $\nu \in \mathcal{M}(\rho,\Delta)$ and showing that the integral in (3.1.4.11) tends to zero.

The last two summands form a harmonic polynomial, the limit of which is also a harmonic polynomial. Comparing the limit with the representation (3.1.4.9), we obtain that $P_{\rho-1}^*(\bullet,v_\epsilon^*)$ tends to zero and $\delta_1(x,v_\epsilon^*,\rho)$ tends to $H_\rho(x)$. $\square$

Theorem 3.1.4.7 (Liouville)** *If $v \in U[\rho]$ satisfies inequality $v(x) \leq 0$ for $z \in \mathbb{R}^m$, then $v(x) \equiv 0$.*

Otherwise it contradicts subharmonicity in 0.

3.1.5 Let us study the connection between $\mathbf{Fr}[u]$ and $\mathbf{Fr}[\mu_u]$.

Note the following properties of the transformations $(\bullet)_t$ and $(\bullet)_{[t]}$.

Theorem 3.1.5.0 (Connection between u_t and μ_t) *One has*

$$(\mu_u)_t = \mu_{u_t}; \ (\mu_v)_{[t]} = \mu_{v_{[t]}}. \tag{3.1.5.0}$$

Proof. By the F. Riesz theorem (Theorem 2.6.4.3) and Theorem 2.5.1.1, GPo3) we have for any $\psi \in \mathcal{D}(\mathbb{R}^m)$,

$$\langle \mu_u, \psi \rangle = \theta_m \langle \Delta u, \psi \rangle = \theta_m \langle u, \Delta \psi \rangle.$$

Using the definition (3.1.3.1), we obtain

$$\langle (\mu_u)_t, \psi \rangle = \langle (\mu_u)_t, \psi((P_t)^{-1}\bullet) \rangle t^{-\rho(t)-m+2}.$$

Thus

$$\langle (\mu_u)_t, \psi \rangle = \theta_m \langle u, \Delta[\psi((P_t)^{-1}\bullet)] \rangle t^{-\rho(t)-m+2}.$$

Since the Laplace operator is invariant with respect to V_t for any t we have

$$\Delta[\psi((P_t)^{-1}\bullet)] = t^{-2}[\Delta\psi]((P_t)^{-1}\bullet).$$

Thus we obtain

$$\begin{aligned}\langle (\mu_u)_t, \psi \rangle &= \theta_m t^{-\rho(t)-m} \langle u, [\Delta\psi]((P_t)^{-1}\bullet) \rangle \\ &= \theta_m \langle u(P_t\bullet) t^{-\rho(t)}, \Delta\psi \rangle = \theta_m \langle u_t, \Delta\psi \rangle = \langle \mu_{u_t}, \psi \rangle.\end{aligned}$$ □

Exercise 3.1.5.1 Do this for $(\bullet)_{[t]}$.

We begin from the case of a non-integer ρ.

Theorem 3.1.5.1 (Connection between Fr's for non-integer ρ) *Let $u \in U(\rho(r))$ and μ_u be its Riesz measure. Then*

$$\mathbf{Fr}[\mu_u] = \{\nu_v : v \in \mathbf{Fr}[u]\}, \tag{3.1.5.1}$$

$$\mathbf{Fr}[u] = \{\Pi(\bullet, \nu, p) : \nu \in \mathbf{Fr}[\mu_u]\}. \tag{3.1.5.2}$$

Proof. Let $\nu \in \mathbf{Fr}[\mu_u]$. There exists $t_n \to \infty$ such that $(\mu_u)_{t_n} \to \nu$ in $\mathcal{D}'$. We can find a subsequence t'_n such that $u_{t'_n} \to v \in \mathbf{Fr}[u]$. Thus $(\mu_u)_{t'_n} \to \nu_v$ and therefore

$\nu = \nu_v$. Hence, $\mathbf{Fr}[\mu_u] \subset \{\nu_v : v \in \mathbf{Fr}[u]\}$. Analogously we can prove that every $\nu_v \in \mathbf{Fr}[\mu_u]$ and hence (3.1.5.1) holds.

Let $\nu \in \mathbf{Fr}[\mu_u]$. We find a sequence $t_n \to \infty$ such that $(\mu_u)_{t_n} \to \nu$ in $\mathcal{D}'$. We find a subsequence t'_n such that $u_{t'_n} \to v \in \mathbf{Fr}[u]$ and $\nu_v = \nu$. By the *Hadamard theorem (Theorem 3.1.4.4) $v = \Pi(\bullet, \nu, p)$. Hence, $\{\Pi(\bullet, \nu, p) : \nu \in \mathbf{Fr}[\mu_u]\} \subset \mathbf{Fr}[u]$. And vice versa, since $\mathbf{Fr}[u] \subset U[\rho]$ (Theorem 3.1.2.2, fr2)), every $v \in \mathbf{Fr}[u]$ is represented as $\Pi(\bullet, \nu_v, p)$ and $\nu_v \in \mathbf{Fr}[\mu]$ by (3.1.5.1). □

Let ρ be integer and $u \in U(\rho(r))$. Let us consider the precompact family of homogeneous polynomials $\delta_t(x, u, \rho)t^{\rho-\rho(t)}$ from Theorem 2.9.4.1. For every $t_n \to \infty$ we can find a subsequence t'_n such that the pair $(\delta_{t'_n}(\bullet, u, \rho){t'_n}^{\rho-\rho(t'_n)}, (\mu_u)_{t'_n})$ tends to a pair (H_ν, ν) where H_ν is a homogeneous harmonic polynomial of degree p. We denote the set of all such pairs as $(\mathcal{H}, \mathbf{Fr})[u]$. Every $v \in U[\rho]$ can be represented in the form (3.1.4.7). Thus for every v the polynomial $H^v := H_p$ is determined.

Theorem 3.1.5.2 (Connection between Fr's for integer ρ) *Let $u \in U(\rho(r))$. Then*

$$(\mathcal{H}, \mathbf{Fr})[u] = \{(H^v, \nu_v) : v \in \mathbf{Fr}[u]\}, \tag{3.1.5.3}$$

$$\mathbf{Fr}[u] = \{v := H_\nu + \Pi_<(\bullet, \nu, \rho) + \Pi_>(\bullet, \nu, \rho) : (H_\nu, \nu) \in (\mathcal{H}, \mathbf{Fr})[u]\}. \tag{3.1.5.4}$$

The proof is clear.

3.1.6 Up to now we supposed that the family of rotations $V_\bullet$ was fixed. Now we take in consideration that it can vary and use the notation $\mathbf{Fr}[u, V_\bullet]$.

Theorem 3.1.6.1 (Dependence of Fr on $V_\bullet$) *Let $\mathbf{Fr}[u, V_\bullet]$ and $\mathbf{Fr}[u, W_\bullet]$ be limit sets of u with respect to rotation families $V_\bullet$ and $W_\bullet$ accordingly. Then for any $v \in \mathbf{Fr}[u, V_\bullet]$ there exist a rotation V^v and $w^v \in \mathbf{Fr}[u, W_\bullet]$ such that*

$$v(x) = w^v(V^v x)$$

for all $x \in \mathbb{R}^m$.

Proof. Let $v \in \mathbf{Fr}[u, V_\bullet]$ and let $t_n \to \infty$ be a sequence such that

$$t_n^{-\rho(t_n)} u(t_n V_{t_n} \bullet) \to v.$$

Since the family V_t is obviously precompact there exists a subsequence (for which we keep the same notation), and a rotation V^v such that $W_{t_n}^{-1} V_{t_n} \to V^v$ and $w \in \mathbf{Fr}[u, W_\bullet]$ such that $t_n^{-\rho(t_n)} u(t_n W_{t_n} \bullet) \to w$.

Now we have

$$\begin{aligned} v(\bullet) &= \mathcal{D}' - \lim t_n^{-\rho(t_n)} u(t_n V_{t_n} \bullet) \\ &= \mathcal{D}' - \lim t_n^{-\rho(t_n)} u(t_n W_{t_n} W_{t_n}^{-1} V_{t_n} \bullet) \\ &= w(V^v \bullet). \end{aligned}$$

□

3.2 Indicators

3.2.1 Let $u \in SH(\rho(r))$ and let $\mathbf{Fr}[u]$ be the limit set. Set

$$h(x,u) := \sup\{v(x) : v \in \mathbf{Fr}[u]\}, \tag{3.2.1.1}$$
$$\underline{h}(x,u) := \inf\{v(x) : v \in \mathbf{Fr}[u]\}. \tag{3.2.1.2}$$

These functionals reflect the asymptotic behavior of u along curves of the form

$$l_{\boldsymbol{x}^0} := \{x = tV_t\boldsymbol{x}^0 : t \in (0,\infty)\} \tag{3.2.1.3}$$

and are called *indicator* of growth of u and *lower indicator* respectively.

Of course, the indicators depend on $\rho(r)$ and V_t, but we will only note that if necessary.

Theorem 3.2.1.1 (Properties of Indicators) *The following holds:*

h1) *$\underline{h}$ is upper semicontinuous, h is subharmonic;*

h2) *they are semiadditive and positively homogeneous, i.e.,*

$$h(x,u_1+u_2) \le h(x,u_1) + h(x,u_2); \tag{3.2.1.4}$$
$$\underline{h}(x,u_1+u_2) \ge \underline{h}(x,u_1) + \underline{h}(x,u_2); \tag{3.2.1.5}$$
$$h,\underline{h}(x,Cu) = Ch,\underline{h}(x,u),\ C \ge 0; \tag{3.2.1.6}$$

h3) *invariance:*

$$h,\underline{h}_{[t]}(x,u) = h,\underline{h}(x,u). \tag{3.2.1.7}$$

Proof. Semicontinuity of $\underline{h}$ follows from Theorem 2.1.2.8 (Commutativity of inf and M(.). Semicontinuity and subharmonicity of h follow from Theorem 2.7.3.4 (Sigurdsson's Lemma). The properties h2) follow from properties of infimum and supremum. The invariance follows from invariance of $\mathbf{Fr}[u]$ (Theorem 3.1.2.2, fr3)). □

Set

$$x^0(x) := P_{|x|}^{-1}(x) \tag{3.2.1.8}$$

where P_t is defined by (3.1.1.1).

This is an intersection of the orbit of P_t that passes through a point x with the unit sphere.

If $V_t \equiv I$,

$$x^0(x) = x/|x| := x^0. \tag{3.2.1.9}$$

Theorem 3.2.1.2 (Homogeneity $h, \underline{h}$**)** *One has*

$$h, \underline{h}(x, \bullet) = |x|^{\rho} h, \underline{h}(x^0(x), \bullet) \tag{3.2.1.10}$$

Thus the indicators are determined uniquely by their values on the unit sphere, i.e., they are "functions of direction". In particular, they are homogeneous for $V_t \equiv I$:

$$h, \underline{h}(x, \bullet) = |x|^{\rho} h, \underline{h}(x^0, \bullet). \tag{3.2.1.11}$$

The proof of (3.2.1.10) follows from h3), Theorem 3.2.1.1 if we set $t := |x|$; $x := P_t^{-1} x$.

3.2.2 In this item we will suppose that $V_t \equiv I$ and study the indicator.

Let $\Delta_{\boldsymbol{x}^0}$ be as defined in Section 2.4.1. Its coefficients depend on a choice of the spherical coordinate system. However, one has

Theorem 3.2.2.1 *Let $\psi(y)$ have continuous second derivatives on the unit sphere S_1. Then the differential form $\Delta_{\boldsymbol{x}^0}\psi(y)dy$ is invariant with respect to the choice of spherical coordinate system.*

Proof. Let $\phi(x)$ be a smooth function in $\mathbb{R}^m$. Then $\Delta\phi(x)dx$ is invariant with respect to the choice of an orthogonal system because Δ (the Laplace operator) and an element of volume are invariant. Set $\phi(x) = \psi(y)$, where $y := x^0 = x/|x|$. Then

$$\Delta\phi dx = \Delta_{\boldsymbol{x}^0}\psi(y)dy\ r^{m-3}dr.$$

Since r is invariant with respect to rotations of the coordinate system, $\Delta_{\boldsymbol{x}^0}\psi(y)dy$ is invariant with respect to the choice of a spherical coordinate system. □

Note that for $m = 2$ this theorem is obvious because

$$\Delta_{\boldsymbol{x}^0} = \frac{d^2}{d\theta^2} \tag{3.2.2.1}$$

and it does not depend on translations with respect to θ.

We define the operator $\Delta_{\boldsymbol{x}^0}$ on $f \in \mathcal{D}'(S_1)$ by

$$\langle \Delta_{\boldsymbol{x}^0} f, \psi \rangle := \langle f, \Delta_{\boldsymbol{x}^0}\psi \rangle, \ \psi \in \mathcal{D}(S_1)$$

in a fixed spherical coordinate system.

The definition is correct. Indeed, suppose in a fixed system

$$\operatorname{supp}\psi \subset S_1 \setminus \{\theta_j = 0; \pi : j = 1, 2, \ldots, m-2\}. \tag{3.2.2.2}$$

Then all the coefficients of $\Delta_{\boldsymbol{x}^0}$ are infinitely differentiable and $\Delta_{\boldsymbol{x}^0}\psi(y) \in \mathcal{D}(S_1)$. By Theorem 3.2.2.1 we obtain that the condition of Theorem 2.3.5.2 ($\mathcal{D}'$ on Sphere) are fulfilled.

Note that for $m = 2$ the operator $\Delta_{\boldsymbol{x}^0}$ is realized by the formula (3.2.2.1) on functions of the form $f = f(e^{i\theta})$, i.e., on 2π-periodic functions.

Theorem 3.2.2.2 (Subsphericality of Indicator) *One has*

$$[\Delta_{\boldsymbol{x}^0} + \rho(\rho + m - 2)]h(y, u) := s > 0 \tag{3.2.2.3}$$

in $\mathcal{D}'(S_1)$.

I.e., s is a measure on S_1.

Proof. It is sufficient to prove this locally, in any spherical system. Let $R(r)$ be finite, infinitely differentiable and nonnegative in $(0; \infty)$ and let $\psi \in \mathcal{D}(S_1)$ be nonnegative and satisfy (3.2.2.2). Set $\phi(x) := R(|x|)\psi(x^0)$. Using the subharmonicity of $h(x, u)$ (h1), Theorem 3.2.1.1 and (3.2.2.2), we have

$$\begin{aligned} 0 &\leq \int h(x,u)\Delta\phi(x)dx \\ &= \int_{(y,r)\in S_1\times(0;\infty)} r^\rho h(y,u)\left[\frac{1}{r^{m-1}}\frac{\partial}{\partial r}r^{m-1}\frac{\partial}{\partial r} + \frac{1}{r^2}\Delta_{\boldsymbol{x}^0}\right]\psi(y)r^{m-1}dydr. \end{aligned}$$

Transforming the last integral we obtain

$$\begin{aligned} \int h(x,u)\Delta\phi(x)dx &= \int_0^\infty r^\rho\left[\frac{1}{r^{m-1}}\frac{\partial}{\partial r}r^{m-1}\frac{\partial}{\partial r}R(r)\right]r^{m-1}dr\int_{S_1} h(y,u)\psi(y)dy \\ &+ \int_0^\infty r^{\rho-2}r^{m-1}R(r)dr\int_{S_1} h(y,u)\Delta_{\boldsymbol{x}^0}\psi(y)dy. \end{aligned} \tag{3.2.2.4}$$

Integrating by parts in the first summand we obtain

$$\int_0^\infty r^\rho\left[\frac{1}{r^{m-1}}\frac{\partial}{\partial r}r^{m-1}\frac{\partial}{\partial r}R(r)\right]r^{m-1}dr = \int_0^\infty R(r)\rho(\rho+m-2)r^{\rho+m-3}dr. \tag{3.2.2.5}$$

Substituting (3.2.2.5) into (3.2.2.4), we have

$$0 \leq \int_0^\infty r^{\rho+m-3}R(r)dr\int_{S_1} h(y,u)[\Delta_{\boldsymbol{x}^0} + \rho(\rho+m-2)]\psi(y)dy.$$

Since $R(r)$ is an arbitrarily nonnegative function,

$$\int_{S_1} h(y,u)[\Delta_{\boldsymbol{x}^0} + \rho(\rho+m-2)]\psi(y)dy \geq 0$$

for arbitrary ψ. □

We will call an upper semicontinuous function which satisfies (3.2.2.3) a ρ-*subspherical* one. Now we are going to study properties of these functions.

3.2.3 We consider the case $m = 2$. A ρ-subspherical function for $m = 2$ is called ρ-*trigonometrically convex* (ρ-t.c.). We will obtain for such a function a representation like in Theorems 3.1.4.4, 3.1.4.5.(*,** Hadamard). Set

$$T_\rho := \frac{d^2}{d\phi^2} + \rho^2.$$

Let us find a fundamental solution of this operator. Let ρ be non-integer. Let us denote as $\widetilde{\cos\rho}(\phi)$ the periodic continuation of $\cos\rho\phi$ from the interval $(-\pi,\pi)$.

Theorem 3.2.3.1 (Fundamental Solution of T_ρ) *One has*

$$\frac{1}{2\rho\sin\pi\rho}T_\rho\widetilde{\cos\rho}(\phi-\pi)=\delta(\phi) \text{ in } \mathcal{D}'(S_1)\ .$$

Proof. Let $f\in\mathcal{D}(S_1)$. We have

$$\int_0^{2\pi}\widetilde{\cos\rho}(\phi-\pi)[f''+\rho^2 f]d\phi=\lim_{\epsilon\to 0}\int_\epsilon^{2\pi-\epsilon}\cos\rho(\phi-\pi)[f''+\rho^2 f]d\phi. \quad (3.2.3.1)$$

Integrating by parts we obtain

$$\int_\epsilon^{2\pi-\epsilon}\cos\rho(\phi-\pi)[f''+\rho^2 f]d\phi$$

$$=\cos\rho(\phi-\pi)\ f'(\phi)|_\epsilon^{2\pi-\epsilon}+\rho\sin\rho(\phi-\pi)\ f(\phi)|_\epsilon^{2\pi-\epsilon}+\int_\epsilon^{2\pi-\epsilon}f(\phi)T_\rho\cos\rho(\phi-\pi)d\phi.$$

However, $T_\rho\cos\rho(\phi-\pi)=0$ for $\phi\in(\epsilon,2\pi-\epsilon)$. Thus the limit in (3.2.3.1) is equal to $f(0)2\rho\sin\pi\rho$. □

Let s be a measure on the circle S_1. Set

$$\Pi(\phi,s):=\int_0^{2\pi}\widetilde{\cos\rho}(\phi-\psi-\pi)s(d\psi).$$

Theorem 3.2.3.2 *One has*

$$T_\rho\Pi(\phi,s)=(2\rho\sin\pi\rho)\ s(\bullet) \text{ in } \mathcal{D}'(S_1).$$

The proof is the same as GPo3) in Theorem 2.5.1.1.

Theorem 3.2.3.3 (Representation of ρ-t.c.f for a non-integer ρ) *Let h be ρ-t.c. on S_1 for non-integer ρ and let $s:=T_\rho h$. Then*

$$h(\phi)=\frac{1}{2\rho\sin\pi\rho}\Pi(\phi,s).$$

The proof is like in Theorem 3.1.4.4 (*Hadamard).

3.2.4 We will suppose in this item that $V_t=I, m=2, \rho$ is integer.

Theorem 3.2.4.1 (Condition on s) *Let ρ be integer, h be ρ-t.c. and $T_\rho h=s$. Then*

$$\int_0^{2\pi}e^{i\rho\phi}s(d\phi)=0. \quad (3.2.4.1)$$

Proof. We have for $f \in \mathcal{D}(S_1)$:

$$\langle s, f\rangle = \langle T_\rho h, f\rangle = \langle h, T_\rho f\rangle.$$

Since $e^{i\rho\phi} \in \mathcal{D}(S_1)$ for integer ρ and $T_\rho e^{i\rho\phi} = 0$, we have for $f := e^{i\rho\phi}$,

$$\langle s, e^{i\rho\bullet}\rangle = \langle h, T_\rho e^{i\rho\bullet}\rangle = 0. \qquad \square$$

Let us denote the periodic continuation of the function $f(\phi) := \phi$ from the interval $[0, 2\pi)$ to $(-\infty, \infty)$ as $\tilde{\phi}$.

Theorem 3.2.4.2 (Generalized Fundamental Solution for T_ρ) *One has*

$$T_\rho\left[-\frac{1}{2\pi\rho}\tilde{\phi}\sin\rho\phi\right] = \delta(\phi) - \frac{1}{\pi}\cos\rho\phi$$

in $\mathcal{D}'(S_1)$.

Proof. Let $\phi \in (\epsilon, 2\pi - \epsilon)$. Then

$$T_\rho\tilde{\phi}\sin\rho\phi = 2\rho\cos\rho\phi$$

because $\tilde{\phi} = \phi$ when $\phi \in (\epsilon, 2\pi - \epsilon)$. We have also

$$(\phi\sin\rho\phi)' = \sin\rho\phi + \phi\rho\cos\rho\phi.$$

Thus

$$\langle T_\rho\tilde{\bullet}\sin\rho\bullet, f\rangle = \int_0^{2\pi}\tilde{\phi}\sin\rho\phi\; T_\rho f d\phi = \lim_{\epsilon\to 0}\int_\epsilon^{2\pi-\epsilon}\phi\sin\rho\phi\; T_\rho f d\phi.$$

Integrating by parts we obtain

$$\int_\epsilon^{2\pi-\epsilon}\phi\sin\rho\phi\; T_\rho f d\phi = \phi\sin\rho\phi f'(\phi)|_\epsilon^{2\pi-\epsilon} - f(\phi)[\sin\rho\phi + \phi\rho\cos\rho\phi|_\epsilon^{2\pi-\epsilon} + \int_\epsilon^{2\pi-\epsilon} T_\rho[\phi\sin\rho\phi]f(\phi)d\phi.$$

Passing to the limit as $\epsilon \to 0$ and taking in account that f is periodic and continuous we obtain

$$\begin{aligned}\langle T_\rho[\tilde{\bullet}\sin\rho\bullet], f\rangle &= -2\pi\rho f(0) + 2\rho\int_0^{2\pi}\cos\rho\phi f(\phi)d\phi\\ &= -2\pi\rho f(0) + 2\rho\langle\cos\rho\bullet, f\rangle.\end{aligned} \qquad \square$$

Set

$$\hat{\Pi}(\phi, s) := \int_0^{2\pi}\widetilde{(\phi - \psi)}\sin\rho(\phi - \psi)s(d\psi).$$

Theorem 3.2.4.3 *One has*

$$T_\rho \hat{\Pi}(\bullet, s) = -2\pi\rho\ s \text{ in } \mathcal{D}'(S_1)$$

for s that satisfies (3.2.4.1).

Proof. Using Theorem 3.2.4.2 we obtain

$$\langle T_\rho \hat{\Pi}(\bullet, s), f\rangle = \langle s, f\rangle - \frac{1}{\pi}\left\langle \int_0^{2\pi} \cos\rho(\bullet - \psi)s(d\psi), f\right\rangle.$$

The last integral is zero because of Theorem 3.2.4.1. □

Theorem 3.2.4.4 (Representation of ρ-t.c.f. for an integer ρ) *Let h be a ρ-t.c.f.for an integer ρ and $T_\rho h := s$. Then*

$$h(\phi) = \Re c e^{i\phi} + \hat{\Pi}(\phi, s)$$

for some complex constant c.

Proof. The function $H(\phi) := h(\phi) - \hat{\Pi}(\phi, s)$ satisfies the equation $T_\rho H = 0$ in $\mathcal{D}'(S_1)$ because of Theorem 3.2.4.3 and it is real. Thus $H(\phi) = \Re c e^{i\phi}$. □

3.2.5 The class TC_ρ of ρ-t.c.functions has a number of properties of subharmonic functions.

The function $\widetilde{\cos\rho\phi}$ is continuous and $\widetilde{\phi}\sin\rho\phi$ is continuous for integer ρ. Therefore any ρ-t.c.f is continuous as follows from Theorem 3.2.3.3 and 3.2.4.4.

Set

$$\mathcal{E}(\phi) := \frac{1}{2\rho}\sin\rho|\phi|.$$

For any interval $I := (\alpha, \beta) \Subset (-\pi, \pi)$ this function satisfies the equality

$$T_\rho \mathcal{E} = \delta$$

in $\mathcal{D}'(\alpha, \beta)$, where δ is the Dirac function in zero.

Let $G_I(\psi, \phi)$ be the Green function of T_ρ for the interval I. By definition it must be symmetric with respect to ϕ, ψ and have the form

$$G_I(\phi, \psi) := \frac{1}{2\rho}\sin\rho|\phi - \psi| + A_I \cos\rho\phi\cos\rho\psi + B_I \sin\rho\phi\sin\rho\psi, \tag{3.2.5.1}$$

where A_I, B_I are chosen such that $G_I(\phi, \psi)$ is equal to zero on $\partial\{I \times I\}$. An explicit form of G_I is given by

$$G_I(\phi, \psi) = \begin{cases} \frac{\sin\rho(\beta-\phi)\sin\rho(\psi-\alpha)}{\rho\sin\rho(\beta-\alpha)}, & \text{for } \psi < \phi; \\ \frac{\sin\rho(\beta-\psi)\sin\rho(\phi-\alpha)}{\rho\sin\rho(\beta-\alpha)}, & \text{for } \phi < \psi. \end{cases}$$

The following assertion is analogous to the Riesz theorem (Theorem 2.6.4.3):

Theorem 3.2.5.1 (Representation on I) *Let $h \in TC_\rho$ and let I be an interval of length* $\operatorname{mes} I < \pi/\rho$. *Then*

$$h(\phi) = Y_\rho(\phi, h) - \int_\alpha^\beta G_I(\phi, \psi) s(d\psi),$$

where $Y_\rho(\phi, h)$ is the only solution of the boundary problem:

$$T_\rho Y = 0, \; Y(\alpha) = h(\alpha), \; Y(\beta) = h(\beta) \tag{3.2.5.2}$$

and $s := T_\rho h$.

Proof. Set

$$\Pi_I(\phi, s) := \int_\alpha^\beta G_I(\phi, \psi) s(d\psi).$$

One can check as in Theorem 3.2.3.2 that $T_\rho \Pi_I = -s$ in $\mathcal{D}'(I)$. Then the function

$$Y_\rho(\phi) := h(\phi) + \Pi_I(\phi, s)$$

satisfies the conditions (3.2.5.2). □

The explicit form of $Y_\rho(\phi)$ is

$$Y_\rho(\phi) = \frac{h(\alpha)\sin\rho(\beta - \phi) + h(\beta)\sin\rho(\phi - \alpha)}{\sin\rho(\beta - \phi)}. \tag{3.2.5.3}$$

Since $\Pi_I(\phi) \geq 0$ we have

Theorem 3.2.5.2 (ρ-Trigonometric Majorant) *Suppose $h \in TC_\rho$ and $Y_\rho(\phi)$ is the solution of* (3.2.5.2). *Then*

$$h(\phi) \leq Y_\rho(\phi), \; \phi \in I$$

if $\beta - \alpha < \pi/\rho$.

This inequality can be written in the symmetric form

$$h(\alpha)\sin\rho(\beta - \phi) + h(\phi)\sin\rho(\alpha - \beta) + h(\beta)\sin\rho(\phi - \alpha) \geq 0 \tag{3.2.5.4}$$

for $\max(\alpha, \phi, \beta) - \min(\alpha, \phi, \beta) < \pi/\rho$. It is called the *fundamental relation of indicator.*

Theorem 3.2.5.3 (Subharmonicity and ρ-t.c.) *A function $h(\phi) \in TC_\rho$ iff the function $u(re^{i\phi}) := h(\phi)r^\rho$ is subharmonic in $\mathbb{R}^2$.*

Proof. Sufficiency follows from Theorem 3.2.2.2. Let us prove necessity. The function $u_1(z) := r^\rho \sin\rho|\phi|$ is subharmonic. Actually, it is harmonic for $\phi \neq 0, \; r \neq 0$ and can be represented in the form

$$u_1(z) = \max(r^\rho \sin\phi, -r^\rho \sin\phi)$$

in a neighborhood of the line $\phi = 0$. Hence, it is subharmonic because of sh2), Theorem 2.6.1.1 (Elementary Properties).

The function $u_2(z) := r^\rho \Pi_I(\phi, s)$ is subharmonic because of sh5) and sh4), Theorem 2.6.1.1. The function $r^\rho Y_\rho(\phi)$ is harmonic for $r > 0$. This can be checked directly. Hence, $u(z)$ is subharmonic for $r > 0$ because of Theorem 3.2.5.1. By Theorem 2.6.2.2 $u(z)$ is also subharmonic for $r = 0$, because it is, obviously, continuous at $z = 0$. □

Theorem 3.2.5.4 (Elementary Properties of ρ-t.c.Functions) *One has*

tc1) *If $h \in TC_\rho$, then $Ah \in TC_\rho$ for $A > 0$;*

tc2) *If $h_1, h_2 \in TC_\rho$, then $h_1 + h_2,\ \max(h_1, h_2) \in TC_\rho$.*

These properties follow from Theorem 3.2.5.3 and properties of subharmonic functions.

Exercise 3.2.5.1 Prove Theorem 3.2.5.4.

Similarly to (usual) convexity, ρ-t.convexity of functions implies several analytic properties.

Theorem 3.2.5.5 *Let $h \in TC_\rho$; then there exist right (h'_+) and left (h'_-) derivatives and they coincide everywhere except, maybe, for a countable set of points.*

Proof. It is enough to prove these properties for the potential

$$\Pi(\phi) := \int_\alpha^\beta \sin \rho|\phi - \psi| s(d\psi),$$

because of (3.2.5.1) and Theorem 3.2.5.1.

We will prove that

$$\Pi'_+(\phi) = \rho \int_\alpha^{\phi-0} \cos \rho(\phi - \psi) s(d\psi) + \rho\mu(\phi) - \rho \int_{\phi+0}^{\beta} \cos \rho(\phi - \psi) s(d\psi); \quad (3.2.5.5)$$

$$\Pi'_-(\phi) = \rho \int_\alpha^{\phi-0} \cos \rho(\phi - \psi) s(d\psi) - \rho\mu(\phi) - \rho \int_{\phi+0}^{\beta} \cos \rho(\phi - \psi) s(d\psi), \quad (3.2.5.6)$$

where $\mu(\phi)$ is the measure, concentrated in the point ϕ.

We have for $\Delta > 0$:

$$\frac{\Pi(\phi + \Delta) - \Pi(\phi)}{\Delta} = \int_\alpha^{\phi-0} \frac{\sin \rho|\phi + \Delta - \psi| - \sin \rho|\phi - \psi|}{\Delta} s(d\psi)$$

$$+ \frac{\sin \rho\Delta}{\Delta}\mu(\phi) + \int_{\phi+0}^{\phi+\Delta} \cdots + \int_{\phi+\Delta}^{\beta} \cdots .$$

Let us estimate the second integral.We have

$$\int_{\phi+0}^{\phi+\Delta} \left| \frac{\sin\rho|\phi+\Delta-\psi| - \sin\rho|\phi-\psi|}{\Delta} \right| s(d\psi) \le \frac{2\sin\rho\Delta}{\Delta}[s(\phi+\Delta)-s(\phi+0)] = o(1)$$

when $\Delta \to +0$.

Passing to the limit, we obtain (3.2.5.5). The equality (3.2.5.6) is obtained in the same way when $\Delta < 0$.

Since $\mu(\phi) \neq 0$ at most in a countable set, for all the other points $\Pi'_+(\phi) = \Pi'_-(\phi)$. □

3.2.6 Now we consider the case $m \geq 3$. We will obtain for the ρ-subspherical function a representation like for the ρ-trigonometrically convex functions.

Theorem 3.2.6.1 (Subharmonicity and Subsphericality) *Let h be subspherical in a neighborhood of $y \in S_1$. Then the function $u(x) := h(y)r^\rho$, $x = ry$ is subharmonic in the corresponding neighborhood of the ray $x = ry : 0 < r < \infty$.*

Proof. Let $f \in \mathcal{D}'(\mathbb{R}^m \setminus 0)$. We can represent it in the form $f := f(rx)$, $x \in S_1$ where $f(\bullet x) \in \mathcal{D}'(0,\infty)$ for any x.

Then

$$\begin{aligned}\langle u, \Delta f\rangle &= \int_0^\infty \int_{S_1} u(rx)\Delta f(rx) r^{m-1} dr dx \\ &= \int_0^\infty \int_{S_1} u(rx)\frac{1}{r^{m-1}}\frac{\partial}{\partial r} r^{m-1}\frac{\partial}{\partial r} f(rx) r^{m-1} dr dx \\ &\quad + \int_0^\infty \int_{S_1} u(rx)\frac{1}{r^2}\Delta_{x^0} f(rx) r^{m-1} dr dx.\end{aligned}$$

Integrating by parts in the first integral, we obtain

$$\int_0^\infty \rho(\rho+m-2) r^{\rho+m-3} \int_{S_1} h(x) f(rx) dr dx.$$

Set

$$\mathcal{S}_\rho := \Delta_{\boldsymbol{x}^0} + \rho(\rho+m-2). \tag{3.2.6.1}$$

Together with the second summand we obtain

$$\langle u, f\rangle = \int_0^\infty \left[\int_{S_1} h(x)\mathcal{S}_\rho f(rx) dx\right] r^{\rho+m-3} dr > 0$$

if $f(rx) \geq 0$. □

Note that the Riesz measure for such u has the form

$$\mu(r^{m-1}drdx) = r^{\rho+m-3}dr\nu_h(dx),$$

where ν_h is a positive measure on S_1, that is equal to $\mathcal{S}_\rho h$ in $\mathcal{D}'(S_1)$.

For a non-integer ρ set

$$\mathcal{E}_\rho(x,y) := \int_0^\infty G_p(x,ry,m)r^{\rho+m-3}dr,$$

where G_p is the primary kernel and $x,y \in S_1$.

Theorem 3.2.6.2 *For non-integer ρ and any ρ-subspherical function h one has*

$$h(x) = \int_{S_1} \mathcal{E}_\rho(x,y)\nu_h(dy).$$

Proof. Set in (3.1.4.6) $v := r^\rho h(x)$. It is clear that $v \in U[\rho]$. We have

$$r^\rho h(x) = \int_{S_1}\int_0^\infty G_p(rx,ty,m)t^{\rho+m-3}dt\nu_h(dx).$$

Now we make the change $t' := t/r$ and use the homogeneity of $G_p(rx,ty,m)$. □

Exercise 3.2.6.1 Show that $\mathcal{E}_\rho(x,y)$ is a fundamental solution of the operator $\mathcal{S}_\rho$.

For an integer $\rho = p$ set

$$\mathcal{E}'_\rho(x,y) := \int_0^1 G_{p-1}(x,ry)r^{\rho+m-3}dr + \int_1^\infty G_p(x,ry)r^{\rho+m-3}dr.$$

Exercise 3.2.6.2 Prove the next

Theorem 3.2.6.3 *For any integer $\rho = p$ and any ρ-subspherical function h one has*

$$h(x) = Y_p(x) + \int_{S_1} \mathcal{E}'_\rho(x,y)\nu_h(dy)$$

where Y_p is some p-spherical function.

For any p-spherical function Y,

$$\int_{S_1} Y(x)\nu_h(dy) = 0.$$

3.2.7 We return to the general case when $x \in \mathbb{R}^m$, V_t is a one-parametric group, $\rho(r)$ is a proximate order and $u \in SH(\rho(r))$. The following theorem represents indicators in a form of limits in the usual topology.

Theorem 3.2.7.1 (Classic Indicators) *One has*

$$h(x,u) = \sup_T [\limsup_{t_j\to\infty} u_{t_j}]^*(x) = [\limsup_{t\to\infty} u_t(x)]^* \tag{3.2.7.1}$$

where $*$ *can be deleted outside a set of zero capacity, and*

$$\underline{h}(x,u) = \inf_T [\limsup_{t_j\to\infty} u_{t_j}]^*(x), \tag{3.2.7.2}$$

where T *is the set of all the sequences that tend to infinity.*

Proof. Let us prove (3.2.7.1). Set

$$h(x,u,\{t_j\}) := \limsup_{t_j\to\infty} u_{t_j}(x). \tag{3.2.7.3}$$

Let $v \in \mathbf{Fr}[u]$ and $u_{t_j} \to v$ in $\mathcal{D}'$. Then

$$h^*(x,u,\{t_j\}) = v(x) \tag{3.2.7.4}$$

by Theorem 2.7.3.3. (H. Cartan+). Thus

$$\sup_T h^*(x,u,\{t_j\}) \geq h(x,u). \tag{3.2.7.5}$$

Let $\epsilon > 0$ be arbitrarily small, and $t_j := t_j(x)$ be a sequence such that

$$h^*(x,u,\{t_j\}) \geq \sup_T h^*(x,u,\{t_j\}) - \epsilon.$$

We can find a subsequence $\{t_j\}$ (we keep the same notation for it) and $v \in \mathbf{Fr}[u]$ such that $u_{t_j} \to v$ in $\mathcal{D}'$. From (3.2.7.4) we obtain

$$h(x,u) \geq v(x) \geq \sup_T h^*(x,u,\{t_j\}) - \epsilon.$$

Thus the reverse inequality to (3.2.7.5) holds. Therefore

$$h(x,u) = \sup_T h^*(x,u,\{t_j\}).$$

Let us prove the second equality in (3.2.7.1). Since

$$\sup_T h(x,u,\{t_j\}) = \limsup_{t\to\infty} u_t(x)$$

we have

$$h(x,u) \geq [\limsup_{t\to\infty} u_t]^*(x). \tag{3.2.7.6}$$

Let us prove the opposite inequality. Let $v \in \mathbf{Fr}[u]$. There exists a sequence $t_j \to \infty$ such that $u_{t_j} \to v$ in $\mathcal{D}'(\mathbb{R}^m)$. By (3.2.7.4)

$$[\limsup_{t\to\infty} u_t]^* \geq h^*(x,u,\{t_j\}) = v(x).$$

Since it holds for every $v \in \mathbf{Fr}[u]$ we have the reverse inequality to (3.2.7.6). Hence, (3.2.7.1) is proved completely.

Let us prove (3.2.7.2). From (3.2.7.4) we have

$$\inf_T h^*(x, u, \{t_j\}) \leq v(x)$$

for all $v \in \mathbf{Fr}[u]$. Therefore

$$\inf_T h^*(x, u, \{t_j\}) \leq \underline{h}(x, u). \tag{3.2.7.7}$$

Let us prove the opposite inequality. Let $\{t_j\}$ be any sequence that tends to ∞. Let us find a subsequence $\{t_{j'}\}$ such that $u_{t_{j'}} \to v$ in $\mathcal{D}'(\mathbb{R}^m)$. Then

$$h(x, u, \{t_j\}) \geq \limsup_{j' \to \infty} u_{t_{j'}}(x).$$

Taking $*$ from the two sides of this inequality and using Theorem 2.7.3.3, we obtain

$$h^*(x, u, \{t_j\}) \geq [\limsup_{j' \to \infty} u_{t_{j'}}]^*(x) = v(x) \geq \underline{h}(x, u).$$

This implies the reverse inequality to (3.2.7.7). Hence (3.2.7.2) holds. □

Corollary 3.2.7.2 *If all the functions* (3.2.7.3) *are upper semicontinuous, then*

$$h(x, u) = \limsup_{t \to \infty} u_t(x), \qquad \underline{h}(x, u) = \liminf_{t \to \infty} u_t(x).$$

Proof. We have $h^*(x, u, \{t_j\}) = h(x, u, \{t_j\})$ and thus

$$h(x, u) = \sup_T [\limsup_{t_j \to \infty} u_{t_j}](x) = \liminf_{t \to \infty} u_t(x),$$
$$\underline{h}(x, u) = \inf_T [\limsup_{t_j \to \infty} u_{t_j}](x) = \liminf_{t \to \infty} u_t(x).$$ □

Theorem 3.2.7.3 (Indicators of Harmonic Function) *Let $u \in SH(\rho(r))$ be harmonic for all the large $|y|$ in a "cone" of the form*

$$Co_\Omega := \{y = P_t x : x \in \Omega,\ t \in (0; \infty)\}$$

where $\Omega \subset S_1$. Then

$$h(x, u) = \limsup_{t \to \infty} u_t(x) \tag{3.2.7.8}$$

and

$$\underline{h}(x, u) = \liminf_{t \to \infty} u_t(x) \tag{3.2.7.9}$$

for $x \in Co_\Omega$.

Proof. The harmonicity of u in Co_Ω implies $[u_t]_\epsilon(x) = u_t(x)$ for large t and sufficiently small ϵ when $x \in Co_\Omega$.

The family $[u_t]_\epsilon$ is uniformly continuous by reg3), Theorem 2.3.4.5 (Properties of Regularizations). Thus the function (3.2.7.5) is continuous. Therefore we can use Corollary 3.2.7.2. □

Theorem 3.2.7.4 (Indicator for $m = 2$) *Let $u \in SH(\mathbb{R}^2)$. Then*

$$h(z, u) = \limsup_{t\to\infty} u_t(x). \tag{3.2.7.10}$$

I.e., the star in (3.2.7.1) can be deleted.

Proof. Let as denote as $h_1(z, u)$ the right part of (3.2.7.10). The "homogeneity" of the indicator (3.2.1.10) and also of $h_1(z, u)$ implies the following property: if the inequality $h_1(z, u) < h(z.u)$ holds for some z_0, it holds on the whole orbit

$$z = \{P_t z_0 : 0 < t < \infty\}$$

that has a positive capacity in $\mathbb{R}^2$. This contradicts Theorem 3.2.7.1. □

3.3 Densities

3.3.1 In the sequel G is an open set, K is a compact set and E a bounded Borel set. Let $\mu \in \mathcal{M}(\rho(r))$ and $\mathbf{Fr}[\mu] := \mathbf{Fr}[\mu, \rho(r), V_t, \mathbb{R}^m]$ be the limit set of μ. Set

$$\begin{aligned}
\overline{\Delta}(G, \mu) &:= \sup\{\nu(G) : \nu \in \mathbf{Fr}[\mu]\};\\
\overline{\Delta}(E, \mu) &:= \inf\{\overline{\Delta}(G, \mu) : G \supset E\};\\
\underline{\Delta}(K, \mu) &:= \inf\{\nu(K) : \nu \in \mathbf{Fr}[\mu]\};\\
\underline{\Delta}(E, \mu) &:= \sup\{\underline{\Delta}(K) : K \subset E\}.
\end{aligned}$$

The quality $\overline{\Delta}(E, \mu)$, ($\underline{\Delta}(E, \mu)$) is called the *upper* (*lower*) *density* of μ relative to the proximate order $\rho(r)$ and the family V_t.

Theorem 3.3.1.1 (Properties of Densities) *The following properties hold:*

dens1) *if $E = \varnothing$, then $\overline{\Delta}(E, \bullet) = \underline{\Delta}(E, \bullet) = 0$;*

dens2) $\forall E,\ \underline{\Delta}(E, \bullet) \le \overline{\Delta}(E, \bullet)$;

dens3) *monotonicity: $\underline{\Delta}, \overline{\Delta}(E_1, \bullet) \le \underline{\Delta}, \overline{\Delta}(E_2, \bullet)$ for $E_1 \subset E_2$;*

dens4) *generalized semi-additivity*[1] *with respect to a set:*

$$\overline{\Delta}(E_1 \cup E_2, \bullet) + \underline{\Delta}(E_1 \cap E_2, \bullet) \le \overline{\Delta}(E_1, \bullet) + \overline{\Delta}(E_2, \bullet), \tag{3.3.1.1}$$

$$\underline{\Delta}(E_1 \cup E_2, \bullet) + \overline{\Delta}(E_1 \cap E_2, \bullet) \ge \underline{\Delta}(E_1, \bullet) + \underline{\Delta}(E_2, \bullet); \tag{3.3.1.2}$$

[1]See Exercise 3.3.1.1

dens5) *continuity from the right and from the left.*

$$E_n \uparrow E \implies \overline{\Delta}(E_n, \bullet) \uparrow \overline{\Delta}(E, \bullet); \; K_n \downarrow K \implies \overline{\Delta}(K_n, \bullet) \downarrow \overline{\Delta}(K, \bullet), \tag{3.3.1.3}$$

$$E_n \downarrow E \implies \underline{\Delta}(E_n, \bullet) \downarrow \underline{\Delta}(E, \bullet); \; G_n \uparrow G \implies \underline{\Delta}(G_n, \bullet) \uparrow \underline{\Delta}(G, \bullet); \tag{3.3.1.4}$$

dens6) *semi-additivity and positive homogeneity with respect to* μ, *i.e.,*

$$\overline{\Delta}(E, \mu_1 + \mu_2) \leq \overline{\Delta}(E, \mu_1) + \overline{\Delta}(E, \mu_2); \tag{3.3.1.5}$$

$$\underline{\Delta}(E, \mu_1 + \mu_2) \geq \underline{\Delta}(E, \mu_1) + \underline{\Delta}(E, \mu_2); \tag{3.3.1.6}$$

$$\overline{\Delta}, \underline{\Delta}(E, \lambda\mu) = \lambda\overline{\Delta}, \lambda\underline{\Delta}(E, \mu) \tag{3.3.1.7}$$

for $\lambda \geq 0$;

dens7) *invariance with respect to the map* $(\bullet)_{[t]}$ (*see,* 3.1.2.4a), *i.e.,*

$$t^{-\rho-m+2}\overline{\Delta}, \underline{\Delta}(P_t E, \bullet) = \overline{\Delta}, \underline{\Delta}(E, \bullet).$$

Proof of Theorem 3.3.1.1. The property dens1) holds because the empty set is open by definition. The properties dens2) and dens3) hold because of the monotonicity of ν.

Let us prove dens4). Since ν is a measure we have

$$\nu(G_1 \cup G_2, \mu) + \nu(G_1 \cap G_2, \mu) = \nu(G_1, \mu) + \nu(G_2, \mu)$$

for any $G_1 \supset E_1$ and $G_2 \supset E_2$.

From this we obtain

$$\nu(G_1 \cup G_2, \mu) + \nu(K_1 \cap K_2, \mu) \leq \nu(G_1, \mu) + \nu(G_2, \mu) \tag{3.3.1.8}$$

for $K_1 \subset E_1$ and $K_2 \subset E_2$.

The right side of (3.3.1.8) is no larger than $\overline{\Delta}(G_1, \bullet) + \overline{\Delta}(G_1, \bullet)$. Now we can take supremum over $\nu \in \mathbf{Fr}\mu$ in the first summand of the left side and infimum in the second summand. Thus we obtain

$$\overline{\Delta}(G_1 \cup G_2, \bullet) + \underline{\Delta}(K_1 \cap K_2, \bullet) \leq \overline{\Delta}(G_1, \bullet) + \overline{\Delta}(G_1, \bullet). \tag{3.3.1.9}$$

Since $\overline{\Delta}(E, \bullet)$ and $\underline{\Delta}(E, \bullet)$ are monotonic with respect to E,

$$\inf\{\overline{\Delta}(G_1 \cup G_2, \bullet) : G_1 \supset E_1, \; G_2 \supset E_2\} = \overline{\Delta}(E_1 \cup E_2, \bullet)$$

and

$$\sup\{\underline{\Delta}(K_1 \cap K_2, \bullet) : K_1 \subset E_1, \; K_2 \subset E_2\} = \underline{\Delta}(E_1 \cup E_2).$$

Thus we obtain the first inequality in dens4) from (3.3.1.9). The second one can be obtained analogously[2].

[2] Exercise 3.3.1.2

Let us prove dens5). For arbitrary $G \supset K$ there exists n_0 such that $K_n \subset G$ for $n > n_0$. According to dens3),

$$\overline{\Delta}(K, \bullet) \leq \overline{\Delta}(K_n, \bullet) \leq \overline{\Delta}(G, \bullet).$$

Hence,

$$\overline{\Delta}(K, \bullet) \leq \lim_{n \to \infty} \overline{\Delta}(K_n, \bullet) \leq \overline{\Delta}(G, \bullet).$$

Taking infimum over all $G \supset K$, we obtain the second assertion in (3.3.1.3).

For $G_n \uparrow G$ we have the equality

$$\lim_{n \to \infty} \overline{\Delta}(G_n, \bullet) = \sup_n \overline{\Delta}(G_n, \bullet) = \overline{\Delta}(G, \bullet) \tag{3.3.1.10}$$

because one can change the order of taking the supremum on n and on $\nu \in \mathbf{Fr}[\mu]$.

Let $E_n \uparrow E$ and let ϵ be arbitrarily small. One can find $G_n \supset E_n$ such that

$$\overline{\Delta}(G_n, \bullet) \leq \overline{\Delta}(E_n, \bullet) + \epsilon.$$

Since $G := \bigcup_1^\infty G_n \supset E$ we have

$$\overline{\Delta}(G_n, \bullet) - \epsilon \leq \overline{\Delta}(E_n, \bullet) \leq \overline{\Delta}(E, \bullet) \leq \overline{\Delta}(G, \bullet).$$

Using (3.3.1.10), we obtain

$$\overline{\Delta}(E, \bullet) - \lim_{n \to \infty} \overline{\Delta}(E_n, \bullet) \leq \epsilon.$$

Since ϵ is arbitrarily small,

$$\overline{\Delta}(E, \bullet) \leq \lim_{n \to \infty} \overline{\Delta}(E_n, \bullet)$$

and hence the first assertion in (3.3.1.3) holds.

The assertion (3.3.1.4) can be proved analogously.[3]

Let us prove dens6). One has

$$\overline{\Delta}(G, \mu_1 + \mu_2) = \sup\{\nu(G) : \nu \in \mathbf{Fr}[\mu_1 + \mu_2]\}.$$

Since

$$\mathbf{Fr}[\mu_1 + \mu_2] \subset \mathbf{Fr}[\mu_1] + \mathbf{Fr}[\mu_2]$$

(see frmu1), Theorem 3.1.3.4 (Properties of $\mu \mapsto \mathbf{Fr}[\mu]$)) one can continue the previous equality as

$$\begin{aligned}&\leq \sup\{\nu(G) : \nu \in \mathbf{Fr}[\mu_1] + \mathbf{Fr}[\mu_2]\}\\ &= \sup\{\nu(G) : \nu \in \mathbf{Fr}[\mu_1]\} + \sup\{\nu(G) : \nu \in \mathbf{Fr}[\mu_2]\} = \overline{\Delta}(G, \mu_1) + \overline{\Delta}(G, \mu_2).\end{aligned}$$

Passing to the infimum over $G \supset E$, we obtain (3.3.1.5). The assertions (3.3.1.6) and (3.3.1.7) can be proved analogously.[4]

The properties dens7) follow from the invariance of $\mathbf{Fr}[\mu]$ (see frm3), Theorem 3.1.3.3. (Properties of $\mathbf{Fr}[\mu]$)). □

[3]See Exercise 3.3.1.3

[4]See Exercise 3.3.1.4

Exercise 3.3.1.1 Prove the *subadditivity* of $\overline{\Delta}(E, \bullet)$:

$$\overline{\Delta}(E_1 \cup E_2, \bullet) \leq \overline{\Delta}(E_1, \bullet) + \overline{\Delta}(E_2, \bullet)$$

and the *superadditivity* of $\underline{\Delta}(E, \bullet)$:

$$\underline{\Delta}(E_1 \cup E_2, \bullet) \geq \underline{\Delta}(E_1, \bullet) + \underline{\Delta}(E_2, \bullet)$$

from Theorem 3.3.1.1.

Exercise 3.3.1.2 Prove (3.3.1.2).

Exercise 3.3.1.3 Prove (3.3.1.4).

Exercise 3.3.1.4 Prove (3.3.1.6) and (3.3.1.7).

Set for $I \subset (0, \infty)$ and $\Omega \subset S_1$,

$$\mathrm{Co}_\Omega(I) := \{x = P_t y : y \in \Omega,\ t \in I\}.$$

Also set $I_t := (0, t)$.

Theorem 3.3.1.2 (Cone's Densities) *One has*

$$\overline{\Delta}, \underline{\Delta}(\mathrm{Co}_\Omega(I_t)) = t^{\rho+m-2}\overline{\Delta}, \underline{\Delta}(\mathrm{Co}_\Omega(I_1)).$$

We obtain this from dens7), Theorem 3.3.1.1, taking $E := \mathrm{Co}_\Omega(I_1)$.

Exercise 3.3.1.5 Show that for $m = 2, S_1 = \{|z| = 1\}, \Omega = \{z = e^{i\phi} : \phi \in (\alpha, \beta)\}$ $\mathrm{Co}_\Omega(I_t)$ is a sector of radius t corresponding to the arc (α, β) on the unit circle.

3.3.2 Let $\delta(E)$ be a *monotonic* function of $E \in \mathbb{R}^m$. A set E is called δ-squarable if

$$\sup_{K \subset E} \delta(K) = \inf_{G \supset E} \delta(G). \tag{3.3.2.1}$$

Example 3.3.2.1 Let $\delta(E)$ be a measure. Then (3.3.2.1) implies $\delta(\partial E) = 0$, i.e., E is δ-squarable in the sense of Section 2.2.3.

Exercise 3.3.2.2 Prove the next

Theorem 3.3.2.1 *If* $\overline{\Delta}(\partial E) = 0$, *then* E *is* $\overline{\Delta}$*-squarable. If* E *is* $\underline{\Delta}$*-squarable, then* $\underline{\Delta}(\partial E) = 0$.

Set

$$E_t := \{x : \exists y \in E : |x - y| < t\}.$$

This is a *t-extension* of E.

A family of sets $\mathcal{A}_1$ is said to be *dense in* a family $\mathcal{A}_2$ if for each set $E_2 \in \mathcal{A}_2$ and an arbitrarily small $\epsilon > 0$ there exists a set $E_1 \in \mathcal{A}_1$ such that

$$\overline{E_1 \Delta E_2} := \overline{(E_1 \setminus E_2) \cup (E_2 \setminus E_1)} \subset (\partial E_2)_\epsilon. \tag{3.3.2.2}$$

Exercise 3.3.2.3 Prove

Theorem 3.3.2.2 *The relation "to be dense in" is reflexive and transitive.*

I.e., $\mathcal{A}_1$ is dense in $\mathcal{A}_1$, and

$$\{\mathcal{A}_1 \text{ is dense in } \mathcal{A}_2\} \wedge \{\mathcal{A}_2 \text{ is dense in } \mathcal{A}_3\} \Longrightarrow \{\mathcal{A}_1 \text{ is dense in } \mathcal{A}_3\}. \tag{3.3.2.3}$$

There are lots of squarable sets.

Theorem 3.3.2.3 *For any monotonic $\delta(E)$ the class of δ-squarable sets is dense in the class of all the subsets of $\mathbb{R}^m$.*

Proof. For any $E \subset \mathbb{R}^m$ set

$$E(t) := E \cup (\partial E)_t. \tag{3.3.2.4}$$

One can check that

$$\overline{E \Delta E(t_1)} \subset (\partial E)_{t_2} \tag{3.3.2.5}$$

and

$$\overline{E(t_1)} \subset \overset{\circ}{E}(t_2) \tag{3.3.2.6}$$

for $t_1 < t_2$.

The function $f(t) := \delta(\overset{\circ}{E}(t))$ is monotonic. Hence, its set of continuity points has a concentration point at $t = 0$.

Suppose $\epsilon > 0$ is arbitrarily small, and $t_0 < \epsilon$ is a continuity point for $f(t)$. From (3.3.2.6) we have

$$\lim_{t \to t_0 - \epsilon} \delta(\overline{E}(t)) \leq \sup_{K \subset E_{t_0}} \delta(K) \leq \inf_{G \supset E_{t_0}} \delta(G) \leq \lim_{t \to t_0 + \epsilon} \delta(\overset{\circ}{E}(t)).$$

Hence, E_{t_0} is δ-squarable. From (3.3.2.5) we have

$$\overline{E \Delta E(t_0)} \subset (\partial E)_\epsilon. \qquad \square$$

Set

$$\overline{\Delta}^{cl}(E) = \limsup_{t \to \infty} \mu_t(E); \ \underline{\Delta}^{cl}(E) = \liminf_{t \to \infty} \mu_t(E).$$

These are classic densities determined without $\mathcal{D}'$-topology. They are monotonic.

The following assertion connects these densities to $\overline{\Delta}$ and $\underline{\Delta}$.

Theorem 3.3.2.4 (Classic Densities) *For any* $\overline{\Delta}^{cl}$*-squarable set* E,

$$\overline{\Delta}^{\mathrm{cl}}(E) = \sup\{\nu(E) : \nu \in \mathbf{Fr}[\mu]\} = \overline{\Delta}(E, \mu). \tag{3.3.2.7}$$

For any $\underline{\Delta}^{\mathrm{cl}}$*-squarable set* E,

$$\underline{\Delta}^{cl}(E) = \inf\{\nu(E) : \nu \in \mathbf{Fr}[\mu]\} = \underline{\Delta}(E, \mu). \tag{3.3.2.7$'$}$$

The theorem follows obviously from the following assertion.

Theorem 3.3.2.5 *One has*

$$\sup_{K\subset E} \overline{\Delta}^{\mathrm{cl}}(K) \le \sup_{\nu\in\mathbf{Fr}} \nu(E) \le \overline{\Delta}(E) \le \inf_{G\supset E} \overline{\Delta}^{\mathrm{cl}}(G); \tag{3.3.2.8}$$

$$\sup_{K\subset E} \underline{\Delta}^{\mathrm{cl}}(K) \le \inf_{\nu\in\mathbf{Fr}} \nu(E) \le \underline{\Delta}(E) \le \inf_{G\supset E} \underline{\Delta}^{\mathrm{cl}}(G). \tag{3.3.2.9}$$

Proof. Let us prove, for example, (3.3.2.9). Let us choose any G and K such that $K \subset E \subset G$. We can find a sequence $t_j \to \infty$ such that

$$\lim_{j\to\infty} \mu_{t_j}(G) = \underline{\Delta}^{\mathrm{cl}}(G).$$

Choose a subsequence t_{j_n} such that $\mu_{t_{j_n}} \to \nu$ in $\mathcal{D}'$ for some $\nu \in \mathbf{Fr}$.

Using Theorems 2.3.4.4.(D$'$and C*) and 2.2.3.1.(C*-limits), we obtain

$$\nu(G) \le \liminf_{n\to\infty} \mu_{t_{j_n}}(G) = \underline{\Delta}^{\mathrm{cl}}(G). \tag{3.3.2.10}$$

By the same theorems

$$\underline{\Delta}^{\mathrm{cl}}(K) \le \limsup_{n\to\infty} \mu_{t_{j_n}}(K) \le \nu(K). \tag{3.3.2.11}$$

From (3.3.2.10) and (3.3.2.11) we obtain

$$\underline{\Delta}^{\mathrm{cl}}(K) \le \nu(E) \le \nu(G) \le \underline{\Delta}^{\mathrm{cl}}(G) \tag{3.3.2.12}$$

because of monotonicity of $\nu(E)$. Taking supremum over all $K \subset E$ and infimum over all $G \supset E$, we obtain (3.3.2.9). □

Exercise 3.3.2.4 Prove (3.3.2.8).

Corollary 3.3.2.6 *The following holds:*

$$\overline{\Delta}^{\mathrm{cl}}(K_t) = \overline{\Delta}(K_t, \mu) = t^{\rho+m-2}\overline{\Delta}(K_1, \mu), t \ge 0, \tag{3.3.2.13}$$

$$\underline{\Delta}^{\mathrm{cl}}(K_t) = \underline{\Delta}(K_t, \mu) = t^{\rho+m-2}\underline{\Delta}(K_1, \mu), t \ge 0. \tag{3.3.2.13$'$}$$

where $K_t = \{x : |x| < t\}$ *is the ball.*

Proof. The right equalities follow from Theorem 3.3.1.2 with $\Omega := S_1$. The left equalities hold at least for one t because of Theorem 3.3.2.4 and hence for all t. □

3.3.3 Let us note generally speaking that values of $\overline{\Delta}$ and $\underline{\Delta}$ on the sets $\mathrm{Co}_\Omega(I_t)$ do not determine their values even on the sets $\mathrm{Co}_\Omega(I)$ for $I = (t_1, t_2)$. However the following assertion holds.

Theorem 3.3.3.1 (Existence of Density) *Let Φ be a dense ring (see, 2.2.3) on S_1. Then the conditions*

$$\overline{\Delta}(\mathrm{Co}_\Omega(I_t)) = \underline{\Delta}(\mathrm{Co}_\Omega(I_t)) \tag{3.3.3.1}$$

for $\Omega \in \Phi$ and some t determine uniquely a measure $\Delta(\Omega)$ on S_1. **Fr**$[\mu]$ *consists of one single measure ν and*

$$\nu(\mathrm{Co}_\Omega(I_t)) = t^{\rho+m-2}\Delta(\Omega) \tag{3.3.3.2}$$

for all the $t \in (0, \infty)$.

To prove this we need an assertion that is valuable by itself. Set

$$\overline{\Delta}(\Omega) := \overline{\Delta}(\mathrm{Co}_\Omega(I_1)); \underline{\Delta}(\Omega) := \underline{\Delta}(\mathrm{Co}_\Omega(I_1)) \text{ for } \Omega \in S_1. \tag{3.3.3.3}$$

We will call them *angular* densities because for $m = 2$ and $V_t \equiv I$, Ω determines an angle in the plane.

Let Ω^G denote an open set in S_1 and Ω^K a closed one.

Theorem 3.3.3.2 (Angular Densities) *One has*

$$\overline{\Delta}(\Omega) = \inf_{\Omega^G \supset \Omega} \overline{\Delta}(\Omega^G); \ \underline{\Delta}(\Omega) = \sup_{\Omega^K \subset \Omega} \underline{\Delta}(\Omega^K). \tag{3.3.3.4}$$

Proof. We need to prove two assertions:

$$\forall \epsilon > 0 \ \exists \Omega^G : \overline{\Delta}(\Omega^G) < \overline{\Delta}(\Omega) + \epsilon; \tag{3.3.3.5}$$

$$\forall \epsilon > 0 \ \exists \Omega^K : \underline{\Delta}(\Omega^K) > \underline{\Delta}(\Omega) - \epsilon. \tag{3.3.3.6}$$

Let us prove (3.3.3.5). Set

$$\Omega^G(\epsilon) := \mathrm{Co}_\Omega(I_{1+\epsilon}) \cup \{|x| < \epsilon\}.$$

This is an open set that contains $\mathrm{Co}_\Omega(I_1)$. One can show the following:

Exercise 3.3.3.1 For every open set $G \supset \mathrm{Co}_\Omega(I_1)$ there exists $\epsilon > 0$ and $\Omega^G \subset S_1$ such that $\Omega^G(\epsilon) \subset G$.

We will show

$$\overline{\Delta}(\Omega^G(\epsilon)) < \overline{\Delta}(\Omega^G) + o(1) \tag{3.3.3.7}$$

uniformly with respect to $\Omega^G \subset S_1$ while $\epsilon \to 0$.

We have from Exercise 3.3.1.1,

$$\overline{\Delta}(\Omega^G(\epsilon)) \leq \overline{\Delta}(\mathrm{Co}_\Omega(I_{1+\epsilon})) + \overline{\Delta}(\{|x| < \epsilon\}). \tag{3.3.3.8}$$

The property dens7), Theorem 3.3.1.1, gives

$$\overline{\Delta}(\mathrm{Co}_{\Omega^G}(I_{1+\epsilon})) = \overline{\Delta}(\mathrm{Co}_{\Omega^G}(I_1))(1+\epsilon)^{\rho+m-2}.$$

Since $\overline{\Delta}(\mathrm{Co}_{\Omega^G}(I_1)) \leq \overline{\Delta}(\{|x|<1\})$ we have

$$\overline{\Delta}(\mathrm{Co}_{\Omega^G}(I_{1+\epsilon})) = \overline{\Delta}(\mathrm{Co}_{\Omega^G}(I_1)) + o(1) \tag{3.3.3.9}$$

uniformly with respect to $\Omega^G \subset S_1$ as $\epsilon \to 0$.

By dens7) we also have

$$\overline{\Delta}(\{|x|<\epsilon\}) = \overline{\Delta}(\{|x|<1\})\epsilon^{\rho+m-2} = o(1). \tag{3.3.3.10}$$

From (3.3.3.10), (3.3.3.9) and (3.3.3.8) we obtain (3.3.3.7). Hence (3.3.3.5) is proved.

Let us prove (3.3.3.6). Set

$$\Omega^K(\epsilon) := \mathrm{Co}_{\Omega^K}(\bar{I}_{1-\epsilon}) \setminus \{|x|<\epsilon\}$$

where $\bar{I}$ is the closure of I.

One can show the following:

Exercise 3.3.3.2 For any compact $K \subset \mathrm{Co}_\Omega(I_1)$ there exist $\Omega^K \subset \Omega$ and $\epsilon > 0$ such that $K \subset \Omega^K(\epsilon) \subset \mathrm{Co}_\Omega(I_1)$.

From the definition of $\underline{\Delta}(\Omega)$ and the monotonicity we obtain (3.3.3.6). □

Proof of Theorem 3.3.3.1. Suppose (3.3.3.1) holds. The property dens7), Theorem 3.3.1.1, implies (3.3.3.1) for all the $t \in (0,\infty)$. Set $\Delta(\Omega) := \overline{\Delta}(\Omega) = \underline{\Delta}(\Omega)$ for $\Omega \in \Phi$. Let us prove that Δ satisfies the conditions Δ1)–Δ3) from Section 2.2.3. The conditions Δ1) and Δ2) follow from dens3) and dens4), Theorem 3.3.1.1, Exercise 3.3.1.1.

Let us prove Δ3). By Theorem 3.3.3.2 for arbitrary $\Omega \in \Phi$ and $\epsilon > 0$ we can choose $\Omega^G \supset \Omega$ such that $\overline{\Delta}(\Omega) > \overline{\Delta}(\Omega^G) - \epsilon$ and $\Omega^K \subset \Omega$ such that $\underline{\Delta}(\Omega) < \underline{\Delta}(\Omega^K) + \epsilon$.

Suppose $\Omega' \in \Phi$ satisfies the condition $\Omega^K \subset \Omega' \subset \Omega^G$. Then

$$\Delta(\Omega') = \overline{\Delta}(\Omega') \leq \overline{\Delta}(\Omega^G) \leq \overline{\Delta}(\Omega) + \epsilon = \Delta(\Omega) + \epsilon$$

and

$$\Delta(\Omega) - \epsilon = \underline{\Delta}(\Omega) - \epsilon \leq \underline{\Delta}(\Omega^K) \leq \underline{\Delta}(\Omega') = \Delta(\Omega'),$$

implying Δ3). □

Chapter 4

Structure of Limit Sets

4.1 Dynamical systems

4.1.1 The most complete and effective description of an arbitrary limit set can be done in terms of dynamical systems (see, [An]).

A family of the form

$$T^t : M \mapsto M, \; t \in \mathbb{R},$$

on a compact metric space (M, d) with a metric $d(\bullet, \bullet)$ is *a dynamical system* $(T^\bullet, M)$ if it satisfies the condition

$$T^{t+\tau} = T^t \circ T^\tau, \; t, \tau \in \mathbb{R}$$

and the map $(t, m) \mapsto T^t m$ is continuous with respect to (t, m), for all $t \in \mathbb{R}$, $m \in M$.

Let $m, m' \in M$, and $\epsilon, s > 0$. An (ϵ, s)*-chain from* m *to* m' is a finite sequence $m_0 = m, m_1, \ldots, m_n = m'$, satisfying the conditions $d(T^{t_j} m_j, m_{j+1}) < \epsilon$, $j = 0, 1, \ldots, n-1$, for some $t_j \geq s$.

A dynamical system $(T^\bullet, M)$ is called *chain recurrent* (see, [HS]), if for an arbitrarily small $\epsilon > 0$ and an arbitrarily large $s > 0$ there exists an (ϵ, s)-chain in M from m to m.

Theorem 4.1.1.1 (Properties of Chain Recurrence) *Let $(T^\bullet, M)$ be a dynamical system on a compact set. Then the following conditions are equivalent:*

cr1) *M is connected and $(T^\bullet, M)$ is chain recurrent;*

cr2) *for every open proper $U \subset M$ satisfying*

$$T^t U \subset U, \; -\infty < t < 0, \tag{4.1.1.1}$$

the boundary ∂U contains a nonempty $T^\bullet$-invariant subset of M;

cr3) *for every closed proper* $K \subset M$ *satisfying*

$$T^t K \subset K, \ t \geq 0, \tag{4.1.1.2}$$

the boundary ∂K *contains a nonempty* $T^\bullet$*-invariant subset of* M;

cr4) *there does not exist any open proper* $V \subset M$ *satisfying* $T^\tau \operatorname{clos} V \subset V$ *for some* $\tau > 0$;

cr5) *for any small* $\epsilon > 0$, *large* $s > 0$, *and every pair of points* m, m' *there exists an* (ϵ, s)*-chain from* m *to* m'.

Proof. The conditions cr2) and cr3) are equivalent. Let us prove, for example, cr2)$\Longrightarrow$ cr3). Set $U := M \setminus K$. It is open. Applying to (4.1.1.2) T^{-t} and, using the invariance of M, we obtain (4.1.1.1) for U. Hence ∂U contains a nonempty invariant subset of M. Since $\partial K = \partial U$ we obtain cr2).

Let us prove the implication cr1)$\Longrightarrow$cr3). Let $K \subset M$ be closed, proper and satisfy (4.1.1.2). Since M is proper ∂K is nonempty.

Let W denote the interior of K in M. The continuity of T and (4.1.1.2) imply

$$T^t W \subset W \tag{4.1.1.3}$$

for $t \geq 0$. Indeed, $T^t W \subset K$. It must be open. Thus it cannot contain any point of ∂K, since else it would contain some neighborhood of this point, contradicting the definition of ∂K.

Suppose that ∂K does not contain any nonempty T-invariant set. Let us show that there exists $s > 0$ such that

$$T^s K \subset W. \tag{4.1.1.4}$$

For any $m \in \partial K$ there exists $t = t(m)$ such that $T^t m \in W$. There exists a neighborhood V_m of m in ∂K that passes to W under $T^{t(m)}$-action because of continuity of $T^t m$ on m.

We also have $T^t V_m \subset W$ for $t > t(m)$ because of (4.1.1.3). Since ∂K is compact we can cover it by a finite number of neighborhoods and obtain s such that

$$T^s \partial K \subset W. \tag{4.1.1.5}$$

(4.1.1.5) and (4.1.1.3) give (4.1.1.4).

Set $\epsilon := 0.5 d(\partial K, T^s K)$. From (4.1.1.2) we see that $T^t K \subset T^s K$ for $t > s$. Therefore there does not exist any (ϵ, s)-chain from a small neighborhood of a point $m \in \partial K$ to itself. This contradicts the chain recurrence of M.

Let us prove cr3)$\Longrightarrow$cr4). Assume that there exists an open proper $V \subset M$ satisfying $T^\tau \operatorname{clos} V \subset V$ for some $\tau > 0$.

We will construct K that does not satisfy cr3). Set $W := \bigcup_{0 \leq t \leq \tau} T^t V$ and $K := \operatorname{clos} W$.

Then

$$T^s W \subset W, \forall s \geq 0. \tag{4.1.1.6}$$

Indeed, let $s = k\tau + s'$, $s' \in [0, \tau)$, $k \in \mathbb{Z}$. Then

$$T^s W = \bigcup_{t \in [0,\tau]} T^{t+s} V. \tag{4.1.1.7}$$

Since $T^\tau V \subset V$ we have $T^{t+k\tau} V \subset T^t V$ for $t > 0$. From (4.1.1.7) we obtain

$$\begin{aligned} T^s W &= \bigcup_{t \in [0,\tau]} T^{t+s'+k\tau} V \subset \bigcup_{t \in [0,\tau]} T^{t+s'} V = \bigcup_{t' \in [s', \tau+s']} T^{t'} V \\ &= \bigcup_{t \in [s',\tau]} T^t V \cup \bigcup_{t \in [\tau, \tau+s']} T^t V := W_1 \cup W_2. \end{aligned}$$

Further we have $W_1 \subset W$ by definition. W_2 can be represented in the form

$$W_2 = \bigcup_{t \in [0,s']} T^{t+\tau} V.$$

Since

$$T^{t+\tau} V = T^t T^\tau V \qquad \text{and} \qquad T^\tau V \subset V$$

by the assumption we get:

$$W_2 \subset W_1 \subset W.$$

This implies (4.1.1.6). The same holds for K because of continuity of T^t, i.e., K satisfies (4.1.1.2).

Let us prove the equality

$$K = \bigcup_{0 \leq t \leq \tau} T^t \operatorname{clos} V. \tag{4.1.1.8}$$

Denote as K' the right side of (4.1.1.8).

The set K' is closed because of compactness of $[0, \tau]$. Indeed, let the sequence $\{T^{t_j} v_j : j = 1, 2, \ldots\} \in T^{t_j}(\operatorname{clos} V)$ converge to w. Choose a subsequence $t_{j_k} \to s \in [0, \tau]$. Then

$$v := \lim_{k \to \infty} v_{j_k} = \lim_{k \to \infty} T^{-t_{j_k}} w = T^{-s} w.$$

Since $\operatorname{clos} V$ is closed, $v \in \operatorname{clos} V$. Thus $w = T^s v$ for some $s \in [0, \tau]$ and some $v \in \operatorname{clos} V$, i.e., $w \in K'$.

Now, $W \subset K'$ because

$$T^t V \subset \operatorname{clos} T^t V = T^t \operatorname{clos} V.$$

Hence,

$$K := \operatorname{clos} W \subset \operatorname{clos} K' = K'.$$

We also have

$$(T^tV \subset W \ \forall t \in [0,\tau]) \Longrightarrow (\operatorname{clos} T^tV = T^t \operatorname{clos} V \subset \operatorname{clos} W = K, \ \forall t \in [0,\tau]).$$

Hence, $K' \subset K$. Therefore $K = K'$, i.e., (4.1.1.8) holds.

From (4.1.1.8) and $T^\tau \operatorname{clos} V \subset V$ we obtain $T^\tau \operatorname{clos} W \subset W$. Hence $T^\tau \partial K \subset W$. This and $\partial K \cap W = \varnothing$ imply

$$T^\tau \partial K \cap \partial K = \varnothing. \tag{4.1.1.9}$$

To obtain a contradiction and complete the proof of cr3) $\Longrightarrow$ cr4) we have to show that K is a proper subset, because both cases: $\partial K = \varnothing$ and $\partial K \neq \varnothing$ will contradict cr3).

Since V is proper T^tV is proper for any $t \in (-\infty,\infty)$. Otherwise $T^tV = M$ implies $V = T^{-t}M = M$, which is a contradiction.

Since V is a neighborhood of the compact set $T^\tau \operatorname{clos} V$ we can find $\alpha > 0$ such that $T^t \circ T^\tau \operatorname{clos} V \subset V$ for $t \in [0,\alpha]$. Then $T^t \operatorname{clos} V \subset T^{-\tau}V$ for $t \in [0,\alpha]$.

By iteration of this inclusion we obtain $T^{jt} \operatorname{clos} V \subset T^{-j\tau}V$ for any integer j. When $j\alpha > \tau$ it follows that $K \subset T^{-j\tau}V$. The last set is proper because we mentioned already that T^tV is proper for any $t \in (-\infty,\infty)$. Hence K is proper.

So K satisfies the conditions of cr3) but ∂K does not contain a nonempty $T^\bullet$-invariant set. This contradiction proves the implication cr3) $\Longrightarrow$ cr4).

Let us prove cr4) $\Longrightarrow$ cr5). Let $\epsilon > 0$ be small and $s > 0$ be large. Let V denote the set of all $m' \in M$ such that there exists an (ϵ, s)-chain from m to m'. This set is open and closed. Indeed, let $m' \in V$. There exists an (ϵ, s)-chain $m = m_0, \dots, m_{n-1}, m_n = m'$ from m to m'. Choose $\epsilon_1 < \epsilon - d(m_n, m_{n-1})$ and consider the closed neighborhood $W := \{m'' : d(m', m'') \le \epsilon_1\}$. Then for any $m'' \in W$ the chain $m = m_0, \dots, m_{n-1}, m_n = m''$ is an (ϵ, s)-chain from m to m''. Hence, with every point, V contains its closed neighborhood. Therefore it is open and closed. Therefore it is a connected component of M.

We also have $T^s m \in V$ because for that case $n = 1, m_0 = m, m_1 = T^s m$. Hence $T^s \operatorname{clos} V \subset V$. If V does not coincide with the whole M the latter contradicts cr4). Hence $V = M$.

Finally, let us prove cr5) $\Longrightarrow$ cr1). If M is a union of two nonempty disjoint sets A and B, then both of them are open and closed. Since M is compact, the distance ϵ between A and B is positive . Hence every $(\epsilon/2, s)$-chain starting at a point of A remains in A, contradicting cr5).

Since for every point $m \in M$ the set V from the proof of cr4)$\Longrightarrow$ cr5) coincides with M, cr1) holds. □

Theorem 4.1.1.2 *Let $T^\bullet$ be chain recurrent on $M_\alpha, \alpha \in A$. Then $T^\bullet$ is chain recurrent on $M = \bigcup_{\alpha \in A} M_\alpha$.*

This is because every (ϵ, s)-chain from m to m' in M_α is also (ϵ, s)-chain in M.

4.1.2 Here we prove two auxiliary assertions that will be used further.

Theorem 4.1.2.1 *Let $T^\bullet$ be chain recurrent on a connected compact M and let $\{q_j\}$ be a sequence in M. Then there exist sequences $\{\alpha_\nu\}$ and $\{\omega_\nu\}$ of real numbers and a sequence $\{p_\nu\}$ in M having $\{q_j\}$ as a subsequence, such that*

$$\alpha_\nu \to -\infty;\ \omega_\nu \to \infty \tag{4.1.2.1}$$

and

$$d(T^{\omega_\nu} p_\nu, T^{\alpha_{\nu+1}} p_{\nu+1}) \to 0 \tag{4.1.2.2}$$

as $\nu \to \infty$.

Proof. In addition to $\{\alpha_\nu\}$, $\{\omega_\nu\}$ and $\{p_\nu\}$ we define, by induction, a sequence $\{\epsilon_\nu\}$ of positive real numbers, tending to zero, and an increasing sequence $\{\nu_j\}$ of positive integers, such that $\{p_{\nu_j}\} = \{q_j\}$ and

$$d(T^{\omega_\nu} p_\nu, T^{\alpha_{\nu+1}} p_{\nu+1}) < \epsilon_\nu, \nu = 1, 2, \dots. \tag{4.1.2.3}$$

We start by setting $\alpha_1 = -1, \epsilon_1 = 1, \nu_1 = 1, \omega_1 = 5$ and $p_1 = q_1$. Assume now that $\alpha_\nu, \epsilon_\nu, \omega_\nu$ and p_ν have been chosen for $\nu = 1, 2, \dots, \nu_j$. Set

$$\alpha = \alpha_{\nu_j} - 1,\ \epsilon = \epsilon_{\nu_j}/2,\ \omega = \omega_{\nu_j}. \tag{4.1.2.4}$$

By Theorem 4.1.1.1, cr5) there exists a sequence $r_0 := T^\omega q_j, r_1, \dots, r_m := T^\alpha q_{j+1}$ such that $d(T^{t_k} r_k, r_{k+1}) < \epsilon$ for $k = 0, 1, \dots, m-1$, where $t_k \geq \omega$. Now we set $\nu_{j+1} = \nu_j + m + 1$. For $\nu = \nu_j + k + 1,\ k = 0, 1, \dots, m-1$, we set $\alpha_\nu = -t_k/2, \omega_\nu = t_k/2, p_\nu = T^{t_k/2} r_k$, and finally, for $\nu = \nu_{j+1}$ we set $\alpha_\nu = \alpha, \epsilon_\nu = \epsilon, \omega_\nu = \omega + 1, p_\nu = q_{j+1}$.

Let us check that with this setting the properties (4.1.2.1) hold . Since $\omega_{\nu_{j+1}} = \omega_{\nu_j} + 1$ we have $\omega_{\nu_j} \to \infty$ as $j \to \infty$. From $t_k \geq \omega = \omega_{\nu_j}$ we obtain $\alpha_\nu \to -\infty$ and $\omega_\nu \to \infty$. Hence (4.1.2.1) holds.

One can see from (4.1.2.4) that $\epsilon_\nu = \epsilon_{\nu_j}/2 \to 0$. To prove (4.1.2.2) it is enough to check (4.1.2.3). For $k = 0$ we have

$$p_\nu = T^{t_0/2} r_0 = T^{(t_0/2)+\omega} q_j = T^{(t_0/2)+\omega} p_{\nu_j}.$$

Hence,

$$T^{\alpha_\nu} p_\nu = T^\omega p_{\nu_j} = T^{\omega_{\nu_j}} p_{\nu_j}.$$

Thus

$$d(T^{\omega_{\nu_j}} p_{\nu_j}, T^{\alpha_\nu} p_\nu) = 0 \tag{4.1.2.5}$$

for this case.

For $k = 1, \dots, m-2$ and the corresponding ν we have

$$T^{\omega_\nu} p_\nu = T^{t_k/2} \circ T^{t_k/2} r_k = T^{t_k} r_k$$

and

$$T^{\alpha_{\nu+1}}p_{\nu+1} = T^{-t_{k+1}/2} \circ T^{t_{k+1}/2} r_{k+1} = r_{k+1}.$$

Hence,

$$d(T^{\omega_\nu}p_\nu, T^{\alpha_{\nu+1}}p_{\nu+1}) = d(T^{t_k}r_k, r_{k+1}) < \epsilon = \epsilon_\nu. \tag{4.1.2.6}$$

Finally, for the last link of the chain we obtain

$$k = m-1,\ \nu = \nu_j + m,\ \nu+1 = \nu_{j+1},\ \alpha_{\nu_{j+1}} = \alpha,$$
$$T^{\alpha_{\nu+1}}p_{\nu+1} = T^{\alpha_{\nu_{j+1}}}p_{\nu_{j+1}} = T^\alpha q_{j+1} = r_m.$$

Thus (4.1.2.6) holds for $k = m-1$. Hence, (4.1.2.3) also holds. Therefore (4.1.2.2) holds. □

Lemma 4.1.2.2 *Let $p_k, q_k \in M$ and $d(p_k, q_k) \to 0$ as $k \to \infty$. Then there exists a sequence $\{\gamma_k \uparrow \infty\}$ such that*

$$d(T^\tau p_k, T^\tau q_k) \to 0 \tag{4.1.2.7}$$

uniformly with respect to $\tau \in [-\gamma_{k+1}, \gamma_k]$.

Proof. Let $[-\gamma, \gamma]$ be a fixed segment. Then $d(T^\tau p_k, T^\tau q_k) \to 0$ uniformly in this segment.

Indeed, suppose there exist sequences τ_j, k_j such that $d(T^{\tau_j}p_{k_j}, T^{\tau_j}q_{k_j}) \geq \epsilon > 0$. Choosing a subsequence we can assume that $\tau_j \to \tau \in [-\gamma, \gamma]$, $p_{k_j} \to p \in M$ and $q_{k_j} \to q = p$. Using continuity of $T^\tau m$ on (τ, m) and continuity of $d(\bullet, \bullet)$ in both arguments we obtain $0 = d(p, p) \geq \epsilon > 0$. This is impossible.

Denote

$$\epsilon(\gamma, k) := \max_{\tau \in [-\gamma, \gamma]} d(T^\tau p_k, T^\tau q_k).$$

This function increases monotonically in γ and tends to zero for any γ as $k \to \infty$.

Choose l_n such that $\epsilon(n, k) \leq 1/n$ for $k \geq l_n$. Set $\gamma_{k+1} := n$ for $l_n < k \leq l_{n+1}$. One can see that $\epsilon(\gamma_{k+1}, k) \to 0$ as $k \to \infty$. Since

$$\max_{\tau \in [-\gamma_{k+1}, \gamma_k]} d(T^\tau p_k, T^\tau q_k) \leq \epsilon(\gamma_{k+1}, k),$$

$\{\gamma_k\}$ satisfies (4.1.2.7). □

4.1.3 We connect the property of being chain recurrent with other well-known characteristics of dynamical systems ([AGL]).

A point $m_0 \in M$ is called *non-wandering* (see [An]) if for any neighborhood $\mathcal{O}$ of m_0 and arbitrarily large number $s \in \mathbb{R}$ there exists $m \in \mathcal{O}$ and $t \geq s$ such that $T^t m \in \mathcal{O}$.

This means that the "returns" take place to an arbitrarily small neighborhood of the point m_0. We shall denote as $\Omega(T^\bullet)$ the set of non-wandering points. It is a closed invariant subset of M.

The set $A \subset M$ is called an *attractor* if it satisfies the following conditions:

attr1) for any neighborhood $\mathcal{O} \supset A$ there exists a neighborhood $\mathcal{O}'$, $A \subset \mathcal{O}' \subset \mathcal{O}$ such that $T^t\mathcal{O}' \subset \mathcal{O}$ $t \in \mathbb{R}$, where $T^t\mathcal{O}'$ is the image of $\mathcal{O}'$;

attr2) there exists a neighborhood $\mathcal{O} \supset A$ such that $T^t m \to A$ when $t \to \infty$ for $m \in \mathcal{O}$.

Theorem 4.1.3.1 *If $\Omega(T^\bullet) = M$, then $(T^\bullet, M)$ is chain recurrent; if $(T^\bullet, M)$ has an attractor $A \neq M$, it is not chain recurrent.*

Proof. The property $\Omega(T^\bullet) = M$ obviously implies the chain recurrence for $m = 1$.Suppose there exists an attractor $A \neq M$. Take a point m_0 that does not belong to A and choose a neighborhood $\mathcal{O} \supset A$ such that $d(m_0, \operatorname{clos}\mathcal{O}) = 2\epsilon > 0$. This is possible because an attractor is closed. Let $\mathcal{O}'$ be chosen by attr1) and s be such that $T^s m \in \mathcal{O}'$. Then there does not exist any (ϵ, s)-chain from a small neighborhood of m_0 itself. By definition (T^t, M) is not chain recurrent. □

Let us give examples of dynamical systems on connected compacts that are chain recurrent.

Theorem 4.1.3.2 *Let M be a connected compact and let T^t $(-\infty < t < \infty)$ be the identity map. Then $(T^\bullet, M)$ is chain recurrent.*

This theorem, of course, is trivial. However, if M consists of a single point this dynamical system determines an important class of subharmonic and entire functions of *completely regular growth* (see [Le, Ch. III]).

Let $m \in M$. Set

$$\mathbb{C}(m) := \operatorname{clos}\{T^t m : -\infty < t < \infty\}. \tag{4.1.3.1}$$

It is closed, connected and invariant.

Exercise 4.1.3.1 Prove this.

Let us denote as $\Omega(m)$ the set of all limits of the form

$$\Omega(m) := \{m' \in M : (\exists t_k \to \infty)(m' = \lim_{k\to\infty} T^{t_k} m\}. \tag{4.1.3.2}$$

This is a limit set as $t \to \infty$. It is the "tangle" at the end of the curve. Denote by $A(m)$ the analogous set for $t \to -\infty$.

Exercise 4.1.3.2 Prove that $A(m)$ and $\Omega(m)$ are invariant.

Theorem 4.1.3.3 $(T^\bullet, \mathbb{C}(m))$ *is chain recurrent iff*

$$A(m) \cap \Omega(m) \neq \varnothing. \tag{4.1.3.3}$$

Proof. Suppose $B := A(m) \cap \Omega(m) = \varnothing$. Then $\Omega(m)$ is an attractor and $(T^\bullet, \mathbb{C}(m))$ is not chain recurrent by Theorem 4.1.3.1.

Suppose $B \neq \varnothing$. We will use cr2) from Theorem 4.1.1.1.

Let U be an open proper subset of $\mathbb{C}(m)$ satisfying (4.1.1.1). Consider two cases:

i) B contains a point of U. Thus U contains a sequence of form $T^{t_k}m$, $t_k \to \infty$. From (4.1.1.1) we obtain that U contains $T^t m$ for all $t \in (-\infty, \infty)$. Thus $U \supset \mathbb{C}(m)$ and $\operatorname{clos} U = \mathbb{C}(m)$. Set $K = \mathbb{C}(m) \setminus U$. One can show that K satisfies (4.1.1.2)(see the beginning of proof of Theorem 4.1.1.1). Hence K contains the set

$$K^* := \bigcap_{t \geq 0} T^t K \tag{4.1.3.4}$$

that is invariant (Exercise 4.1.3.3).

Therefore $K^* \subset K \subset \operatorname{clos} U \setminus U = \partial U$. By cr2) $(T^t, \mathbb{C}(m))$ is chain recurrent.

ii) B contains no point of U. Then $B \subset A(m) \subset \partial U$. By cr2) $(T^\bullet, \mathbb{C}(m))$ is chain recurrent. □

Exercise 4.1.3.3 Let U satisfy (4.1.1.1) and $K := M \setminus U$. Prove that K^* from (4.1.3.4) is invariant.

4.1.4 The connectedness of M is a necessary condition for a dynamical system to be chain recurrent.

Let M be a subset of a linear space. The set M is called *polygonally connected* if every pair of points m_1, m_2 can be connected by a polygonal path.

Of course, polygonal connectedness implies connectedness and even arcwise connectedness.

Theorem 4.1.4.1 *Let $(T^\bullet, M)$ be a dynamical system such that M is a polygonally connected set. Then $(T^\bullet, M)$ is chain recurrent.*

Proof. Let U be an open proper subset of M, satisfying (4.1.1.1). We choose $m_1 \in U$ and m_2 in an invariant subset K^* of $K := M \setminus U$. Then there exists a polygonal path from m_1 to m_2 :

$$m_\theta := (j + 1 - \theta)m'_j + (\theta - j)m'_{j+1}, \text{ for } \theta \in [j, j+1],$$

$$j = 0, 1, \ldots, l-1;\ m'_0 := m_1,\ m'_l := m_2.$$

Now M is invariant, so for each t the continuous path $\theta \mapsto T^t m_\theta$ lies in M.

If $t \in (-\infty, 0)$ its initial point $T^t m_1$ belongs to U and its endpoint $T^t m_2$ belongs to $K^* \subset K$.

For each $t \in (-\infty, 0)$ we set

$$\theta(t) := \min[\theta \in [0; l] : T^t m_\theta \in K].$$

Then $\theta(t) > 0$, $T^t m_{\theta(t)} \in \partial U$ and (4.1.1.1) implies that $t \mapsto \theta(t)$ is a decreasing function. Hence the limit

$$\theta(-\infty) := \lim_{t\to-\infty} \theta(t)$$

exists and is positive.

Set $m_3 := m_{\theta(-\infty)}$. We claim that $A(m_3) \subset \partial U$ ($A(\cdot)$ is a set defined before Theorem 4.1.3.3). If $\theta(-\infty) \in (j, j+1]$ for some $j \in [0, l]$ then $\theta(t) \in (j, j+1]$ for t that is near to $-\infty$, and

$$T^t m_3 = T^t m_{\theta(t)} + (\theta(t) - \theta(-\infty))T^t m'_j + (\theta(-\infty) - \theta(t))T^t m'_{j+1}.$$

The first term in the right-hand side lies in ∂U. The set M is compact and invariant so the other terms tend to zero as $t \to -\infty$. Hence $A(m_3) \subset \partial U$.

Thus ∂U contains this invariant subset and $(T^\bullet, M)$ is chain recurrent by cr2), Theorem 4.1.1.1. □

We have the obvious

Corollary 4.1.4.2 *Let $(T^\bullet, M)$ be a dynamical system such that M is a compact convex set. Then $(T^\bullet, M)$ is chain recurrent.*

This is because the polygonal path can be taken as a line segment connecting every pair of points.

4.1.5 Let $U[\rho, \sigma]$ be a set of subharmonic functions defined in (3.1.2.4). It is invariant with respect to the transformation $(\bullet)_{[t]}$ defined in (3.1.2.4a).

Set (subindex!)

$$T_t v := v_{[e^t]}. \tag{4.1.5.1}$$

Since $(\bullet)_{[t]}$ has the property (3.1.2.4b)

$$T_{t+\tau} v = (T_t \circ T_\tau)v, \ \forall t, \tau \in \mathbb{R}. \tag{4.1.5.2}$$

By Theorem 3.1.2.3, $T_\bullet$ is continuous in the appropriate topology and hence $(T_\bullet, U[\rho, \sigma])$ is a dynamical system.

Theorem 4.1.5.1 (Universality of $U[\rho, \sigma]$) *Let $(T^\bullet, M)$ be a chain recurrent dynamical system on a compact set M. Then for any ρ, σ there exists $U \subset U[\rho, \sigma]$ and a homeomorphism* $\mathrm{imb} : M \mapsto U$ *such that* $\mathrm{imb} \circ T^t = T_t \circ \mathrm{imb}$, $t \in (-\infty, \infty)$.

I.e., any dynamical system can be imbedded in $(T_\bullet, U[\rho, \sigma])$.

This plot is developed in [Az(2008)].

It is sufficient to prove the theorem by supposition $P_t x = tx$ because $(T^P_\bullet, U[\rho, \sigma])$ is a dynamical system for any P_t and

$$\mathrm{imb} : (T_\bullet, U[\rho, \sigma]) \mapsto (T^P_\bullet, U[\rho, \sigma])$$

where $\mathrm{imb} : u(x) \mapsto T_{-t} T^P_t u(x)$ is also a homeomorphism of dynamical systems.

Exercise 4.1.5.1 Consider Theorem 3.1.6.1 from this point of view.

We need some auxiliary definitions and results. Let us denote as $\mathcal{M}(S^{m-1})$ the set of measures ν with bounded full variation on the unit sphere S^{m-1}. Introduce the metric $d(\nu, 0) := \text{Var } \nu$ and consider the set

$$K := \{\nu : \nu > 0,\ d(\nu, 0) \leq 1\},$$

i.e., the intersection of the cone of positive measures with the unit ball.

The following assertion is a corollary of Keller's theorem (see, e.g., [BP, Thm. 3.1, p. 100]).

Theorem 4.1.5.2 (Imbedding) *Every metric compact set can be homeomorphically imbedded to K.*

Thus we can assume below that for any $m \in M$ there exists a positive measure

$$Y(\bullet, m) = Y(dx^0, m) \in K$$

such that

$$(Y(\bullet, m_1) = Y(\bullet, m_2)) \Longrightarrow (m_1 = m_2) \tag{4.1.5.3}$$

and $Y(\bullet, m)$ is continuous with respect to the metrics.

We also introduce a new coordinate system. For $x := e^y x^0 \in \mathbb{R}^m \setminus 0$ set $\text{Pol}(x) = (y, x^0)$. This formula gives a one-to-one map from $\mathbb{R}^m \setminus 0$ onto the cylinder $Cyl := (-\infty, \infty) \times S^{m-1}$. Thus, for any $(y, x^0) \in Cyl$, $\text{Pol}^{-1}(y, x^0) = e^y x^0$.

For $m = 2$ this is a common cylinder.

4.1.6

Proof of Theorem 4.1.5.1. We consider separately the cases of integer and non-integer ρ.

Let ρ be non-integer and $\sigma > 0$. For any $v \in U[\rho, \sigma]$, one has the representation of Theorem 3.1.4.4 (*Hadamard),

$$v(x) = \Pi(x, \mu, \rho) \tag{4.1.6.1}$$

where $\mu \in \mathcal{M}[\rho, \Delta]$ and Δ depends only on σ (Theorem 2.8.3.3).

Vice versa, every $\mu \in \mathcal{M}[\rho, \Delta]$ generates v by (4.1.6.1) and

$$v_{[t]}(x) = \Pi(x, \mu_{[t]}, \rho).$$

Let us "transplant" μ in Cyl. For μ that has a dense $f_\mu(rx^0)$, we set

$$\nu(dy \otimes dx^0) := f_\mu(e^y x^0) e^{(-\rho-2)y}(dy \otimes dx^0),$$

i.e., the density f_ν of ν is defined by

$$f_\nu(x^0, y) := f_\mu(e^y x^0) e^{(-\rho-2)y}.$$

Respectively,

$$f_\mu(x^0, r) = f_\nu(x^0, \log r) r^{\rho+2}.$$

We can extend this equality for all $\mu \in \mathcal{M}[\rho, \Delta]$ by using a limit process in $\mathcal{D}'$ topology.

Exercise 4.1.6.3 Do that using, for example, Theorem 2.3.4.5 (Properties of Regularization).

We can also define ν as a distribution in $\mathcal{D}'(Cyl)$. Namely, for $\psi \in \mathcal{D}(Cyl)$ we set

$$\psi^*(x^0, r) := \psi(\mathrm{Pol}^{-1}(x^0, \log r))r^{-\rho-m+2}$$

and

$$\langle \nu, \psi \rangle := \int \psi^*(x^0, r)\mu(dx^0 \otimes r^{m-1}dr).$$

Exercise 4.1.6.4 Check that this definition gives the same ν.

The transformation $P_t x = (x^0, tr),\ rx^0 \in \mathbb{R}^m \setminus 0$ passes to

$$\mathrm{Pol} \circ P_t \circ \mathrm{Pol}^{-1}(x^0, y) = (x^0, y + \log t).$$

Thus $T_{e^\tau}\mu$ gives a transformation $S_\tau \nu$ defined by

$$S_t f_\nu(x^0, y) := f_\nu(x^0, y + t)$$

for densities or by

$$\langle S_t \nu, \psi \rangle := \int \psi(x^0, y - t)\nu(dx^0 \otimes dy) \tag{4.1.6.2}$$

for distributions ($\psi \in \mathcal{D}(Cyl)$.)

Exercise 4.1.6.5 Check the equivalence.

From $\mu \in \mathcal{M}[\rho, \Delta]$ we obtain

$$\int\limits_{y \le 0} e^{(\rho+m-2)y} S_t \nu(dy \otimes dx^0) \le \Delta,\ t \in \mathbb{R}. \tag{4.1.6.3}$$

Exercise 4.1.6.6 Check this.

Let $X(t)$ be a positive function satisfying the condition

$$\int\limits_{-\infty}^{\infty} X(t)dt = 1$$

and such that the linear hull of its translations are dense in $L^1(-\infty, \infty)$. We can choose, for example, the function

$$X(t) := \frac{1}{\sqrt{2\pi}} e^{-\frac{t^2}{2}}$$

because its Fourier transformation does not vanish in $\mathbb{R}$ (it is $e^{-\frac{s^2}{2}}$).

Exercise 4.1.6.7 Check these properties.

Let us define $\nu(\bullet, m)$ by

$$\langle \nu(\bullet, m), \psi \rangle := \int\limits_{(x^0,y)\in Cyl} \psi(x^0, y) \left(\rho \int\limits_{-\infty}^{\infty} Y(dx^0, T^{y-t}m) X(t)dt \right) dy. \quad (4.1.6.4)$$

Now we check the property

$$S_\tau \nu(\bullet, m) = \nu(\bullet, T^\tau m).$$

Using (4.1.6.2), we obtain

$$\begin{aligned}
\langle S_\tau \nu(\bullet, m), \psi \rangle &= \int \psi(x^0, y) \left(\rho \int\limits_{-\infty}^{\infty} Y(dx^0, T^{y+\tau-t}m) X(t)dt \right) dy \\
&= \int \psi(x^0, y) \left(\rho \int\limits_{-\infty}^{\infty} Y(dx^0, T^{y-t}(T^\tau m)) X(t)dt \right) dy \\
&= \langle \nu(\bullet, m), T^\tau m \rangle.
\end{aligned}$$

We also check the condition (4.1.6.3).

$$\begin{aligned}
\int\limits_{y\le 0} e^{\rho y} S_t \nu(dy \otimes dx^0) &= \int\limits_{-\infty}^{\infty} X(t)dt \int\limits_{y\le 0} e^{\rho y})\rho dy \int\limits_{S^{m-1}} Y(dx^0, T^{y+\tau-t}m \\
&\le \sup_{\tau\in\mathbb{R}} Y(S^{m-1}, T^\tau m) \int\limits_{-\infty}^{\infty} X(t)dt \le 1,
\end{aligned}$$

since $Y(\bullet, \bullet) \in K$.

Now we should "transplant" ν back to $\mathbb{R}^m \setminus 0$ such that S_τ passes to $(\bullet)_{[e^\tau]}$. Define $\mu(\bullet, m)$ by

$$\langle \mu(\bullet, m), \psi^* \rangle := \langle \nu(\bullet, m), \psi \rangle, \quad (4.1.6.5)$$

where $\psi^*(rx^0) \in \mathcal{D}(\mathbb{R}^m \setminus 0)$ and

$$\psi(x^0, y) := \psi^*(e^y x^0) e^{-(\rho-m+2)y} \in \mathcal{D}(Cyl).$$

Then

$$\langle (\mu)_{[e^\tau]}, \psi^* \rangle = \langle (\mu), T_{-\tau}\psi^* \rangle = \langle \nu, S - \tau\psi \rangle = \langle S_\tau \nu, \psi \rangle.$$

The condition $\mu(\bullet, m) \in \mathcal{M}[\rho, \sigma]$ is also satisfied.

Exercise 4.1.6.8 Check these properties.

Now we use (4.1.6.1) to transplant the dynamical system to $U[\rho, \sigma]$. This completes a construction of a homomorphism $(T^t, M) \mapsto (T_t, U[\rho, \sigma])$.

Let us check that it is an imbedding, i.e., we must check the one-to-one correspondence. One-to-one correspondence of $v(\bullet, m)$ and $\mu(\bullet, m)$ is known (Theorem 3.1.4.4). One-to-one correspondence of $\mu(\bullet, m)$ and $\nu(\bullet, m)$ can be also checked easily.

Exercise 4.1.6.9 Check this in detail.

So we should check the one-to-one correspondence of $\nu(\bullet, m)$ and $Y(\bullet, m)$. Suppose

$$\nu(\bullet, m_1) = \nu(\bullet, m_2).$$

Then

$$\langle \nu(\bullet, m_1), \psi \rangle = \langle \nu(\bullet, m_2), \psi \rangle \ \forall \psi \in \mathcal{D}(Cyl).$$

In particular, set

$$\psi(x^0, y) = \phi(x^0) R(y), \ \phi \in \mathcal{D}(S^{m-1}), \ R \in \mathcal{D}(-\infty, \infty).$$

Then

$$\langle \nu(\bullet, m_1), \psi \rangle = \int R(y) dy \int_{-\infty}^{\infty} \langle Y(\bullet, T^{y-t} m_1), \phi \rangle_{S^{m-1}} X(t) dt \tag{4.1.6.6}$$

$$= \langle \nu(\bullet, m_2), \psi \rangle = \int R(y) dy \int_{-\infty}^{\infty} \langle Y(\bullet, T^{y-t} m_2), \phi \rangle_{S^{m-1}}$$

where

$$\langle Y(\bullet), \phi \rangle_{S^{m-1}} := \int_{S^{m-1}} \phi(x^0) Y(dx^0).$$

Set

$$F_j(y) := \langle Y(\bullet, T^y m_j), \phi \rangle_{S^{m-1}}, \ j = 1, 2.$$

From (4.1.6.6) we obtain for the convolutions

$$(F_1 * X)(y) \equiv (F_2 * X)(y), \ y \in (-\infty, \infty).$$

Thus

$$F_1(y) \equiv F_2(y), \ y \in (-\infty, \infty)$$

because of the property of X.

Hence

$$Y(\bullet, T^y m_1) \equiv Y(\bullet, T^y m_2), \ y \in (-\infty, \infty).$$

In particular, for $y = 0$ we have

$$Y(\bullet, m_1) = Y(\bullet, m_2).$$

Hence $m_1 = m_2$ because of (4.1.5.3), and this completes the proof of one-to-one correspondence.

Consider the case of an integer ρ. For this case we can use $v \in U[\rho, \sigma]$ of the form

$$v(x) = \Pi_<(x, \mu, \rho) + \Pi_>(x, \mu, \rho)$$

instead of (4.1.6.1).

□

Exercise 4.1.6.10 Check this.

4.1.7 The most simple set satisfying the conditions of Theorem 4.1.3.3 is the set that is generated by a function $v \in U[\rho]$ that has the property

$$v_{[te^P]} = v_{[t]}, \ t \in (0, \infty)$$

for some P.

Then

$$T_{t+P}v = T_t v, \ t \in (-\infty, \infty),$$

i.e., the dynamical system $T_\bullet$ is *periodic* with the period P on the set

$$\mathbb{C}(v) = \{T_t v : 0 \le t \le P\}.$$

Theorem 4.1.7.1 (Periodic Limit Set) *For all $P > 0, \rho > 0, \sigma > 0$, there exists $v \in U[\rho, \sigma]$ such that the dynamical system $(T_\bullet, \mathbb{C}(v))$ is periodic with the period P.*

Proof. Suppose ρ is non-integer. Let us take $\mu \in \mathcal{M}[\rho, \Delta]$ such that the canonical potential $\Pi(x, \mu, [\rho])$ belongs to $U[\rho, \sigma]$. This is possible because of Theorem 3.1.4.2 (*Brelot-Borel).

Denote as μ_P^* the restriction of μ on the spherical ring $\{x : 1 < |x| < e^P\}$ and set

$$\mu_P := \sum_{k=-\infty}^{\infty} T_{kP}\mu_P^*.$$

We have $\mu_P \in \mathcal{M}[\rho, \Delta]$ and

$$T_{t+P}\mu_P = T_t\Big(\sum_{k=-\infty}^{\infty} T_{(k+1)P}\mu_P^*\Big) = T_t\mu_P.$$

Then $v := \Pi(x, \mu_P, [\rho]) \in U[\rho, \sigma]$ and $T_{t+P}v = T_t v$ because of (3.1.5.0).

For an integer ρ we use the function

$$v(x) := \Pi_<(x, \mu_P, \rho) + \Pi_>(x, \mu_P, \rho).$$

□

4.2 Subharmonic function with prescribed limit set

4.2.1 The following two theorems describe structure of limit sets in terms of dynamical systems.

Theorem 4.2.1.1 (Necessity) *Let $u \in SH(\mathbb{R}^m, \rho, \rho(r))$. Then the dynamical system $(T_\bullet, \mathbf{Fr}[u, \bullet])$ is chain recurrent.*

The chain recurrence is also sufficient.

Theorem 4.2.1.2 (Sufficiency) *Let U be a compact connected and $T_\bullet$-invariant subset of $U[\rho, \sigma]$ for some $\sigma > 0$, such that the dynamical system $(T_\bullet, U)$ is chain recurrent. Then for any proximate order $\rho(r) \to \rho$ there exists $u \in SH(\mathbb{R}^m, \rho, \rho(r))$ such that*

$$\mathbf{Fr}[u, \rho(r), V_t, \mathbb{R}^m] = U.$$

Proof of Theorem 4.2.1.1. We need the curve u_t, $t \geq 1$, and $\mathbf{Fr}[u, \bullet]$ to be contained in a common metric space X. Thus we set

$$X := \{v \in SH(\mathbb{R}^m) : \sup_{r \geq 1} M(r, v) r^{-\rho-1} \leq \sup_{r \geq 1} M(r, u) r^{-\rho-1}\}.$$

We want to use Theorem 4.1.1.1 cr 2). Let U be an open proper subset of $\mathbf{Fr}[u, \bullet]$ satisfying (4.1.1.1) and let F be a $T_\bullet$-invariant subset of $K := \mathbf{Fr}[u, \bullet] \setminus U$.

Such F exists. Indeed, K is closed and $T_t K \subset K$ for $t > 0$ (see proof of Theorem 4.1.1.1, cr2)$\Longleftrightarrow$cr3)). Thus $\Omega(K) \subset K$ where $\Omega(\bullet)$ was defined in (4.1.3.2). The set $\Omega(K)$ is invariant with respect to T_t (see Exercise 4.1.3.1). So the set of such sets F is not empty.

If F intersects ∂U at a point v, then $A(v) \subset F \cap \partial U$. Since $A(v)$ is invariant (Exercise 4.1.3.2) ∂U contains a nonempty $T_\bullet$-invariant set. So we obtain the assertion of the theorem using Theorem 4.1.1.1, cr2).

Suppose F does not intersect ∂U. Let U_0 be an open set in X such that

$$U_0 \cap \mathbf{Fr}[u, \bullet] = U, \ \operatorname{clos} U_0 \cap \mathbf{Fr}[u, \bullet] = \operatorname{clos} U \tag{4.2.1.1}$$

(see Exercise 4.2.1.1). Since $\operatorname{clos} U_0 \cap F = \varnothing$ we can take a sequence of open neighborhoods $U_1, U_2, \dots$ of F in X such that all sets $\operatorname{clos} U_j$, $j = 1, 2, \dots$ do not intersect $\operatorname{clos} U_0$ and $U_j \downarrow F$.

By definition of $\mathbf{Fr}[u, \bullet]$ we can find intervals $a_j \leq t \leq b_j$ with $a_j \to \infty$ such that $u_{e^{a_j}} \in \partial U_j$, $u_{e^{b_j}} \in \partial U_0$, and $u_{e^t} \notin \operatorname{clos} U_0 \cup \operatorname{clos} U_j$ for $a_j < t < b_j$. We can pass to a subsequence and assume that

$$u_{e^{a_j}} \to w \in F. \tag{4.2.1.2}$$

Let us use the identity

$$u_{e^{t+a_j}} = (u_{e^{a_j}})_{e^t} \frac{\rho(e^t)\rho(e^{a_j})}{\rho(e^{t+a_j})}. \tag{4.2.1.3}$$

By (4.2.1.2), (4.2.1.3) and the property (3.1.2.2) of a proximate order we obtain

$$u_{e^{t+a_j}} \to T_t w \in F$$

uniformly for any bounded set of t. Thus $b_j - a_j \to \infty$.

Passing to a subsequence we may assume that $u_{e^{b_j}} \to v \in \mathbf{Fr}[u, \bullet] \cap \partial U_0 = \partial U$. Since $u_{e^{t+b_j}} \to T_t v$ and $u_{e^{t+b_j}} \notin U_0$ when $a_j - b_j < t < 0$ we obtain that $T_t v \notin U$ when $t < 0$.

Hence the whole backward orbit $\{T_t v : t < 0\}$ lies in ∂U, which must therefore contain the $T_\bullet$-invariant set $A(v)$. □

Exercise 4.2.1.1 Prove that the set

$$U_0 := \bigcup_{v \in U} \{w \in X : \operatorname{dist}(v, w) < \operatorname{dist}(v, K)/2\}$$

satisfies the conditions (4.2.1.1).

Proof. We have

$$U_0 \supset U \Longrightarrow U_0 \cap \mathbf{Fr}[u, \bullet] \supset U \cap \mathbf{Fr}[u, \bullet] = U.$$

Thus

$$U_0 \cap \mathbf{Fr}[u, \bullet] \supset U. \tag{4.2.1.4}$$

From (4.2.1.4) we have

$$\operatorname{clos} U_0 \cap \mathbf{Fr}[u, \bullet] = \operatorname{clos} U_0 \cap \operatorname{clos} \mathbf{Fr}[u, \bullet] = \operatorname{clos}(U_0 \cap \mathbf{Fr}[u, \bullet]) \supset \operatorname{clos} U. \tag{4.2.1.5}$$

Finally (4.2.1.4) $\wedge$ (4.2.1.5) $\Longrightarrow$ (4.2.1.1). □

4.2.2 To prove Theorem 4.2.1.2 we need some preparation. Theorems of the next Sections form the basis of the construction that we will use in the proof.

Let β be an infinitely differentiable function on $\mathbb{R}$ such that $0 \le \beta(x) \le 1$, $\beta(x) = 0$ for $x \le 0$ and $\beta(x) = 1$ for $x \ge 1$. We can set, for example,

$$\beta(x) := A \int_{-\infty}^{x} \alpha(y+1) dy$$

where α is taken from (2.3.1.1) and

$$A = \int_{-\infty}^{\infty} \alpha(y+1) dy.$$

Suppose that the sequences $\{r_k, \sigma_k,\ k = 0, 1, \dots\}$ satisfy the following conditions:

$$r_0 = 1;\ r_k < r_k\sigma_k < r_{k+1}/\sigma_{k+1} < r_{k+1},\ k = 1, 2, \dots, \tag{4.2.2.1}$$

$$\sigma_k \uparrow \infty;\ \frac{r_{k+1}}{\sigma_{k+1} r_k \sigma_k} \uparrow \infty. \tag{4.2.2.2}$$

Set

$$\psi_k(r) := \beta\left(\frac{\log r - \log(r_k/\sigma_k)}{\log(\sigma_k r_k) - \log(r_k/\sigma_k)}\right) - \beta\left(\frac{\log r - \log(r_{k+1}/\sigma_{k+1})}{\log(\sigma_{k+1} r_{k+1}) - \log(r_{k+1}/\sigma_{k+1})}\right),$$

$$\psi_0(r) := 1 - \beta\left(\frac{\log r - \log(r_1/\sigma_1)}{\log(\sigma_1 r_1) - \log(r_1/\sigma_1)}\right).$$

The sequence $\{\psi_k\}$, $k = 0, 1, \dots$ forms a *partition of unity* with the following properties:

Theorem 4.2.2.1 (Partition of Unity) *One has*

$$\sum_{k=0}^{\infty} \psi_k = 1; \tag{prtu1}$$

$$\operatorname{supp} \psi_k \subset (r_k/\sigma_k, r_{k+1}\sigma_{k+1}); \tag{prtu2}$$

$$\psi_k(r) = 1,\ \textit{for}\ r \in (r_k\sigma_k, r_{k+1}/\sigma_{k+1}); \tag{prtu3}$$

$$\operatorname{supp} \psi_k \cap \operatorname{supp} \psi_l = \varnothing\ \textit{for}\ |k - l| > 1; \tag{prtu4}$$

$$\lim_{k\to\infty} \max_r \psi_k'(r) r = \lim_{k\to\infty} \max_r \psi_k''(r) r^2 = 0. \tag{prtu5}$$

Moreover

$$\max_r |\psi_k'(r) r|, \max_r |\psi_k''(r) r^2| \le \gamma_k \tag{prtu6}$$

where γ_k can be made to tend to zero arbitrarily fast by choosing the sequences $\{\sigma_k\}$ and $\{r_k\}$.

Proof. Set

$$\beta_k(r) := \beta\left(\frac{\log r - \log(r_k/\sigma_k)}{\log(\sigma_k r_k) - \log(r_k/\sigma_k)}\right).$$

The functions $\beta_k(r)$ and $\beta_{k+1}(r)$ vanish for $r < r_k/\sigma_k$ because $\beta(x) = 0$ for $x \le 0$, and both of them are equal to 1 for $r \ge \sigma_k r_k$ because $\beta(x) = 1$ for $x \ge 1$. Hence, (prtu2) holds.

One has for any $r \in (0, \infty)$,

$$\sum_{k=0}^{n} \psi_k = 1 - \beta_{n+1}(r).$$

As mentioned, $\beta_{n+1}(r) = 0$ for n such that $r_{n+1}/\sigma_{n+1} > r$. Thus (prtu1) holds.

Counting derivatives of ψ_k, we have:

$$\max_r |r\psi_k'(r)| \leq$$

$$\left[(\log(\sigma_k r_k) - \log(r_k/\sigma_k))^{-1} + (\log(\sigma_{k+1} r_{k+1}) - \log(r_{k+1}/\sigma_{k+1}))^{-1}\right] \max_x |\beta'|(x).$$

Thus we can take the right side of the inequality as γ_k and regulate its vanishing by choice of the ratio in (4.2.2.2). The same holds for $r^2\psi''(r)$. Hence (prtu5) and (prtu6) are proved.

Exercise 4.2.2.1 Check (prtu4). □

4.2.3 Now we construct a function which is of zero type but has a "maximal possible" mass density.

Theorem 4.2.3.1 (Maximal Mass Density Function) *Let $\rho(r) \to \rho$, $\rho > 0$ be a smooth proximate order (i.e., having properties (2.8.1.8)), and let $\gamma(r)$, $r \in [0, \infty)$, satisfy the conditions: $\gamma(r) > 0$ and $\gamma(r) \to 0$, as $r \to \infty$.*

Then there exists an infinitely differentiable subharmonic function $\Phi(x)$ such that

$$\Delta\Phi(x) \geq \gamma(x)|x|^{\rho(r)-2} \tag{4.2.3.1}$$

and

$$(\Phi)_t \to 0 \tag{4.2.3.2}$$

in $\mathcal{D}'$ as $t \to \infty$.

To prove Theorem 4.2.3.1 we need an elementary lemma.

Theorem 4.2.3.2 (Convex Majorization) *Let $a(s)$, $s \in [s_0, \infty)$ be a function such that $a(s) \to -\infty$ as $s \to \infty$. Then there exists an infinitely differentiable, convex function $k(s)$ such that:*

k1) $k(s) \geq a(s)$;

k2) $k(s) \downarrow -\infty$ *as* $s \to \infty$;

k3) $k^{(n)}(s) \to 0$ *for all* $n = 1, 2, \ldots$.

Proof. Set

$$a^*(s) := \sup\{a(t) : t \geq s\}.$$

Then $a^*(s) \downarrow -\infty$ as $s \to \infty$.

Set $b_0 := -a^*(s_0)$ and denote as $s(b)$, $b \in [b_0, +\infty)$ the function inverse to the function $-a^*(s)$. Let us construct a convex function that majorates $s(b)$ and tends to infinity monotonically with all its derivatives. It can be done in the following way.

First we construct a piecewise linear convex function. Set

$$s_1(b) := s_0 + 1 + \alpha_0(b - b_0), \ b \in [b_0, b_0 + 1],$$

and choose α_0 such that the inequality $s_1(b) > s(b)$ holds for $b \in [b_0, b_0 + 1]$.

For this we choose

$$\alpha_0 \geq \sup_{b \in [b_0, b_0+1]} \frac{s(b) - s_0 - 1}{b - b_0}.$$

Since $s(b) - s_0 - 1 < 0$ the right side is finite.

For all the following intervals we set

$$s_1(b) := s_1(b_0 + j) + \alpha_j(b - b_0 - j), \ b \in [b_0 + j, b_0 + j + 1],$$

where $\alpha_j \geq \alpha_{j-1}$ and satisfies the condition

$$\alpha_j \geq \sup_{b \in [b_0+j, b_0+j+1]} \frac{s_1(b) - s_1(b_0 + j)}{b - b_0 - j}.$$

To obtain a smooth function, set

$$s_2(b) := \int \alpha(b - x) s_1(x) dx, \tag{4.2.3.3}$$

where $\alpha(x)$ is defined by (2.3.1.1). Then $s_2(b)$ is infinitely differentiable, monotonic and convex.

Exercise 4.2.3.1 Check this.

Set

$$k(s) := -s_2^{-1}(s), \tag{4.2.3.4}$$

where $s_2^{-1}(s)$ is the inverse function to s_2. One can check that $k(s)$ satisfies the properties k1), k2), k3). □

Exercise 4.2.3.2 Check that $k(s)$ satisfies k1), k2), k3).

Proof of Theorem 4.2.3.1. We are going to show that Φ can be taken in the form

$$\Phi(x) := c e^{k(\log |x|^2)} |x|^{\rho(|x|)} \tag{4.2.3.5}$$

where c and $k(s)$ will be chosen later.

Note that $\Phi(x) = \Phi(|x|)$ depends only on $r = |x|$ and pass to the variable $s := \log r^2$. Then for $\phi(s) := \Phi(e^{s/2})$ we have

$$\begin{aligned} \Delta\Phi(x) &= r^{1-m} \frac{\partial}{\partial r} r^{m-1} \frac{\partial}{\partial r} c e^{k(\log r^2)} r^{\rho(r)} \\ &= c e^{-s} \left(\frac{\partial^2}{\partial s^2} + \frac{m-2}{2} \frac{\partial}{\partial s} \right) \phi(s) \geq c m e^{-s} \min[\phi''(s), \phi'(s)]. \end{aligned} \tag{4.2.3.6}$$

Let us chose k as in Theorem 4.2.3.2 with $a(s) := \log \gamma(r) = \log \gamma(e^{\frac{s}{2}})$. Now we estimate the derivatives from below.

$$\phi'(s) = \phi(s)\left[k'(s) + \frac{1}{4}s\rho'(e^{\frac{s}{2}}) + \frac{1}{2}\rho(e^{\frac{s}{2}})\right].$$

By k3) and k1), $k'(s) \to 0$ and $k(s) \geq a(s)$. Also $s\rho'(e^{\frac{s}{2}}) \to 0$ and $\rho(e^{\frac{s}{2}}) \to \rho$ by properties of proximate order (Theorem 2.8.1.4). Thus we can chose c such that

$$\phi'(s) > \frac{1}{m}e^{\log \gamma(e^{\frac{s}{2}}) + \frac{s}{2}\rho(e^{\frac{s}{2}})}. \tag{4.2.3.7}$$

Differentiating once again, we obtain

$$\phi''(s) = \phi(s)\left[k'(s) + \frac{1}{4}s\rho'(e^{\frac{s}{2}}) + \frac{1}{2}\rho(e^{\frac{s}{2}})\right]^2 + \left[k''(s) + \frac{1}{2}\rho'(e^{\frac{s}{2}}) + \frac{1}{8}s\rho''(e^{\frac{s}{2}})\right].$$

From here we obtain by choosing c:

$$\phi''(s) > \frac{1}{m}e^{\log \gamma(e^{\frac{s}{2}}) + \frac{s}{2}\rho(e^{\frac{s}{2}})}. \tag{4.2.3.8}$$

Using (4.2.3.6), (4.2.3.7) and (4.2.3.8) we obtain:

$$\Delta\Phi(s) > e^{\log \gamma(e^{\frac{s}{2}}) + \frac{s}{2}\rho(e^{\frac{s}{2}})}.$$

Returning to the variable r we obtain (4.2.3.1). Correctness of (4.2.3.2) can be checked directly using k2) and properties of the proximate order (Theorem 2.8.1.3).

Exercise 4.2.3.3 Check this. □

4.2.4 We have already approximated distributions and subharmonic functions by infinitely differentiable functions (Theorems 2.3.4.5 and 2.6.2.3). Now we need to make more precise this approximation. Namely, we are going to make it uniform with respect to $v \in U[\rho, \sigma]$. We will denote

$$\partial^l := \frac{\partial^{|l|}}{(\partial x_1)^{l_1}(\partial x_2)^{l_2}\cdots(\partial x_m)^{l_m}} \tag{4.2.4.1}$$

where $l = (l_1, l_2, \ldots, l_m)$, $|l| = l_1 + l_2 + \cdots + l_m$.

Set for $v \in U[\rho, \sigma]$

$$R_\epsilon v(x) := \int \alpha_\epsilon(x-y)v(y)dy \tag{4.2.4.2}$$

where α_ϵ is taken from (2.3.1.3).

We have changed the notation from 2.3.1 and 2.6.2 because a subindex of v was already engaged for t.

For a fixed $0 < \delta \leq 0.5$, set

$$\mathrm{Str}(\delta) := \{x : \delta \leq |x| \leq \delta^{-1}\}. \tag{4.2.4.3}$$

Theorem 4.2.4.1 (Estimation of R_ϵ) *Let $v \in U[\rho, \sigma]$. Then*

R1. *for a fixed $g \in \mathcal{D}(\mathbb{R}^m \setminus 0)$ with $\operatorname{supp} g \subset \operatorname{Str}(\delta)$,*

$$|\langle R_\epsilon v - v, g\rangle| \leq o(1, g) 2\sigma\delta^{-\rho} \tag{4.2.4.4}$$

where $o(1, g) \to 0$ as $\epsilon \to 0$;

R2. *the inequality*

$$|\partial^l R_\epsilon v(x)| \leq A(m)\sigma\epsilon^{-|l|-m+1}|x|^{-|l|+\rho}, \tag{4.2.4.5}$$

with $A(m)$ depending only on m, holds for $\epsilon < |x|/2$.

Proof. One has

$$\langle R_\epsilon v, g\rangle = \langle v, R_\epsilon g\rangle. \tag{4.2.4.6}$$

Thus

$$\langle R_\epsilon v - v, g\rangle = \langle v, R_\epsilon g - g\rangle. \tag{4.2.4.7}$$

Exercise 4.2.4.1 Check (4.2.4.6) and (4.2.4.7).

Now

$$|\langle v, R_\epsilon g - g\rangle| \leq \max_{\operatorname{Str}(\delta)} |R_\epsilon g - g|(x) \int\limits_{\operatorname{Str}(\delta)} |v|(x)dx. \tag{4.2.4.8}$$

The first factor is $o(1)$ because g is smooth. For the second one we have

$$\int\limits_{\operatorname{Str}(\delta)} |v|(x)dx \leq 2 \int\limits_{\operatorname{Str}(\delta)} v^+(x)dx \leq 2\sigma\delta^{-\rho}. \tag{4.2.4.9}$$

This and (4.2.4.8) imply R1).

Differentiating the equality

$$R_\epsilon v(x) := C_m \int \epsilon^{-m}\alpha(|x-y|/\epsilon)v(y)dy,$$

we have

$$|\partial^l R_\epsilon v(x)| \leq C_m\epsilon^{-|l|-m} \max_{\{|y|<\epsilon\}} |\partial^l\alpha(|y|)| \int\limits_{\{|y|<\epsilon\}} |v(x-y)|dy.$$

Suppose $|x| = 1$. Then for $0 < \epsilon \leq 0.5$, we have

$$\int\limits_{|y|<\epsilon} |v|(x-y)dy \leq \int\limits_{1-\epsilon<|x|<1+\epsilon} |v|(x) \leq 2 \int\limits_{1-\epsilon<|x|<1+\epsilon} v^+(x)dx$$
$$\leq \sigma_m 2 \cdot 2\epsilon\sigma(1+\epsilon)^\rho \leq \sigma_m 6\sigma\epsilon$$

where σ_m is the square of the unit sphere. Hence for $|x| = 1$

$$|\partial^l R_\epsilon v(x)| \leq A(m)\sigma\epsilon^{-|l|-m+1} \tag{4.2.4.10}$$

with

$$A(m) = 6\sigma_m \max_{y\in\mathbb{R}^m} |\partial^l \alpha(|y|)|.$$

Set $t = |x|$. Apply the inequality (4.2.4.10) to $v := v_{[t]}(y)$ with $y := x/|x|$. Then

$$|\partial^l R_\epsilon v_{[t]}(y)| \leq A(m)\sigma\epsilon^{-|l|-m+1}.$$

Computing the derivatives, we obtain

$$\partial^l R_\epsilon v_{[t]}(x) = t^{-\rho} t^{|l|} \partial^l R_\epsilon v(x)|_{x=ty}.$$

Thus one has R2. □

4.2.5 In this section we describe the main part of a construction that will be used in the proof of Theorem 4.2.1.2.

Let $\{v_j \in U[\rho, \sigma],\ j = 1, 2, \ldots\}$ and $\{\psi_j,\ j = 1, 2\ldots\}$ be the partition of unity from Theorem 4.2.2.1. Let us chose $\epsilon_j \downarrow 0$ such that the condition

$$\gamma_j \epsilon_j^{-m} \to \infty \tag{4.2.5.1}$$

holds for γ_j taken from Theorem 4.2.2.1, (prtu 6). Set

$$v(x|t) := \sum_{j=0}^{\infty} \psi_j(t)(v_j)_{[t]}(x), \tag{4.2.5.2}$$

where $(\cdot)_{[t]}$ is defined by (3.1.2.4a).

One can see that $v(x|t) \in U[\rho, 3\sigma]$ for all t.

Exercise 4.2.5.1 Show this, using properties of $\{\psi_j\}$ and invariance of $U[\rho, \sigma]$ with respect to $(\cdot)_{[t]}$.

We can consider $v(x|t)$ as a curve (a *pseudo-trajectory*) in $U[\rho, 3\sigma]$.

Set

$$u(x) := \sum_{j=0}^{\infty} \psi_j(|x|) R_{\epsilon_j}(v_j)(x)|x|^{\rho(|x|)-\rho}. \tag{4.2.5.3}$$

where R_ϵ is defined by (4.2.4.2).

It is an infinitely differentiable function in $\mathbb{R}^m$.

Theorem 4.2.5.1 (Construction) *One has*

$$u_t - v(\bullet|t) \to 0 \tag{4.2.5.4}$$

in $\mathcal{D}'(\mathbb{R}^m)$, *and*

$$\Delta u(x) = f(x) + \gamma(x)|x|^{\rho(|x|)-2} \tag{4.2.5.5}$$

with $f(x) \geq 0$ *and* $\gamma(x) = o(1)$ *as* $|x| \to \infty$.

Let us note that the function $u(x)$ is "almost-subharmonic" and can be made subharmonic by summing with the function Φ from Theorem 4.2.3.1.

Exercise 4.2.5.2 Prove this.

So we have

Theorem 4.2.5.2 (Pseudo-Trajectory Asymptotics) *For any* $v(x|t)$ *of the form* (4.2.5.2) *there exists an infinitely differentiable function* $u \in SH(\rho(r))$ *that satisfies* (4.2.5.4).

Proof of Theorem 4.2.5.1. One has

$$u_t(x) := \sum_{j=0}^{\infty} \psi_j(t|x|)(R_{\epsilon_j}(v_j))_{[t]}(x)a(x,t),$$

where

$$a(x,t) := \frac{|tx|^{\rho(t|x|)-\rho}}{t^{\rho(t)-\rho}}.$$

For any $0 < \delta < 0.5$ and $x \in \mathrm{Str}(\delta)$, $a(x,t) \to 1$ uniformly in $|x|$as $t \to \infty$. This follows from Theorem 2.8.1.3, ppo3).

Exercise 4.2.5.3 Check this in detail.

We have

$$u_t(x) - v(x|t) = \sum_{j=0}^{\infty} [\psi_j(t|x|)(R_{\epsilon_j}(v_j))_{[t]}(x)a(x,t) - \psi_j(t)(v_j)_{[t]}(x)], \quad (4.2.5.6)$$

and there are no more than three summands in the sum for sufficiently large $t = t(\delta)$ because of Theorem 4.2.2.1, prtu4. Let us estimate every summand. One has

$$\begin{aligned} b_j(x,t) &:= [\psi_j(t|x|)(R_{\epsilon_j}(v_j))_{[t]}(x)a(x,t) - \psi_j(t)(v_j)_{[t]}(x)] \\ &= [\psi_j(t|x|) - \psi_j(t)](R_{\epsilon_j}(v_j)))_{[t]}a(x,t) + \psi_j(t)(R_{\epsilon_j}(v_j))_{[t]}(x)[a(x,t) - 1] \\ &\quad + \psi_j(t)[(R_{\epsilon_j}(v_j))_{[t]}(x) - (v_j)_{[t]}(x)] \\ &:= (a_1 + a_2 + a_3)(x,t). \end{aligned}$$

Let us estimate $\langle b_j(\bullet,t), g\rangle$ for every $g \in \mathcal{D}(\mathbb{R}^m \setminus 0)$.

We can assume that supp $g \subset \mathrm{Str}(\delta)$. Set

$$M(g) := \max_{x \in \mathrm{Str}(\delta)} |g|(x).$$

We have

$$|\langle a_1(\bullet,t), g\rangle| \le M(g) \max_{r \in (0,\infty)} |r\psi_j'(r)|\delta^{-1} \int_{\mathrm{Str}(\delta)} |(R_{\epsilon_j}(v_j))_{[t]}|(x)dx.$$

One can check that

$$\int_{\mathrm{Str}(\delta)} |(R_{\epsilon_j}(v_j))_{[t]}|(x)dx \leq 3\sigma\delta^{-\rho}.$$

Exercise 4.2.5.4 Check this using (4.2.4.9) and the invariance of $U[\rho,\sigma]$ with respect to $(\bullet)_{[t]}$ (see (3.1.2.4)).

Hence

$$|\langle a_1(\bullet,t),g\rangle| \leq C_1(g)\gamma_j. \tag{4.2.5.7}$$

Let us estimate $a_2(x,t)$. We have

$$\langle a_2(\bullet,t),g\rangle \leq \max_{\mathrm{Str}(\delta)} |a(x,t)-1|\psi_j(t)M(g)3\sigma\delta^{-\rho} = C_2(g)o(1) \tag{4.2.5.8}$$

where $o(1)\to 0$ as $t\to\infty$.

For estimating $a_3(x,t)$, we use Theorem 4.2.4.1 (Estimation of R_ϵ), (4.2.4.4):

$$|\langle a_3(\bullet,t),g\rangle| \leq o(\epsilon_j,g)2\sigma\delta^{-\rho} \tag{4.2.5.9}$$

where $o(\epsilon_j,g)\to 0$ as $j\to\infty$.

Hence (4.2.5.7), (4.2.5.8) and (4.5.5.9) imply

$$\langle b_j(\bullet,t),g\rangle \to 0 \tag{4.2.5.10}$$

as $t\to\infty$ and $j\to\infty$.

Suppose, for a large fixed t, the sum (4.2.5.6) contains $b_j(x,t)$ for $j=j(t), j=j(t)+1$ and $j=j(t)+2$. This implies that $j(t)\to\infty$ as $t\to\infty$.

Since

$$\langle u_t(\bullet)-v(\bullet|t),g\rangle = \langle b_{j(t)}(\bullet,t),g\rangle + \langle b_{j(t)+1}(\bullet,t),g\rangle + \langle b_{j(t)+2}(\bullet,t),g\rangle$$

we obtain from (4.2.5.10) that $\langle u_t(\bullet)-v(\bullet|t),g\rangle \to 0$ as $t\to\infty$ for any $g\in\mathcal{D}(\mathbb{R}^m\setminus 0)$. This is (4.2.5.4).

Let us prove (4.2.5.5). We have

$$\Delta u = \sum_{j=0}^{\infty}[\Delta(R_{\epsilon_j}v_j)(x)\psi_j(x)|x|^{\rho(|x|)-\rho} + \sum_{l,m.k}\partial^l(R_{\epsilon_j}v_j)(x)\partial^n\psi_j(x)\partial^k|x|^{(\rho|x|)-\rho}], \tag{4.2.5.11}$$

where l,m,k are multi-indexes that satisfy the condition: in any summand there are derivatives in the same variable, the derivatives of ψ_j and $|x|^{\rho(|x|)-\rho}$ have no more than second order and the derivatives of $R_{\epsilon_j}v_j(x)$ have no more than first order.

Exercise 4.2.5.5 Check this.

As usual, the derivative of zero order is the function itself. For any $x \in \mathrm{Str}(\delta)$, the outside sum contains no more then three summands. First we consider only the terms in the square brackets. The first term is nonnegative because of subharmonicity of $R_{\epsilon_j} v_j$ and non-negativity of all the other factors. Set

$$f(x) := \sum_{j=0^\infty} [\Delta(R_{\epsilon_j} v_j)(x)\psi_j(x)|x|^{\rho(|x|)-\rho} \geq 0. \tag{4.5.2.12}$$

Using Theorem 4.2.4.1, R2) we obtain

$$|\partial^l (R_{\epsilon_j} v_j)(x)| \leq A(m)\sigma\epsilon_j^{-|l|-m+1}|x|^{-|l|+\rho} \tag{4.2.5.13}$$

for $|l| = 0$ or $|l| = 1$.

From Theorem 4.2.2.1, prtu6), and inequality $|\partial_{x_i}|x|| \leq 1$ we obtain

$$|\partial^n \psi_j(|x|) \leq |\psi^{(n)}(r)||_{r=|x|} \leq \gamma_j |x|^{-|n|} \tag{4.2.5.14}$$

for $|n| = 1, 2$.

Using properties of the smooth proximate order (Theorem 2.8.1.4), one can obtain

$$|\partial^{|k|}|x|^{\rho(|x|)-\rho}| = (|x|^{\rho(|x|)-\rho-|k|})|_{r=|x|}(1 + o(1)), \tag{4.2.5.15}$$

as $|x| \to \infty$.

Exercise 4.2.5.6 Check in detail (4.2.5.13), (4.2.5.14) and (4.2.5.15).

Thus, for every term of the inner sum, we have

$$\begin{aligned}
&|\partial^l (R_{\epsilon_j} v_j)(x)\partial^n \psi_j(x)\partial^k |x|^{\rho(|x|)-\rho}| \\
&\leq A(m)\sigma\gamma_j\epsilon_j^{-|l|-m+1}|x|^{-2+\rho}|x|^{\rho(|x|)-\rho} \\
&\leq \beta_j |x|^{\rho(|x|)-2},
\end{aligned} \tag{4.2.5.16}$$

where $\beta_j \to 0$ because of the condition (4.2.5.1).

Recall that for every large x the outside sum contains no more than three summands, say, $j = j(x), j = j(x) + 1$ and $j = j(x) + 2$. Thus $j(x) \to \infty$ as $|x| \to \infty$. Hence (4.2.5.12) and (4.2.5.16) imply (4.2.5.5). □

4.2.6

Proof of Theorem 4.2.1.2. Let $v(\bullet|t)$ have the form (4.2.5.2). We denote as $\Omega(v)$ a set of the $\mathcal{D}'$-limits of the form

$$w := \lim_{t_k \to \infty} v(\bullet|t_k).$$

We are going to construct some $v(\bullet|t)$ for which

$$\Omega(v) = U, \tag{4.2.6.1}$$

and at the next step to use Theorem 4.2.5.2 to obtain a subharmonic function with the same limit set.

First we describe the construction of the function $v(\bullet|t)$. Let $\{r_k, t_k,\ k = 1, 2, \dots\}$ be an alternating sequence $r_0 = 1,\ r_k < t_k < r_{k+1}$ such that

$$\lim_{k\to\infty} \frac{t_k}{r_k} = \lim_{k\to\infty} \frac{r_{k+1}}{t_k} = \infty.$$

Let us chose in U a countable, dense set $\{g_j\}$ and form from it a sequence $\{w_k\}$ such that every element g_j is repeated infinitely often. For example,

$$w_1 := g_1, w_2 := g_1, w_3 := g_2, w_4 := g_1, w_5 := g_2, w_6 := g_3, \dots .$$

Set

$$q_k := (w_k)_{[1/t_k]} = T_{-\log t_k} w_k$$

in the notation (4.1.5.1).

Now we use that $(T_\bullet, U)$ is chain recurrent. Set

$$\alpha_k := \log \frac{r_k}{t_k}; \quad \omega_k := \log \frac{r_{k+1}}{t_k}$$

and find, by Theorem 4.1.2.1, a sequence $\{v_j\} \supset \{q_k\}$ such that the condition (4.1.2.1) holds, i.e.,

$$T_{\omega_k} v_k - T_{\alpha_{k+1}} v_{k+1} \to 0 \tag{4.2.6.2}$$

as $k \to \infty$.

Set in Theorem 4.1.2.3

$$p_k := T_{\alpha_{k+1}} v_{k+1},\ q_k := T_{\omega_k} v_k$$

and find γ_k such that the condition

$$T_\tau \circ T_{\omega_k} v_k - T_\tau \circ T_{\alpha_{k+1}} v_{k+1} \to 0 \tag{4.2.6.3}$$

holds uniformly for $\tau \in [-\gamma_{k+1}, \gamma_k]$.

Set

$$\sigma_k := \min \left[e^{\gamma_k}, \sqrt{\frac{t_k}{r_k}} \right].$$

These σ_k satisfy the conditions (4.2.2.1) and (4.2.2.2).

Exercise 4.2.6.1 Check this.

We define $v(\bullet|t)$ by (4.2.5.2) with described v_j and with ψ_j from Theorem 4.2.2.1, corresponding to the chosen r_j and σ_j. Let us prove (4.2.6.1).

Consider for fixed k the following three cases.

1. $t \in [r_k \sigma_k, r_{k+1}/\sigma_{k+1})$;
2. $t \in [r_{k+1}/\sigma_{k+1}, r_{k+1})$;
3. $t \in [r_k, r_k \sigma_k)$.

For the first case we have

$$v(\bullet|t) = (v_k)_{[t/t_k]} = T_{\log(t/t_k)} v_k.$$

For the second one

$$\begin{aligned} v(\bullet|t) &= \psi_k(t)(v_k)_{[t/t_k]} + \psi_{k+1}(t)(v_{k+1})_{[t/t_{k+1}]} \\ &= (v_k)_{[t/t_k]} + \psi_{k+1}(t)[(v_{k+1})_{[t/t_{k+1}]} - (v_k)_{[t/t_k]}. \end{aligned}$$

We transform the expression in the square brackets

$$\begin{aligned} (v_{k+1})_{[t/t_{k+1}]} &= T_{\log(t/t_{k+1})} v_{k+1} = T_{\log(t/r_{k+1})} \circ T_{\log(r_{k+1}/t_{k+1})} v_{k+1} \\ &= T_{\log(t/r_{k+1})} \circ T_{\alpha_{k+1}} v_{k+1}. \end{aligned}$$

For the second term, we obtain

$$(v_k)_{[t/t_k]} = T_{\log(t/r_{k+1})} \circ T_{\omega_k} v_k.$$

Exercise 4.2.6.2 Check this.

Setting $\tau := \log(t/r_{k+1})$, we have

$$v(\bullet|t) = (v_k)_{[t/t_k]} + \psi_{k+1}(t)[T_\tau \circ T_{\alpha_{k+1}} v_{k+1} - T_\tau \circ T_{\omega_k} v_k], \tag{4.2.6.4}$$

where $\tau \in [-\log \sigma_{k+1}, 0) \subset [-\gamma_{k+1}, \gamma_k]$. For the third case, set $\tau := \log(t/r_k)$. Then

$$v(\bullet|t) = (v_k)_{[t/t_k]} + \psi_k(t)[T_\tau \circ T_{\omega_{k-1}} v_{k-1} - T_\tau \circ T_{\alpha_k} v_k], \tag{4.2.6.5}$$

where $\tau \in [0, \log \sigma_k) \subset [-\gamma_k, \gamma_{k-1}]$.

Let $t_N \to \infty$ be an arbitrary sequence. Choosing a subsequence, we may suppose that there exist the limits $(v_{k(t_N)})_{[t_N/t_{k(t_N)}]} \to v^* \in U$ and $v(\bullet|t_N) \to v_\infty$.

Choosing a subsequence, we may suppose that t_N satisfies either 1 or 2 or 3 For case 1, we obtain at once $v_\infty = v^* \in U$.

For case 2, from (4.2.6.4), (4.2.6.2) and Theorem 4.1.1.3 we obtain that the superfluous addends tend to zero, and hence $v_\infty \in U$.

The same holds for case 3. Hence $\Omega(v) \subset U$.

Further, for $t = t_k$, we have $v(\bullet|t) = w_k$. The sequence $\{w_k\}$ contains the set $\{g_j\}$ that is dense in U. Thus $\Omega(v) \supset U$. Thus equality (4.2.6.1) has been proved.

As already said, the application of Theorem 4.2.5.2 concludes the proof. □

4.3 Further properties of limit sets

4.3.1 Let as mark the following property of the pseudo-trajectory $v(\bullet|t)$ defined in (4.2.5.2):

Theorem 4.3.1.1 *One has*

$$T_\tau v(\bullet|e^t) - v(\bullet|e^{t+\tau}) \to 0 \tag{4.3.1.1}$$

as $t \to \infty$ *uniformly with respect to* $\tau \in [a,b]$ *for any* $[a,b] \subset (-\infty, \infty)$.

Proof. Using the definition of $(\bullet)_t$ (see (3.1.2.1)) and (4.2.5.4), the remainder in (4.3.1.1) can be represented in the form

$$b(t,\tau,\bullet) := T_\tau v(\bullet|e^t) - v(\bullet|e^{t+\tau}) = T_\tau(u_{e^t}) - u_{e^{t+\tau}} + o(1)$$

where $o(1) \to 0$ uniformly with respect to $\tau \in [a,b]$ for any $[a,b] \subset (-\infty,\infty)$.

Exercise 4.3.1.1 Check this in detail.

Then we obtain

$$b(t,\tau,\bullet) = u_{e^{t+\tau}}[e^{\rho(e^t)-\rho(e^{t+\tau})} - 1] + o(1) \to 0$$

uniformly in the same sense due to precompactness of the family $\{u_{e^t}\}$ and properties of the proximate order.

Exercise 4.3.1.2 Check this in detail. □

The property (4.3.1.1) shows that the pseudo-trajectory $v(\bullet|t)$ behaves asymptotically like the dynamical system $T_\bullet$. Thus a pseudo-trajectory with this property is called an *asymptotically dynamical* pseudo-trajectory with *dynamical asymptotics* $T_\bullet$ (a.d.p.t.).

Theorem 4.2.5.1 shows that for any a.d.p.t. of the form (4.2.5.2) there exists $u \in SH(\rho(r))$ that satisfies the condition

$$u_{e^t} - v(\bullet|e^t) \to 0 \tag{4.3.1.2}$$

as $t \to \infty$.

The following assertion shows that we can suppose $v(\bullet|\bullet)$ to be an arbitrary, in some sense, a.d.p.t.

We call a pseudo-trajectory $w(\bullet|\bullet)$ *piecewise continuous* if the property

$$w(\bullet|t+h) - w(\bullet|t) \to 0 \tag{4.3.1.3}$$

as $h \to 0$ holds for all t except perhaps a countable set without points of condensation.

Let $U \subset U[\rho, \sigma]$ for some $\sigma > 0$. A pseudo-trajectory $w(\bullet|\bullet)$ is called *ω-dense* in U if $\Omega(w) = U$ (see (4.1.3.2)), i.e.,

$$\{v \in U[\rho] : (\exists t_j \to \infty)\ v = \mathcal{D}' - \lim w(\bullet|e^{t_j})\} = U. \tag{4.3.1.4}$$

We have proved already that $v(\bullet|\bullet)$ defined by (4.2.5.2) has this property (see (4.2.6.1)).

Now we consider again the dynamical system $(T_\bullet, U)$ where $U \subset U[\rho, \sigma]$ for some $\sigma > 0$ and T_t is defined by (4.2.1.1).

Theorem 4.3.1.2 (A.D.P.T. and Chain Recurrence) $(T_\bullet, U)$ *is chain recurrent iff there exists an a.d.p.t. that is piecewise continuous and ω-dense in U.*

Necessity has been proved already, because the pseudo-trajectory (4.2.6.2) possesses this property. Sufficiency will be proved later.

The claim of piecewise continuity can be justified by

Theorem 4.3.1.3 *For any $u \in SH(\rho(r))$ there exists a piecewise continuous pseudo-trajectory $w(\bullet|\bullet)$ such that*

$$u_t - w(\bullet|t) \to 0 \tag{4.3.1.5}$$

as $t \to \infty$.

Of course, $w(\bullet|\bullet)$ is a.d.p.t.

Exercise 4.3.1.2 Check this.

4.3.2

Proof of Theorem 4.3.1.3. Let $\{t_n\}$ be any sequence such that

$$t_n \to \infty,\ t_{n+1}/t_n \to 1, \tag{4.3.2.1}$$

for example, $t_n = n$.

There exists a sequence $\{v_n\} \subset \mathbf{Fr}[u]$ such that

$$u_{t_n} - v_n \to 0. \tag{4.3.2.2}$$

Set

$$w(\bullet|t) := v_n,\ \text{for } t_n < t \le t_{n+1}. \tag{4.3.2.3}$$

This is a piecewise continuous function.

Let us prove that

$$u_t - w(\bullet|t) \to 0. \tag{4.3.2.4}$$

Assume the opposite; i.e., there exists a sequence $\{t'_k\}$ such that it is not true. We can suppose that

$$u_{t'_k} \to w_1 \in \mathbf{Fr}[u],\ w(\bullet|t'_k) \to w_2 \in \mathbf{Fr}[u],\ w_1 \ne w_2. \tag{4.3.2.5}$$

Let us find a sequence $\{n_k\}$ such that $t_{n_k} < t'_k < t_{n_k+1}$. Then

$$t_{n_k}/t'_k \to 1. \tag{4.3.2.6}$$

From (4.3.2.5), (4.3.2.3) and (4.3.2.2) we have

$$u_{t_{n_k}} \to w_2. \tag{4.3.2.7}$$

Then we have, using properties of $(\bullet)_t$ and the proximate order,

$$u_{t'_k} = (u_{t_{n_k}})_{[t'_k/t_{n_k}]}(1 + o(\log(t'_k/t_{n_k})) \to w_2 \tag{4.3.2.8}$$

because of (4.3.2.6) and the continuity of $u_{[t]}$ on (u, t).

However (4.3.2.8) contradicts (4.3.2.5). Thus (4.3.2.4) holds. □

4.3.3 Now we will prepare the proof of Theorem 4.3.1.2.

Let $\{v_k,\ k = 1, 2, \dots\} \subset U[\rho, \sigma]$ for some σ be a sequence of functions and $\{r_k,\ k = 1, 2, \dots\}, \{t_k\ k = 1, 2, \dots\}$ be two sequences such that

$$0 < r_k < t_k < r_{k+1},\ k = 1, 2, \dots \tag{4.3.3.1}$$

and

$$\lim_{k\to\infty} t_k/r_k = \lim_{k\to\infty} r_{k+1}/t_k = \infty. \tag{4.3.3.2}$$

Set

$$w^*(\bullet|t) := (v_k)_{t/t_k},\ \text{for } t \in [r_k, r_{k+1}) \tag{4.3.3.3}$$

where $k = 1, 2, \dots$.

Theorem 4.3.3.1 *Let $w(\bullet|\bullet) \subset U$ be an arbitrary ω-dense a.d.p.t. and $\{p_j,\ j = 1, 2, \dots\} \subset U$ an arbitrary sequence. Then there exists a sequence $\{v_k,\ k = 1, 2, \dots\} \supset \{p_j,\ j = 1, 2, \dots\}$ and sequences $\{r_k,\ k = 1, 2, \dots\}$ and $\{t_k,\ k = 1, 2, \dots\}$ satisfying* (4.3.3.1) *and* (4.3.3.2) *such that for $w^*(\bullet|\bullet)$ determined by* (4.3.3.3) *the condition*

$$w^*(\bullet|t) - w(\bullet|t) \to 0 \tag{4.3.3.4}$$

as $t \to \infty$ is fulfilled.

This proposition shows that any ω-dense a.d.p.t. is equivalent to one constructed of long pieces of trajectories of the dynamical system $T_\bullet$.

Proof of Theorem 4.3.3.1. We can take sequences $\{\epsilon_j \downarrow 0,\ j = 1, 2, \dots\}$ and $\{b_j \uparrow \infty,\ j = 1, 2, \dots\}$ and choose a sequence $\{\tau_j,\ j = 1, 2, \dots\}$ such that the inequalities

$$d(T_\tau p_j - T_\tau w(\bullet|\tau_j)) < \epsilon_j/2 \tag{4.3.3.5}$$

and

$$d(T_\tau w(\bullet|t) - w(\bullet|e^\tau t)) < \epsilon_j/2,\ t > \tau_j \tag{4.3.3.6}$$

are fulfilled uniformly with respect to $\tau \in [b_{j+1}^{-2}, b_j^2]$.

Indeed, $w(\bullet|\bullet)$ is ω-dense in U, and hence we can find $\tau_n \to \infty$ such that

$$p_n - w(\bullet|\tau_n) \to 0.$$

Set in Lemma 4.1.2.2,

$$p_n := p_n,\ q_n := w(\bullet|\tau_n),\ \gamma_n := 2\log b_j.$$

Then for any ϵ_j we can find $\tau_j := \tau_{n_j}$ such that (4.3.3.5) holds uniformly with respect to $\tau \in [b_{j+1}^{-2}, b_j^2]$.

The inequality (4.3.3.6) holds, because $w(\bullet|\bullet)$ is asymptotically dynamical (see (4.3.1.1)).

We can also suppose without loss of generality that $\tau_j > \tau_{j-1}b_{j-1}^2$, i.e., that the sequence $\{\tau_j\}$ is rather thin.

The inequality (4.3.3.5) shows that for intervals of t that are determined by the inequality $b_{j+1}^{-2} \le t/\tau_j \le b_j^2$ our pseudo-trajectory is already close to some trajectories.

Now we divide the spaces between such intervals into equal parts in the logarithmic scale such that their logarithmic lengths would be between $\log b_j$ and $\log b_{j+1}$, so that they tend to infinity.

To this end, set

$$n_j := \left[\frac{\log \tau_{j+1} - \log \tau_j}{b_j}\right]$$

where $[\cdot]$ means the entire part, and

$$\gamma_j := (\tau_{j+1}/\tau_j)^{\frac{1}{2n_j}}.$$

It is clear that $b_j < \gamma_j < b_j^2$. As centers of new intervals we take the points

$$\tau_{j,l} := \tau_j\gamma_j^{2l},\ l = 0, 1, \ldots, n_j.$$

Thus $\tau_{j,0} = \tau_j$ and $\tau_{j,n_j} = \tau_{j+1}$. The ends of the intervals are $\tau_{j,l}/\gamma_j$ and $\tau_{j,l}\gamma_j$. Now we complete the sequence $\{p_j\}$ by the values of the pseudo-trajectory $w(\bullet|t)$ in the centers of the intervals, i.e., we set

$$p_{j,l} := w(\bullet|\tau_{j,l}),\ l = 1, \ldots, n_j - 1$$

For $t \in (\tau_{j,l}/\gamma_j, \tau_{j,l}\gamma_j),\ l = 1, \ldots, n_j - 1$ we have

$$d((p_{j,l})_{t/\tau_{j,l}} - w(\bullet|t)) < \epsilon_j/2 \tag{4.3.3.7}$$

because of (4.3.3.6).

For $l = 0$ and $l = n_j$ we set accordingly

$$p_{j,0} := p_j;\ p_{j,n_j} := p_{j+1}.$$

Using (4.3.3.5) and (4.3.3.6) we have an inequality like (4.3.3.7) for $l = 0, l = n_j$ but with ϵ_j instead of $\epsilon_j/2$.

We complete the proof, re-denoting all the centers $\tau_{j,l}$ as t_k, all the ends as r_k and all the $p_{j,l}$ as v_k. □

4.3.4

Proof of sufficiency in Theorem 4.3.1.2. A direct corollary of the previous Theorem 4.3.3.1 is

$$w^*(\bullet|r_k - 0) - w^*(\bullet|r_k) \to 0 \tag{4.3.4.1}$$

as $k \to \infty$.

Really, $w^*(\bullet|\bullet)$ is an a.d.p.t.

Exercise 4.3.4.1 Check this as in Theorem 4.3.1.1 using that $w(\bullet|\bullet)$ is asymptotically dynamical.

For $\tau \in [-\epsilon, 0]$ and $t = r_k$ we have uniformly on τ,

$$T_\tau w^*(\bullet|t) - w^*(\bullet|t) = T_\tau(v_k)_{r_k/t_k} - (v_{k+1})_{r_k/t_k} \to 0.$$

Setting $\tau = 0$ we obtain (4.3.4.1).

Let $V \subset U$ be an arbitrary open set, $\epsilon > 0$ arbitrarily small and $s > 0$ arbitrarily large. We show that there exists an (ϵ, s)-chain from V to V.

Choose s_1 such that

i. for $r_k > s_1$, $d(w^*(\bullet|r_k - 0), w^*(\bullet|r_k)) < \epsilon$. This is possible by virtue of (4.3.4.1).

ii. $w(\bullet|s_1) \in V$. This is possible because $w(\bullet|\bullet)$ is ω -dense.

iii. $d(w^*(\bullet|t), w(\bullet|t)) < d(w(\bullet|s_1), \partial V)$ for $t > s_1$. This is possible because of Theorem 4.3.3.1.

Choose $s_2 > s_1$ such that $w(\bullet|s_2) \in V$. This is possible because $w(\bullet|\bullet)$ is ω-dense. Then the pseudo-trajectory $w^*(\bullet|e^t)$ for $s_1 \le e^t \le s_2$ is an (ϵ, s)-chain connecting $w^*(\bullet|s_1)$ and $w^*(\bullet|s_2)$ that belong to V.

Exercise 4.3.4.2 Check this in detail.

Hence $(T_\bullet, U)$ is chain recurrent.

4.3.5 We will prove one more existence theorem that is a corollary of Theorem 4.2.1.2.

Theorem 4.3.5.1 *Let $\Lambda \subset U[\rho]$ be a compact connected and $T_\bullet$-invariant subset of $U[\rho]$. Then for any proximate order $\rho(r) \to \rho$ there exists $u \in SH(\mathbb{R}^m, \rho, \rho(r))$ such that*

$$h(x, u) = \sup\{v(x) : v \in \Lambda\}, \tag{4.3.5.1}$$

$$\underline{h}(x, u) = \inf\{v(x) : v \in \Lambda\}. \tag{4.3.5.2}$$

Proof. Let $U := Conv\Lambda$ be the convex hull of Λ. It is linearly connected and hence polygonally connected (see 4.1.4). By Theorem 4.1.4.1 it is chain recurrent and by Theorem 4.2.1.2 for any proximate order $\rho(r) \to \rho$ there exists $u \in SH(\mathbb{R}^m, \rho, \rho(r))$

such that

$$\mathbf{Fr}[u, \rho(r), V_t, \mathbb{R}^m] = U.$$

Since every $v \in U$ can be represented in the form $v = av_1 + (1-a)v_2$ for $0 \leq a \leq 1$, $v_1, v_2 \in \Lambda$ we obtain (4.3.5.1) and (4.3.5.2) from Theorem 3.2.1.1 (Properties of Indicators), h2).

Exercise 4.3.5.1 Check this.

4.3.6 In applications we need the following

Theorem 4.3.6.1 *Let $p \in P \subset \mathbb{R}^m$ and let P be a connected closed set. Let $U_P := \{v(z,p) : p \in P \subset \mathbb{R}^m\}$ be a family of functions with parameter p such that for every $p \in P$, $v(\bullet, p) \in U[\rho]$ and satisfy the condition* (4.1.3.3). *Then there exists $u \in SH(\rho(r))$ such that $\mathbf{Fr}[u] = U_P$.*

This is a direct corollary of Theorems 4.1.1.2, 4.1.3.3 and 4.2.1.2.

Exercise 4.3.6.1 Explain this in detail.

4.3.7 In the next three sections we return to the periodic limit sets (see Theorem 4.1.7.1). We show that the limit set $\mathbf{Fr}[u, \rho(r), V_\bullet, \mathbb{R}^m]$ of every subharmonic function $u \in SH(\rho(r), \mathbb{R}^m)$, $\rho(r) \to \rho$ for non-integer ρ can be approximated in some sense by periodic limit sets ([Gi], [GLO, Ch. 3, § 2, Thm. 10]).

Here we give some definitions. Let $X_n \subset U[\rho], n = 1, 2, \ldots$ be a sequence of compact sets. We say that X_n *converges to a compact set* $Y \subset U[\rho]$, i.e.,

$$\mathcal{D}' - \lim_{n \to \infty} X_n = Y \qquad (4.3.7.1)$$

if the following two conditions hold:

converg1) $\forall x_n \in X_n$, $n = 1, 2, \ldots$ $\exists x_{n_j} \in X_{n_j}, j = 1, 2, \ldots$ and $y \in Y$ such that $\mathcal{D}' - \lim_{j \to \infty} x_{n_j} = y$;

converg2) $\forall y \in Y$ $\exists x_n \in X_n$, $n = 1, 2, \ldots$, such that $x_n \to y$.

On every compact set K in $\mathcal{D}'$-topology one can introduce a metric $d(\bullet, \bullet)$ such that the topology generated by this metric is equivalent to $\mathcal{D}'$-topology (see, e.g., [AG(1982)]).

Denote by

$$X_\epsilon := \{y \in K : \exists x \in X \text{ such that } d(y, x) < \epsilon\}$$

the *ϵ-neighborhood* of X.

Let X, Y be two compact sets. Set

$$d(X, Y) := \inf\{\epsilon : X \subset Y_\epsilon, Y \subset X_\epsilon\}.$$

Exercise 4.3.7.1 Prove the assertion

$$(4.3.7.1) \iff \{d(X_n, Y) \to 0\}. \tag{4.3.7.2}$$

We prove the following

Theorem 4.3.7.1 (Approximation by Periodic Limit Sets) *Let $u \in SH(\rho(r), \mathbb{R}^m)$, $\rho(r) \to \rho$ for non-integer ρ. Then for every $V_\bullet$ there exists a sequence $u_n \in SH(\rho(r), \mathbb{R}^m)$ with periodic limit sets $\mathbf{Fr}[u_n, \rho(r), V_\bullet, \mathbb{R}^m]$ such that $\mathbf{Fr}[u_n, \bullet] \to \mathbf{Fr}[u, \bullet]$.*

This theorem is a corollary of the following

Theorem 4.3.7.2 *Let $\mu \in \mathcal{M}(\rho(r), \mathbb{R}^m)$, $\rho(r) \to \rho$ for non-integer ρ. Then for every $V_\bullet$ there exists a sequence $\mu_n \in SH(\rho(r), \mathbb{R}^m)$ with periodic limit sets $\mathbf{Fr}[\mu_n, \rho(r), V_\bullet, \mathbb{R}^m]$ such that $\mathbf{Fr}[\mu_n, \bullet] \to \mathbf{Fr}[\mu, \bullet]$.*

Proof of Theorem 4.3.7.1. The canonical potential $u(x):=\Pi(x,\mu,p)$ (see (2.9.2.1)) of a measure $\mu \in \mathcal{M}(\rho(r), \mathbb{R}^m)$ belongs to $SH(\rho(r), \mathbb{R}^m)$ by Theorem 2.9.3.3 and has a limit set

$$\mathbf{Fr}[u, \bullet] = \{\Pi(\bullet, \nu, p) : \nu \in \mathbf{Fr}[\mu_u]\}$$

by Theorem 3.1.4.4 (*Hadamard). The potentials $u_n(x) := \Pi(x, \mu_n, p)$ have periodic limit sets

$$\mathbf{Fr}[u_n, \bullet] = \{\Pi(\bullet, \nu, p) : \nu \in \mathbf{Fr}[\mu_{u_n}]\}$$

by Theorem 3.1.5.0. Let us prove that

$$\mathbf{Fr}[u_n, \bullet] =: X_n \to Y := \mathbf{Fr}[u, \bullet].$$

If $v_n \in \mathbf{Fr}[u_n, \bullet]$ then from the corresponding sequence of $\nu_n := \nu_{v_n} \in \mathbf{Fr}[\mu_n, \bullet]$ we can find a subsequence ν_{n_j} and $\nu \in \mathbf{Fr}[\mu, \bullet]$ such that $\nu_n \to \nu$ (by Theorem 2.2.3.2 (Helly)). It is easy to check, using Theorem 3.1.4.3 (*Liouville), that $v^* = \mathcal{D}' - \lim_{j\to\infty} \Pi(\bullet, \nu_{n_j})$ exists and coincides with $v = \Pi(\bullet, \nu, p) \in \mathbf{Fr}[u, \bullet]$.

So the condition converg1) is verified. In the same way one can check converg2). □

Exercise 4.3.7.2 Prove this in detail.

4.3.8 Now we are going to prove Theorem 4.3.7.2. We begin from

Proposition 4.3.8.1 *For any $\mu \in \mathcal{M}(\rho(r), \bullet)$ there exists $\hat{\mu} \in \mathcal{M}(\rho, \bullet)$ such that*

$$\mathbf{Fr}[\hat{\mu}, \rho, \bullet] = \mathbf{Fr}[\mu, \rho(r), \bullet]. \tag{4.3.8.0}$$

In other words we can suppose further that $\rho(r) \equiv \rho$.

Proof. Set $L(r) = r^{\rho(r)-\rho}$ and

$$\hat{\mu}(dx) := L^{-1}(|x|)\mu(dx). \tag{4.3.8.1}$$

Using properties of proximate order (Section 2.8.1., po1)–po4)), it is easy to check that

$$[L^{-1}(r)]' = L^{-1}(r)o(1) \text{ and } [L(r)]' = L(r)o(1), \text{ as } \to 0. \tag{4.3.8.2}$$

Exercise 4.3.8.1 Prove this.

Let us show that $\hat{\mu} \in \mathcal{M}[\rho, \Delta]$ for some Δ. Indeed

$$\frac{\hat{\mu}(R)}{R^{\rho+m-2}} = R^{-\rho-m+2}\int_0^R \frac{\mu(dr)}{L(r)} = \frac{\mu(r)}{R^{\rho+m-2}L(r)}|_0^R + R^{-\rho-m+2}\int_0^R \mu(r)(L^{-1})'dr.$$

We suppose that $\mu(r) = 0$ in some neighborhood of zero. Using (4.3.8.2) we obtain further for the last expression,

$$\mu(R)R^{-\rho(r)} + R^{-\rho-m+2}\int_0^R \mu(r)(L^{-1})o(1)dr.$$

Using the l'Hospital rule, we obtain

$$\lim_{R\to\infty} R^{-\rho-m+2}\int_0^R \mu(r)(L^{-1}(r))o(1/r)dr = (-\rho-m+2)\lim_{R\to\infty}\mu(R)R^{-\rho(R)}o(1/R).$$

Thus

$$\limsup_{R\to\infty}\frac{\hat{\mu}(R)}{R^{\rho+m-2}} \leq \limsup_{R\to\infty}\mu(R)(L^{-1}(R))[1+o(1/R)] = \overline{\Delta}[\mu,\rho(r)] < \infty.$$

Let us note that $\mu_t = L(t)\hat{\mu}_{[t]}$. This implies equality (4.3.8.0) because $L(t) \to 1$ as $t \to \infty$.

Exercise 4.3.8.2. Prove this in detail. □

Proof of Theorem 4.3.7.2. As we already said we can also suppose that $\mu \in \mathcal{M}[\rho]$. Let $\nu \in \mathbf{Fr}[\mu]$. We can suppose that

$$\nu(\{|x| = 1\}) = 0. \tag{4.3.8.2a}$$

Otherwise we can find τ such that $\nu_{[\tau]}(\{|x| = 1\}) = 0$ and if $\nu_n \to \nu_\tau$ and are periodic, then $(\nu_n)_{[1/\tau]}$ are also periodic and $(\nu_n)_{[1/\tau]} \to \nu$.

Let $r_n \to \infty$ be such that $\mu_{[r_n]} \to \nu$. By passing to subsequences we can make $r_{n+1}/r_n > r_n$.

Denote $K_n := \{x : r_n \leq |x| < r_{n+1}\}$. Set for every $E \subset K_n$

$$\mu_n|_E := \mu(E)$$

and continue it periodically with the period $P_n = r_{n+1}/r_n$ by the equality

$$\mu_n(P_n^k E) = P_n^{k\rho}\mu(E), \ k = \pm 1, \pm 2, \dots. \tag{4.3.8.3}$$

Since every $X \in \mathbb{R}^m$ can be represented in the form

$$X = \bigcup_{k=-\infty}^{\infty} \{X \cap P_n^k K_n\},$$

we can define

$$\mu_n(X) := \sum_{k=-\infty}^{\infty} \mu_n(\{X \cap P_k^n K_n\}).$$

It is easy to check that μ_n is periodic with period P_n and $\mu_n \in \mathcal{M}[\rho, \Delta]$ with Δ independent of n.

Exercise 4.3.8.3 Check this.

Let us prove that

$$\mathbf{Fr}[\mu] = \lim_{n\to\infty} \mathbf{Fr}[\mu_n]. \tag{4.3.8.4}$$

Check the condition convergl). Let $\nu_{n_j} \in \mathbf{Fr}[\mu_{n_j}]$ and suppose $\mathcal{D}' - \lim_{j\to\infty} \nu_{n_j} := \nu$. Let us prove that $\nu \in \mathbf{Fr}[\mu]$.

Since $\mathbf{Fr}[\mu_{n_j}]$ is a periodic limit set,

$$\nu_{n_j} = (\mu_{n_j})_{[\tau_j]}.$$

Take k_j such that

$$\tau_j' := \tau_j P_{n_j}^{k_j} \in [r_{n_j}, r_{n_j+1}).$$

From periodicity μ_{n_j} we obtain

$$\nu_{n_j} = (\mu_{n_j})_{[\tau_j']}.$$

Passing to a subsequence if necessary, we can consider three cases:

i) $\lim_{j\to\infty} \tau_j'/r_{n_j} = \infty$, $\lim_{j\to\infty} \tau_j'/r_{n_j+1} = 0$;

ii) $\lim_{j\to\infty} \tau_j'/r_{n_j} = \tau$; $1 \leq \tau < \infty$; In this case we have also $\lim_{j\to\infty} \tau_j'/r_{n_j+1} = 0$.

iii) $\lim_{j\to\infty} \tau_j'/r_{n_j+1} = \tau$; $0 < \tau \leq 1$; In this case we have also $\lim_{j\to\infty} \tau_j'/r_{n_j} = \infty$.

Consider the case i). Let $\phi \in \mathcal{D}(\mathbb{R}^m \setminus O)$. Then $\operatorname{supp} \phi(x/\tau_j') \subset (r_{n_j}, r_{n_j+1})$ for $j \geq j_0$. It is easy to see that, for $j \geq j_0$,

$$\langle (\mu_{n_j})_{[\tau_j']}, \phi \rangle = \langle \mu_{[\tau_j']}, \phi \rangle.$$

Exercise 4.3.8.4 Check this.

Since $\mu_{[\tau_j']} \to \nu \in \mathbf{Fr}[\mu]$ by definition the condition converg1) holds for the case i).

Consider the case ii). Recall that $O \notin \operatorname{supp}\phi$. Then there exists $1 \leq c < \infty$ such that

$$\operatorname{supp}\phi \subset \{x : |x| \in (1/c, c)\}.$$

Define

$$\phi_t(x) := \phi(x/t)(1/t)^\rho.$$

Represent τ_j' in the form

$$\tau_j' := e_j \tau r_{n_j} \qquad \text{where} \qquad e_j := \frac{\tau_j'}{r_{n_j}\tau}.$$

The condition ii) means that

$$e_j \to 1. \tag{4.3.8.4a}$$

Compute

$$\langle \nu_j, \phi \rangle := \langle (\mu_{n_j})_{[\tau_j']}, \phi \rangle = \langle \mu_{n_j}, ((\phi_\tau)_{e_j})_{r_{n_j}} \rangle.$$

Note that

$$\operatorname{supp}\phi_\tau \subset \{x : |x| \in (\tau/c, \tau c)\}.$$

We can increase c so that $1 \in (\tau/c, \tau c)$.

Consider the following partition of unity. Choose the functions $\eta_k \in \mathcal{D}(\mathbb{R}^m)$, $k = 1, 2, 3$ so that

$$\eta_1(t) + \eta_2(t) + \eta_3(t) = 1$$

for $t \geq 1$ and

$$\begin{aligned}
&\operatorname{supp}\eta_1 \subset \{x : |x| < 1 - \epsilon\},\\
&\operatorname{supp}\eta_2 \subset \{x : |x| \in (1 - 2\epsilon, 1 + 2\epsilon)\},\\
&\operatorname{supp}\eta_3 \subset \{x : |x| > 1 + \epsilon\},
\end{aligned}$$

where ϵ is an arbitrary number, satisfying

$$\tau/c < 1 - 2\epsilon < 1 + 2\epsilon < \tau c.$$

Represent ϕ_τ in the form

$$\phi_\tau = \psi_1 + \psi_2 + \psi_3, \qquad \text{where} \qquad \psi_k = \phi_\tau \eta_k,\ k = 1, 2, 3.$$

In this notation

$$\langle \nu_j, \phi \rangle = \sum_{k=1}^{3} \langle \mu_{n_j}, (\psi_k)_{e_j r_{n_j}} \rangle. \tag{4.3.8.5}$$

Choose j_ϵ such that for $j \geq j_\epsilon$ the following inclusions hold:

$$\begin{aligned}
\operatorname{supp}(\psi_1)_{e_j r_{n_j}} &\subset \{x : |x| \in (r_{n_j}\tau/c, r_{n_j}(1-\epsilon))\};\\
\operatorname{supp}(\psi_2)_{e_j r_{n_j}} &\subset \{x : |x| \in ((1-\epsilon)r_{n_j}, r_{n_j}(1+\epsilon))\};\\
\operatorname{supp}(\psi_3)_{e_j r_{n_j}} &\subset \{x : |x| \in ((1+\epsilon)r_{n_j}, \tau r_{n_j}\}.
\end{aligned}$$

Thus for ψ_3 we have

$$\begin{aligned}
\langle \mu_{n_j}, (\psi_3)_{e_j r_{n_j}} \rangle &= \int (e_j r_{n_j})^{-\rho}\psi_3(|x|/(e_j r_{n_j}))\mu_{n_j}(dx)\\
&= \int (e_j r_{n_j})^{-\rho}\psi_3(|x|/(e_j r_{n_j}))\mu(dx) = \langle \mu_{[r_{n_j}]}, (\psi_3)_{e_j} \rangle.
\end{aligned}$$

Since $\mu_{[r_{n_j}]} \xrightarrow{\mathcal{D}'} \nu$ and $(\psi_3)_{e_j} \xrightarrow{\mathcal{D}} \psi_3$ we have (see Theorem 2.3.4.6)

$$\lim_{j\to\infty} \langle \mu_{n_j}, (\psi_3)_{e_j r_{n_j}} \rangle = \langle \nu, \psi_3 \rangle. \tag{4.3.8.6}$$

Consider the addend with ψ_1. Because of periodicity μ_{n_j} we have

$$\langle \mu_{n_j}, (\psi_1)_{e_j r_{n_j}} \rangle = \langle (\mu_{n_j})_{[P_{n_j}]}, (\psi_1)_{e_j r_{n_j}} \rangle.$$

Transforming the RHS we obtain

$$\langle (\mu_{n_j})_{[P_{n_j}]}, (\psi_1)_{e_j r_{n_j}} \rangle = \langle \mu_{n_j}, ((\psi_1)_{P_{n_j}})_{e_j r_{n_j}} \rangle.$$

Since $P_{n_j} = r_{n_j+1}/r_{n_j}$ the following inclusion holds for $j \geq j_\epsilon$:

$$\begin{aligned}
\operatorname{supp}(\psi_1)_{P_{n_j} e_j r_{n_j}} &\subset \{x : |x| \in (P_{n_j} r_{n_j}\frac{\tau}{c}, P_{n_j} r_{n_j}(1-\epsilon))\}\\
&= \{x : |x| \in (r_{n_j+1}\frac{\tau}{c}, r_{n_j+1}(1-\epsilon))\} \subset \{x : |x| \in (r_{n_j}, r_{n_j+1})\}.
\end{aligned}$$

Thus

$$\langle \mu_{n_j}, ((\psi_1)_{P_{n_j}})_{e_j r_{n_j}} \rangle = \langle \mu, (\psi_1)_{P_{n_j} e_j r_{n_j}} \rangle = \langle \mu, (\psi_1)_{e_j r_{n_j}} \rangle = \langle \mu_{[r_{n_j}]}, (\psi_1)_{e_j} \rangle.$$

Hence

$$\lim_{j\to\infty} \langle \mu_{n_j}, (\psi_1)_{e_j r_{n_j}} \rangle = \langle \nu, \psi_1 \rangle \tag{4.3.8.7}$$

because $e_j \to 1$ and $\mu_{[r_{n_j}]} \xrightarrow{\mathcal{D}'} \nu$ (see Theorem 2.3.4.6).

From (4.3.8.5), (4.3.8.6) and (4.3.8.7) we obtain

$$\lim_{j\to\infty} \langle \nu_j, \phi \rangle = \langle \nu, \psi_1 + \psi_3 \rangle + \lim_{j\to\infty} \langle \mu_{n_j}, ((\psi_2)_{e_j r_{n_j}} \rangle.$$

Let us estimate the last limit. We have

$$\lim_{j\to\infty}\langle \mu_{n_j}, (\psi_2)_{e_j r_{n_j}}\rangle = \langle (\mu_{n_j})_{[e_j r_{n_j}]}, \psi_2\rangle.$$

Define

$$E_1(\epsilon) := \{x : |x| \in (1-2\epsilon, 1)\}; E_2(\epsilon) := \{x : |x| \in [1, 1+2\epsilon)\}.$$

Suppose ϵ is chosen so that

$$\nu(\partial E_k) = 0,\ \ k = 1, 2. \tag{4.3.8.7a}$$

Recall that ν satisfies the condition (4.3.8.2a), hence E_1, E_2 are ν-squarable and hence (see Theorem 2.2.3.7)

$$\lim_{n\to\infty}\mu_{[r_n]}(E_k(\epsilon)) = \nu(E_k(\epsilon)),\ \ k = 1, 2.$$

Define

$$C_\phi := \max\{\phi(x) : x \in \mathbb{R}^m\}.$$

Then for $j \geq j_\epsilon$,

$$\begin{aligned}|\langle (\mu_{n_j})_{[e_j r_{n_j}]}, \psi_2\rangle| &\leq C_\phi (\mu_{n_j})_{[e_j r_{n_j}]}(E_1(\epsilon)\cup E_2(\epsilon))\\ &= C_\phi((\mu_{n_j})_{[e_j r_{n_j}]}(E_1(\epsilon)) + (\mu_{n_j})_{[e_j r_{n_j}]}(E_2(\epsilon)).\end{aligned}$$

By definition

$$(\mu_{n_j})_{[e_j r_{n_j}]}(E_2(\epsilon)) = \mu_{[e_j r_{n_j}]}(E_2(\epsilon)).$$

Because of (4.3.8.4a) we obtain

$$\lim_{j\to\infty}\mu_{[e_j r_{n_j}]}(E_2(\epsilon)) = \nu(E_2).$$

Exercise 4.3.8.5 Check in detail.

To compute the limit of the first addend we use periodicity of μ_{n_j} :

$$\mu_{n_j}(E_1(\epsilon)) = P_{n_j}^{-\rho}\mu_{n_j}(P_{n_j}E_1(\epsilon)) = (\mu_{n_j})_{[P_{n_j}]}(E_1(\epsilon)),$$

where

$$r_{n_j}P_{n_j}E_1(\epsilon) = \{x : |x| \in (r_{n_j+1}(1-2\epsilon), r_{n_j+1})\}.$$

Thus

$$\begin{aligned}(\mu_{n_j})_{[e_j r_{n_j}]}(E_1(\epsilon)) &= (\mu_{n_j})_{[e_j r_{n_j}P_{n_j}]}(E_1(\epsilon))\\ &= (\mu_{n_j})_{[e_j r_{n_j+1}]}(E_1(\epsilon)) = \mu_{[e_j r_{n_j+1}]}(E_1(\epsilon)).\end{aligned}$$

From this we obtain

$$\lim_{j\to\infty}(\mu_{n_j})_{[e_j r_{n_j}]}(E_1(\epsilon)) = \nu(E_1(\epsilon)).$$

Therefore

$$\lim_{j\to\infty} |\langle(\mu_{n_j})_{[e_j r_{n_j}]}, \psi_2\rangle| \leq C_\phi \nu(E_1(\epsilon) \cup E_2(\epsilon)). \tag{4.3.8.8}$$

Because of (4.3.8.2a) we have

$$\nu(\{x : |x| \in (1-2\epsilon, 1+2\epsilon)\}) \to 0 \tag{4.3.8.9}$$

as $\epsilon \to 0$ over the set of ϵ satisfying (4.3.8.7a). From (4.3.8.8) and (4.3.8.9) we obtain

$$\lim_{j\to\infty} \left\langle (\mu_{n_j})_{[e_j r_{n_j}]}, \psi_2 \right\rangle \to 0 \tag{4.3.8.10}$$

and

$$\langle \nu, \psi_2 \rangle \to 0 \tag{4.3.8.11}$$

when $\epsilon \to 0$. Hence if $\epsilon \to 0$ satisfying (4.3.8.7a) we have

$$\begin{aligned} &\lim_{j\to\infty} \langle \nu_j, \phi\rangle - \langle \nu_{[\tau]}, \phi\rangle \\ &\quad = \lim_{\epsilon\to 0} [\langle \nu, \psi_1 + \psi_3\rangle + \lim_{j\to\infty} \langle(\mu_{n_j})_{[e_j r_{n_j}]}, \psi_2\rangle - \langle \nu, \psi_1 + \psi_2 + \psi_3\rangle] \\ &\quad = \lim_{\epsilon\to 0} [\lim_{j\to\infty} \langle(\mu_{n_j})_{[e_j r_{n_j}]}, \psi_2\rangle - \langle \nu, \psi_2\rangle] = 0. \end{aligned}$$

The last equality holds because every addend tends to zero.

The case iii) can be considered in an analogous way.

Exercise 4.3.8.6 Consider it.

Thus the condition converg1) was checked. □

4.3.9 Now we should check the condition converg2). We need

Lemma 4.3.9.1 *Let $\mu \in \mathcal{M}[\rho]$, $\nu \in \mathbf{Fr}[\mu]$ and $r_n \to \infty$, $n = 1, 2, \ldots$ be a sequence such that*

$$\mathcal{D}' - \lim_{n\to\infty} \mu[r_n] = \nu_0. \tag{4.3.9.1}$$

Then passing if necessary to a subsequence, we can find $\{r_n\}$ such that for arbitrarily $\nu \in \mathbf{Fr}[\mu]$ a sequence $t_j \to \infty$ exists such that

$$\mathcal{D}' - \lim_{j\to\infty} \mu[t_j] = \nu \tag{4.3.9.2}$$

and for every n we can find $t_j \in [r_n, r_{n+1}]$.

Proof. Note that if the assertion of the lemma is satisfied for the sequence $\{r_n,\ n = 1, 2, \ldots\}$ it is satisfied for every subsequence of $\{r_n,\ n = 1, 2, \ldots\}$.

Let M be a countable set that is dense in $\mathbf{Fr}[\mu]$. Since reduction $\mathcal{D}'$-topology on $\mathcal{M}[\rho]$ is metrizable, it is sufficient to prove that we can choose a subsequence r_n for which assertion of the lemma is satisfied for all $\nu \in M$. We can do it using a diagonal process.

Let $r_n^0 \to \infty$ be an arbitrary sequence such that

$$\mathcal{D}' - \lim_{n\to\infty} \mu_{[r_n^0]} = \nu_0$$

and let a sequence t_j^1 satisfy the condition

$$\mathcal{D}' - \lim_{j\to\infty} \mu_{[t_j^1]} = \nu_1.$$

Omitting in the sequence $\{r_n^0\}$ the ends of the segments $[r_n^0, r_{n+1}^0]$ that do not contain elements $\{t_j\}$ we obtain a subsequence $\{r_n^1\}$. Continuing in such a way we obtain a subsequence $\{r_n^m\}$ satisfying (4.3.9.1) and the sequences $\{t_j^1\}, \{t_j^2\}, \dots, \{t_j^m\}$, $j = 1, 2\dots$ satisfying

$$\mathcal{D}' - \lim_{j\to\infty} \mu_{[t_j^l]} = \nu_l, \; l = 1, 2, \dots m. \tag{4.3.9.3}$$

Taking a diagonal sequence $\{r_n^n\}$, $n = 1, 2, \dots$ we observe that it is a subsequence of every subsequence $\{r_n^m\}$ and hence satisfies the assertion of the lemma. □

Proof of converg2). We can suppose that $\{r_n\}$ from the construction of μ_n with periodic limit sets satisfies the assertion of Lemma 4.3.9.1. Let $\nu \in \mathbf{Fr}[\mu]$ and $\mu_{t_j} \to \nu$ under condition $t_j \in [r_n, r_{n+1}]$. We should consider as in the proof of converg1) three cases i), ii) and iii). But all these cases were already considered and hence it was proved that

$$\nu_n := (\mu_n)_{r_n} \to \nu.$$

Exercise 4.3.9.1 Check this. □

4.4 Subharmonic curves. Curves with prescribed limit sets

4.4.1 In this paragraph we consider subharmonic functions $u \in SH(\rho(r))$ in the plane of finite type with respect to some proximity order $\rho(r) \to \rho$.

The pair $\boldsymbol{u} := (u_1, u_2)$, $u_1, u_2 \in SH(\rho(r))$ is called a *subharmonic curve* (which for brevity we will refer to simply as a *curve*).

The family

$$(\boldsymbol{u})_t := ((u_1)_t, (u_2)_t)$$

is precompact in the topology of convergence in $\mathcal{D}'$-topology on every component. The set of all limits

$$\mathbf{Fr}[\boldsymbol{u}] := \{\boldsymbol{v} = (v_1, v_2) : \exists t_j \to \infty, \boldsymbol{v} = \mathcal{D}' - \lim_{j\to\infty} \boldsymbol{u}_{t_j}\}$$

is called the limit set of the curve $\boldsymbol{u}$.

Actually this set describes coordinated asymptotic behavior of pairs of subharmonic functions.

Theorem 4.4.1.1 $\mathbf{Fr}[\boldsymbol{u}]$ *is closed, connected, invariant with respect to* $(\bullet)_{[t]}$ *(see* 3.1.2.4a) *and is contained in the set*

$$\boldsymbol{U}[\rho, \boldsymbol{\sigma}] := \{\boldsymbol{v} = (v_1, v_2) : v_n(z) \leq \sigma_n |z|^\rho, v_n(0) = 0, n = 1, 2.\}$$

where $\boldsymbol{\sigma} := (\sigma_1, \sigma_2)$

Exercise 4.4.1.1 Prove this by using Theorem 3.1.2.2.

Let us define $\boldsymbol{\sigma} > 0$ as $\sigma_n > 0,\ n = 1, 2$. Set

$$\boldsymbol{U}[\rho] := \bigcup_{\boldsymbol{\sigma}>0} \boldsymbol{U}[\rho, \boldsymbol{\sigma}].$$

We will write $\boldsymbol{U} \subset \boldsymbol{U}[\rho]$ if $\boldsymbol{U} \subset \boldsymbol{U}[\rho, \boldsymbol{\sigma}]$ for some $\boldsymbol{\sigma}$.

Since $(\boldsymbol{T}_\bullet, \boldsymbol{U}[\rho])$ is a dynamical system we have two theorems analogous to Theorems 4.2.1.1 and 4.2.1.2.

Exercise 4.4.1.2 Formulate and prove these theorems.

All the other assertions and definitions of Sections 4.2,4.3 can be repeated for subharmonic curves.

Let $\boldsymbol{U} \subset \boldsymbol{U}[\rho]$. Set

$$U' := \{v' : \exists v'' : (v', v'') \in \boldsymbol{U}\}.$$

This is a projection of $\boldsymbol{U}$. Set for $v' \in U'$,

$$U''(v') := \{v'' : (v', v'') \in \boldsymbol{U}\}.$$

This is the fibre over v'.

Theorem 4.4.1.2 *Let* $\boldsymbol{U} \Subset \boldsymbol{U}[\rho]$ *be closed and invariant and assume that every fiber* $U''(v')$ *is convex. Let* $U' = \mathbf{Fr}[u']$ *for some* $u' \in U(\rho(r))$. *Then there exists* $u'' \in U(\rho(r))$ *such that* $\mathbf{Fr}(u', u'') = \boldsymbol{U}$.

We construct a pseudo -trajectory asymptotics in the form (4.2.5.2) replacing u with $\boldsymbol{u}$ and v with $\boldsymbol{v}$. We can directly check that this curve satisfies the assertion of the theorem.

Exercise 4.4.1.3 Check this.

Theorem 4.4.1.3 (Concordance Theorem) *Let* $u \in U(\rho(r))$ *and* $v^0 \in \mathbf{Fr}[u]$, *and suppose* $v \in U[\rho]$ *has the property*

$$\lim_{\tau \to -\infty} T_\tau v = \lim_{\tau \to +\infty} T_\tau v = \tilde{v}.$$

Then there exists a function $w \in U(\rho(r))$ *such that the limit set of the curve* $\boldsymbol{u} = (u, w)$ $\mathbf{Fr}[\boldsymbol{u}] = (\mathbf{Fr}[u], \mathbb{C}(v))$ *and for every sequence* $t_n \to \infty$ *such that* $\lim_{n\to\infty} w_{t_n} = v$,

$$\lim_{n\to\infty} \boldsymbol{u}_{t_n} = (v^0, v). \tag{4.4.1.3.}$$

For proving this theorem we should use a.d.p.t. (4.2.5.2). If $v_j = v^0$ we replace v_j by $\boldsymbol{v}_j := (v^0, v)$. If $v_j \neq v^0$ we replace v_j by $\boldsymbol{v}_j := (v_j, \tilde{v})$.

Exercise 4.4.1.3 Do that and exploit Theorem 4.3.1.2 and Theorem 4.2.1.2.

Corollary 4.4.1.4 *Under conditions of Theorem* 4.4.1.3, *if* $\lim_{n\to\infty} w_{t_n} = T_\tau v$, *then* $\lim_{n\to\infty} u_{t_n} = T_\tau v^0$.

We should apply T_τ to (4.4.1.3) and use its continuity in $\mathcal{D}'$-topology. □

Chapter 5

Applications to Entire Functions

5.1 Growth characteristics of entire functions

5.1.1 Let $f(z)$ be an entire function. The function $u(z) := \log|f(z)|$ is subharmonic in $\mathbb{R}^2(= \mathbb{C})$. Hence the scale of growth subharmonic functions considered in Section 2.8 is transferred completely to entire functions. We will mark passing to entire function by changing index u for index f. For example,

$$M(r, f) := M(r, \log|f|), T(r, f) := T(r, \log|f|).$$

If $u(z) := \log|f(z)|$ has order $\rho[u] = \rho$, then $f(z)$ has order $\rho[f] := \rho$ and so on.

We will write $f \in A(\rho, \rho(r))$ and say "*f is an entire function of order ρ and normal type with respect to proximate order $\rho(r)$*" if $\log|f|$ is a subharmonic function of order ρ and normal type with respect to the same proximate order. Shortly, if $\log|f| \in SH(\rho, \rho(r), \mathbb{R}^2)$, then $f \in A(\rho, \rho(r))$.

Exercise 5.1.1.1 Give definitions of

$$T(r, f),\ M(r, f),\ \rho_T[f],\ \rho_M[f],\ \sigma_T[f, \rho(r)],\ \sigma_M[f, \rho(r)]$$

and reformulate all the assertions of Section 2.8 in terms of entire and meromorphic functions.

5.1.2 A divisor of zeros of an entire function can be represented as an integer mass distribution n on a discrete set $\{z_j\} \subset \mathbb{C}$. The multiplicity of a zero z_j is the mass concentrated at the point z_j.

The notation for characteristics of the behavior of zeros will mimic that of the behavior of masses, replacing μ for n. For example, $n(K_r), n(r)$ is the number of zeros (with multiplicities) in the disk K_r, $\rho[n]$ is the convergence exponent, $\overline{\Delta}[n]$ is the upper density and so on.

Exercise 5.1.1.2 Give definitions of $N(r,n)$, $\rho_N[n]$, $\overline{\Delta}_N[n]$, $p[n]$.

5.1.3 The limit set $\mathbf{Fr}[f]$ of an entire function $f \in A(\rho,\rho(r))$ is defined as the limit set of the subharmonic function $u(z) := \log|f(z)| \in SH(\rho,\rho(r),\mathbb{R}^2)$ (see Section 3.1), i.e.,

$$\mathbf{Fr}[f] := \mathbf{Fr}[\log|f|]. \tag{5.1.3.1}$$

It possesses, of course, all the properties described in Chapters 3, 4 but it is not clear now if there exists an *entire* function with prescribed limit set, i.e., whether the subharmonic function in Theorem 4.2.1.2 can be chosen to be $\log|f(z)|$ where $f \in A(\rho,\rho(r))$. It turns out that this is possible and we prove this in Section 5.3.

As it was mentioned in 3.1.1 the general form of $V_\bullet$ for the case of the plane is

$$V_t z = ze^{i\gamma\log t},$$

where γ is real.

The limit set $\mathbf{Fr}[n]$ of a divisor n is the limit set of the corresponding mass distribution n (see 3.1.3).

Of course generally speaking n_t (see (3.1.3.2)) is not an integer mass distribution.

Exercise 5.1.3.1 Give a complete definition of $\mathbf{Fr}[f]$ and $\mathbf{Fr}[n]$, and reformulate all the theorems of Sections 3.1.2, 3.1.3 in terms of entire functions and their zeros.

The connection between $\mathbf{Fr}[f]$ and $\mathbf{Fr}[n]$ is preserved completely (see Section 3.1.5).

Exercise 5.1.3.2 Reformulate the theorems of Section 3.1.5 for entire functions.

5.1.4 Let $f = f_1/f_2$ be a meromorphic function, where f_1, f_2 have no common zeros. If $f_2(0) = 1$, $f_1(0) \neq 0$ and $f_1, f_2 \in A(\rho,\rho(r))$, then $u := \log|f_1| - \log|f_2| \in \delta SH(\rho,\rho(r))$, and we write $f \in \mathrm{Mer}(\rho,\rho(r))$ and say "*f is a meromorphic function of order ρ and normal type with respect to the proximate order $\rho(r)$*". For $f \in \mathrm{Mer}(\rho,\rho(r))$ we use the following characteristics: $T(r,f), \rho_T[f], \sigma_T[f,\rho(r)]$. The charge of $\log|f|$ consists of integer positive and negative masses.

5.2 $\mathcal{D}'$-topology and topology of exceptional sets

5.2.1 Let $\alpha-$mes be the Carleson measure defined in Section 2.5.4. Set for $C \subset \mathbb{R}^2$,

$$\alpha - \overline{\mathrm{mes}}C := \limsup_{R\to\infty}[\alpha - \mathrm{mes}(C \cap K_R)]R^{-\alpha}. \tag{5.2.1.1}$$

It is called the *relative* Carleson α-measure.

Theorem 5.2.1.1 (Properties of the Relative Carleson Measure) *One has*

rCm1) *If C is bounded $\alpha - \overline{\text{mes}}C = 0$;*

rCm2)
$$\alpha - \overline{\text{mes}}(C_1 \cap C_2) \leq \alpha - \overline{\text{mes}}C_1 + \alpha - \overline{\text{mes}}C_2,$$

i.e., the relative Carleson measure is sub-additive;

rCm3)
$$C_1 \subset C_2 \Rightarrow \alpha - \overline{\text{mes}}(C_1) \leq \alpha - \overline{\text{mes}}C_2,$$

i.e., the relative Carleson measure is monotonic with respect to sets;

rCm4)
$$\alpha_1 > \alpha_2 \Rightarrow \alpha_1 - \overline{\text{mes}}C \leq \alpha_2 - \overline{\text{mes}}C,$$

i.e., the relative Carleson measure is monotonic with respect to α.

Exercise 5.2.1.1 Prove this.

A set $C \subset \mathbb{R}^2$ for which $\alpha - \overline{\text{mes}}C = 0$ is called a $C_0^\alpha - set$. If $\alpha - \overline{\text{mes}}C = 0$ for all $\alpha > 0$, C is called a $C_0^0 - set$.

Let us recall that if $u_1, u_2 \in SH(\rho, \rho(r), \mathbb{R}^2)$, then $u = u_1 - u_2 \in \delta SH(\rho, \rho(r), \mathbb{R}^2)$ (see Section 2.8.2).

Theorem 5.2.1.2 ($\mathcal{D}'$-topology and Exceptional sets) *Let $u \in \delta SH(\rho, \rho(r), \mathbb{R}^2)$. In order that*

$$u_t \to 0 \tag{5.2.1.2}$$

in $\mathcal{D}'$ as $t \to \infty$ it is sufficient that

$$u(z)|z|^{-\rho(|z|)} \to 0 \tag{5.2.1.3}$$

as $z \to \infty$ outside some C_0^2-set.

If (5.2.1.2) *holds, then* (5.2.1.3) *holds outside some C_0^0-set.*

5.2.2 To prove Theorem 5.2.1.2 we need some auxiliary assertions. Recall that dz is an element of area following the notation of the previous chapters.

Proposition 5.2.2.1 *Let $u \in SH(\rho, \rho(r), \mathbb{R}^2)$, and $C_{0,R}^2 := C_0^2 \cap K_R$. Then*

$$\int_{C_{0,R}^2} |u|(z)dz = o(R^{\rho(R)+2}) \tag{5.2.2.1}$$

as $R \to \infty$.

Proof. Suppose (5.2.2.1) does not hold. Then there exists a sequence $R_j \to \infty$ such that

$$\lim_{R_j \to \infty} R_j^{-\rho(R_j)-2} \int_{C_{0,R_j}^2} |u|(z)dz = A > 0. \tag{5.2.2.2}$$

Consider the following family of δ-subharmonic functions:

$$u_j(\zeta) := R_j^{-\rho(R_j)} u(\zeta R_j). \tag{5.2.2.2a}$$

It can be represented as a difference $u_j = u_{1,j} - u_{2,j}$ of subharmonic functions of the same form.

Thus it is precompact in L_{loc} (Theorem 2.7.1.3). Let us choose a convergent subsequence for which we keep the same notation. Its limit v is a locally summable function.

Now let χ_j be the characteristic functions of the sets

$$E_j := R_j^{-1} C^2_{0,R_j}.$$

Since $\text{mes}\, E_j \to 0$ it is possible to choose a sequence (for which we keep the same notation) such that $\chi_j \to 0$ almost everywhere. We will also suppose that R_j are the same for χ_j and u_j. Thus

$$\int\limits_{|\zeta|\leq 1} |\chi_j(\zeta)u_j(\zeta) - 0 \cdot v(\zeta)| d\zeta = \int\limits_{|\zeta|\leq 1} |\chi_j(\zeta)u_j(\zeta)| d\zeta \to 0.$$

By change of variables $z = R_j\zeta$ we obtain that

$$R_j^{-\rho(R_j)-2} \int\limits_{C^2_{0,R_j}} |u|(z)dz = \int\limits_{|\zeta|\leq 1} |\chi_j(\zeta)u_j(\zeta)| d\zeta \to 0.$$

Hence the limit in (5.2.2.2) is equal to zero. Contradiction. □

Proposition 5.2.2.2 *Under condition* (5.2.1.2) *the set*

$$C := \{z : |u(z)||z|^{-\rho(|z|)} > \epsilon\}$$

is a C^0_0-set for arbitrary ϵ.

Proof. Assume the contrary; that is , $\exists \alpha > 0$ such that

$$\alpha - \overline{\text{mes}} C = 2\delta > 0. \tag{5.2.2.3}$$

One can see that for some $\eta > 0$,

$$\limsup_{R\to\infty} (\alpha - \text{mes}\, K_{\eta R}) R^{-\alpha} \leq \delta/2. \tag{5.2.2.4}$$

Exercise 5.2.2.1 Check this.

(5.2.2.3) and (5.2.2.4) imply that there exists a sequence $R_j \to \infty$ such that

$$\lim_{R_j \to \infty} \alpha - \text{mes}[C \cap (K_{R_j} \setminus K_{\eta R_j})] R_j^{-\alpha} \geq \frac{3}{2}\delta.$$

Set

$$E_j := R_j^{-1} C \cap (K_{R_j} \setminus K_{\eta R_j}).$$

It is clear that $E_j \subset K_1 \setminus K_\eta$ and for sufficiently large j,

$$\alpha - \text{mes}\, E_j \geq \delta. \tag{5.2.2.5}$$

Set u_j as in (5.2.2.2a). We claim that for large j and $\zeta \in E_j$,

$$|u_j|(\zeta) \geq \frac{\epsilon}{2}|\zeta|^\rho. \tag{5.2.2.6}$$

Indeed,

$$|u_j|(\zeta) = \frac{|u|(R_j\zeta)}{R_j^{\rho(R_j)}} = \frac{|u|(z)}{|z|^{\rho(|z|)}}(1 + o(1))|\zeta|^\rho \geq \frac{\epsilon}{2}|\zeta|^\rho.$$

We used here properties of the proximate order and the equivalence

$$z = R_j\zeta \in C \cap (K_{R_j} \setminus K_{\eta R_j}) \Leftrightarrow \zeta \in E_j.$$

Exercise 5.2.2.2 Check this in detail.

Now we will show that the condition (5.2.1.2) contradicts (5.2.2.6). Since $u \in \delta SH(\rho, \rho(r), \mathbb{R}^2)$ it is a difference of $u_1, u_2 \in SH(\rho, \rho(r), \mathbb{R}^2)$. The corresponding sequences $u_{1,j}$ and $u_{2,j}$ are precompact in $\mathcal{D}'$ and there exist subsequences (with the same notation) that converge to v_1 and v_2, respectively.

By Theorem 2.7.5.1 these sequences converge to v_1 and v_2 with respect to $\alpha-$mes on $K_1 \setminus K_\eta$. Since $u_t \to 0$ in $\mathcal{D}'$, it follows that $v_1 = v_2$. Thus $u_j \to 0$ with respect to $\alpha-$mes on $K_1 \setminus K_\eta$. However, this contradicts (5.2.2.5) and (5.2.2.6). □

Proposition 5.2.2.3 *Let $\{C_j\}_1^\infty$ be a sequence of C_0^0-sets. There exists a sequence $R_j \to \infty$ such that the set*

$$C = \bigcup_{j=1}^{\infty} \{C_j \cap (K_{R_{j+1}} \setminus K_{R_j})\} \tag{5.2.2.7}$$

is a C_0^0-set.

Proof. Choose $\epsilon_j \downarrow 0$ and $\alpha_j \downarrow 0$. Set $R_0 := 1$. Suppose R_{j-1} was already chosen. Take R_j such that

$$\alpha_j - \operatorname{mes}[C \cap K_{R_{j-1}}] < \epsilon_j R^{\alpha_j} \tag{5.2.2.8}$$

for $R > R_j$.

It is possible because of property rC1) Theorem 5.2.1.1. We can also increase R_j so that

$$\alpha_j - \operatorname{mes}[C_j \cap K_R] < \epsilon_j R^{\alpha_j} \tag{5.2.2.9}$$

and

$$\alpha_j - \operatorname{mes}[C_{j+1} \cap K_R] < \epsilon_j R^{\alpha_j} \tag{5.2.2.10}$$

for $R > R_j$.

It is possible because C_j and C_{j+1} are C_0^0-sets.

Let us estimate $\alpha_j - \operatorname{mes}[C \cap K_R]$ for $R_j \leq R < R_{j+1}$. From (5.2.2.8), (5.2.2.9) and (5.2.2.10) we obtain

$$\alpha_j - \operatorname{mes}[C \cap K_R] \leq 3\epsilon_j R^{\alpha_j}. \tag{5.2.2.11}$$

Let $\alpha > 0$ be arbitrarily small. Find $\alpha_j < \alpha$. For $R_{j+1} \geq R > R_j$ we have

$$\alpha - \operatorname{mes}[C \cap K_R]R^{-\alpha} \leq \alpha_j - \operatorname{mes}[C \cap K_R]R^{-\alpha_j} \leq 3\epsilon_j.$$

Hence $\alpha - \overline{\operatorname{mes}} C = 0$. □

5.2.3

Proof of Theorem 5.2.1.2. Let $\phi \in \mathcal{D}(\mathbb{C})$ and $\operatorname{supp} \phi \subset K_R$. Then for any $\epsilon > 0$,

$$J(t) := \int \phi(z) u_t(z) dz = \left(\int\limits_{K_R \setminus K_\epsilon} + \int\limits_{K_\epsilon} \right) \phi(z) u_t(z) dz := J_1(t) + J_2(t). \tag{5.2.3.1}$$

We have for J_2 (see 2.8.2.3):

$$|J_2|(t) \leq \max_{|z| \leq \epsilon} |\phi(z)| \times \text{const} \int_0^\epsilon T(r, |u_t|) r dr \leq \text{const}\, T(\epsilon, |u_t|)\epsilon^2. \tag{5.2.3.2}$$

Further (see Theorem 2.8.2.1)

$$\begin{aligned} T(r, |u_t|) &\leq 2T(r, u_t) + O(t^{-\rho(t)}) \leq 2[T(r, u_{1,t}) + T(r, u_{2,t})] + O(t^{-\rho(t)}) \\ &\leq 2[M(r, u_{1,t}) + M(r, u_{2,t})] + O(t^{-\rho(t)}). \end{aligned} \tag{5.2.3.3}$$

Using (5.2.3.2), (5.2.3.3) and (3.1.2.3) we obtain

$$\limsup_{t \to \infty} |J_2(t)| \leq \text{const}\, \epsilon^{\rho+2}. \tag{5.2.3.4}$$

Exercise 5.2.3.1 Check this using the change of variable $z = t\zeta$.

To estimate $J_1(t)$ write

$$|J_1(t)| \leq \text{const}\Bigg(\int\limits_{\tilde{K}_t \setminus C^2_{0,Rt}} |u(z)|dz + \int\limits_{C^2_{0,Rt}} |u(z)|dz \Bigg) t^{-\rho(t)-2} := J_{1,1}(t) + J_{1,2}(t), \tag{5.2.3.5}$$

where $\tilde{K}_t := \{z : \epsilon t \leq |z| \leq Rt\}$.

The summand $J_{1,1}$ is $o(1)$ as $t \to \infty$ by (5.2.1.3).

Exercise 5.2.3.2 Check this using the properties of the proximate order (Theorem 2.8.1.3, ppo3).

The summand $J_{1,2}$ is $o(1)$ by Theorem 5.2.2.1. Thus

$$\limsup_{t\to\infty} |J(t)| \leq \text{const}\, \epsilon^{\rho+2}$$

for any ϵ. Hence it is equal to zero and the sufficiency of (5.2.1.3) has been proved.

Let us prove sufficiency of (5.2.1.2). Let $\epsilon_j \downarrow 0$. By Theorem 5.2.2.2 we choose a C^0_0-set C_j outside which $|u(z)||z|^{-\rho(|z|)} < \epsilon_j$.

We construct the set C by (5.2.2.7). Outside C we have (5.2.1.3). And by Theorem 5.2.2.3 it is a C^0_0-set. □

5.3 Asymptotic approximation of subharmonic functions

5.3.1 One of the widely applied methods of constructing entire functions with a prescribed asymptotic behavior is the following: First construct a subharmonic function behaving asymptotically as the logarithm of modulus of the entire function,and then approximate it in some sense by the logarithm of modulus of entire function such that the asymptotic is preserved.

Various queries about the a precision of the approximation and about the metric in which it was implemented generated a spectrum of theorems of such kind that we will demonstrate.

Historically the first theorems of this kind were proved for concrete functions, the masses of which were concentrated on sufficiently smooth curves (in particular, on lines, see, e.g., [BM, Ev, Kj, Ar], ...)

In such cases the approximation was very precise and exceptional sets where the approximation failed were small and determined.

The first general case was proved in [Az(1969)]. Next it was developed in [Yu(1982)], and vastly improved in [Yu(1985)]. It is the following

Theorem 5.3.1.1 (Yulmukhametov) *Let $u \in SH(\rho)$. Then there exists an entire function f such that for every $\alpha \geq \rho$,*

$$|u(z) - \log|f(z)|| < C_\alpha \log|z|$$

for $z \notin E_\alpha$, where E_α is an exceptional set that can be covered by discs $D_{z_j,r_j} := \{z : |z - z_j| < r_j\}$ satisfying the condition

$$\sum_{|z_j|>R} r_j = o(R^{\rho-\alpha}),\ \ R \to \infty.$$

This theorem is precise in the following sense: If

$$||z| - \log|f(z)|| = o(\log|z|),\ \ z \to \infty,\ \ z \notin E,$$

then for every covering of E by discs D_{z_j,r_j} and every $\epsilon > 0$

$$\sum_{|z_j|<R} r_j \geq R^{1-\epsilon},\ \ R \to \infty,$$

i.e., in any case this sum is not even bounded.

However it is necessary to remark that the construction from [Yu(1985)] "rigidly" fastens zeros of the entire function, whereas the construction of [Az(1969)] and [Yu(1982)] gives some possibilities to move them, which is needed in some constructions.

Let us also mention that such approximation generates an approximation of a plurisubharmonic function by the logarithm of the modulus of an entire function in $\mathbb{C}^p$ (see [Yu(1996)]).

It is also useful to approximate subharmonic functions in an integral metric, for example L^p, as was done in [GG].

Set

$$\|g\|_p := \left(\int_0^{2\pi} |g(t)|^p dt\right)^{1/p}.$$

Denote by $Q(r,u)$ a function that satisfies the conditions:

1) if u is of finite order, then $Q(r,u) = O(\log r)$;
2) if u is of infinite order, then $Q(r,u) = O(\log r + \log \mu_u(r))$.

Theorem 5.3.1.2 (Girnyk, Gol′dberg) *For every subharmonic function in $\mathbb{C}$ u there exists an entire function f such that $\|u(re^{i\cdot}) - \log|f|(re^{i\cdot})\|_p = Q(r,u)$.*

This theorem also considers functions of infinite order. In this case, it is possible to replace $\mu_u(r)$ by $T(r,u)$ or $M(r,u)$ in $Q(r,u)$ outside an exceptional set $E \subset \mathbb{R}^+$ of finite measure. This theorem is also unimprovable for subharmonic functions of finite order, because, for example, $u = \frac{1}{2}\log|z|$ gives, as it is possible to prove:

$$\liminf_{r\to\infty} \frac{\|u(re^{i\bullet}) - \log|f|(re^{i\bullet})\|_p}{\log r} > 0.$$

However it was found [LS], [LM] that the remainder term $O(\log|z|)$ that was regarded the best possible is not precise and in some "regular" cases can be replaced with $O(1)$ outside a bigger (but still "small") set .

Set, for $E \subset \mathbb{C}$:

$$\Delta(E) := \limsup_{r\to\infty} \frac{\operatorname{mes} E \cap D_{0,R}}{R^2}.$$

Theorem 5.3.1.3 (Lyubarskii, Malinnikova) *Let u be a subharmonic function in $\mathbb{C}$ with μ_u satisfying the conditions: $\mu_u(\mathbb{C}) = \infty$ and there exists $\alpha > 0,\ q > 1,\ R_0 > 0$ such that*

$$\mu_u(D_{0,qR} \setminus D_{0,R}) > \alpha$$

for all $R > R_0$.

Then there exists an entire function f such that for every $\epsilon > 0$,

$$|u(z) - \log|f(z)|| < C_\epsilon$$

for $z \in \mathbb{C} \setminus E_\epsilon$ with $\Delta(E_\epsilon) < \epsilon$.

So if μ_u has no "Hadamard's gaps" such approximation is possible.

In this book we restrict ourself to a weaker and simply proved theorem that is sufficient for our aim

Theorem 5.3.1.4 (Approximation Theorem) *For every $u \in SH(\rho, \rho(r))$ there exists an entire function f such that*

$$D' - \lim_{t\to\infty} (u - \log|f|)_t = 0.$$

Nevertheless this theorem has an important

Corollary 5.3.1.5 *For every $u \in SH(\rho, \rho(r))$ there exists an entire function f such that*

$$\mathbf{Fr}[u] = \mathbf{Fr}[f].$$

5.3.2 Now we prove Theorem 5.3.1.4. We can suppose, because of Theorem 3.1.6.1 (Dependence **Fr** on $V_\bullet$), that in the definition of $(\bullet)_t$ (see 3.1.2.1) $V_t \equiv I$

We prove this theorem for the case non-integer ρ. For proving this theorem we need

Lemma 5.3.2.1 *Let $u \in \delta SH(\rho, \rho(r))$, for ρ non-integer, and ν is its charge. Then $u_t \to 0$ iff $\nu_t \to 0$ in $\mathcal{D}'$ as $t \to \infty$.*

Proof. Sufficiency. Suppose $u_t := (u_1)_t - (u_2)_t \not\to 0$. There exists a subsequence $t_j \to \infty$ and subharmonic functions v_1 and v_2 such that

$$u_{t_j} = (u_1)_{t_j} - (u_2)_{t_j} \to v_1 - v_2 := v \neq 0. \tag{5.3.2.1}$$

Applying to (5.3.2.1) the continuity of Δ in $\mathcal{D}'$ and using the conditions of the theorem, we obtain

$$\nu_{t_j} \to \frac{1}{2\pi}\Delta v = 0.$$

Hence v is harmonic. Since $v_1, v_2 \in U[\rho,]$ also $v \in U[\rho]$ (see Theorem 2.8.2.1, t3), t4) and Theorem 2.8.2.3).

Exercise 5.3.2.1 Prove this in detail.

By Theorem 3.1.4.3 we obtain $v = 0$. Contradiction.

Necessity. Since the Laplace operator is continuous in $\mathcal{D}'$-topology, the assertion $u_t \to 0$ implies $\nu_t := \frac{1}{2\pi}\Delta u_t \to 0$. □

Now we describe a construction of the zero distribution of the future entire function. Let $u \in SH(\rho)$ and μ be its mass distribution. Set

$$R_{j+1} := R_j (j+1)^{4/\kappa} \tag{5.3.2.2}$$

where $\kappa := \min(\rho - [\rho], [\rho] + 1 - \rho)$.

Let us divide all the plane by circles of the form $S_{R_j} := \{|z| = R_j\}$ such that $R_{j+1}/R_j \to \infty$ and $\mu(S_{R_j}) = 0$.

Exercise 5.3.2.2 Prove that it is possible.

Choose a sequence $\delta_j \downarrow 0$. Divide every annulus $K_j := \{z : R_j \leq |z| < R_{j+1}\}$ by circles $S_{R_{j,n}}$ for

$$R_{j,n} := \left(\frac{1+\delta_j}{1-\delta_j}\right)^n R_j, \; n = 0, 1, 2, \ldots, n_j,$$

where

$$n_j := \left[\frac{\log \frac{R_{j+1}}{R_j}}{\log \frac{1+\delta_j}{1-\delta_j}}\right],$$

and by rays

$$L_k := z : \arg z = k\delta_j, \; k = 0, 1, \ldots, [2\pi/\delta_j].$$

They divide all the plane into sectors $K_{j,n,k}$. We can choose δ_j in such a way that $\mu(\partial K_{j,n,k}) = 0$ because $\mu(K_{j,n,k})$ is a monotonic function of δ_j and has only a countable set of jumps.

Exercise 5.3.2.3 Explain this in detail.

Choose a point $z_{j,n,k}$ in every sector $K_{j,n,k}$ and concentrate all the mass of the sector at this point. In other words we consider a new mass distribution $\hat{\mu}$ that has masses concentrated in the points $z_{j,n,k}$ and $\hat{\mu}(z_{j,n,k}) = \mu(K_{j,n,k})$.

The next lemma shows that $\hat{\mu}$ is close to μ.

Lemma 5.3.2.2 *One has*

$$\hat{\mu}_t - \mu_t \to 0$$

in $\mathcal{D}'$ as $t \to \infty$.

Proof. Assume the contrary, i.e., $\hat{\mu}_t - \mu_t \not\to 0$. Choose a sequence $t_l \to \infty$ such that $\hat{\mu}_{t_l} \to \hat{\nu}$ and $\mu_{t_l} \to \nu$, $\nu, \hat{\nu} \in \mathcal{M}[\rho]$, $\nu \neq \hat{\nu}$. Then there exists a disc $K_{z_0,r_0} := \{z : |z - z_0| < r_0\}$ such that $\nu(K_{z_0,r_0}) \neq \hat{\nu}(K_{z_0,r_0})$. We can assume that this disc does not contain zero since for all the $\nu \in \mathcal{M}[\rho]$ the condition $\nu(K_r) \leq \Delta r^\rho, \forall r > 0$ is fulfilled.

Suppose, for example,

$$\nu(K_{z_0,r_0}) > \hat{\nu}(K_{z_0,r_0}). \tag{5.3.2.3}$$

Set $a := \nu(K_{z_0,r_0}) - \hat{\nu}(K_{z_0,r_0}) > 0$. Choose ϵ such that

$$\nu(K_{z_0,r_0}) < \nu(\overline{K_{z_0,r_0-\epsilon}}) + a/3. \tag{5.3.2.4}$$

This is possible because the countable additivity of $\hat{\nu}$ implies $\lim\limits_{r'\uparrow r} \nu(K_{z_0,r'}) = \nu(K_{z_0,r})$.

Consider now the sets $t_l K_{z_0,r_0}, t_l K_{z_0,r_0-\epsilon}$. For sufficiently large t_l they are contained in the union of the annuluses $K_{j_l} \cup K_{j_l+1}$.

As $j_l \to \infty$ the diameters of all the sectors $K_{j_l,n,k}$ are $o(R_{j_l})$ uniformly. Thus they are $o(t_l)$. Hence for such t_l's we can find a union Γ_l of sectors covering $t_l K_{z_0,r_0-\epsilon}$ that does not intersect the circle of $t_l K_{z_0,r_0}$.

We have $\hat{\mu}(\Gamma_l) = \mu(\Gamma_l)$ by definition of $\hat{\mu}$. Using the monotonicity of measures, we obtain $\mu(t_l K_{z_0,r_0-\epsilon}) \leq \hat{\mu}(t_l K_{z_0,r_0})$, whence

$$\mu_{t_l}(K_{z_0,r_0-\epsilon}) \leq \hat{\mu}_{t_l}(K_{z_0,r_0}).$$

Passing to the limit as $l \to \infty$ and using Theorems 2.2.3.1 and 2.3.4.4, we obtain $\nu(\overline{K_{z_0,r_0-\epsilon}}) \leq \hat{\nu}(K_{z_0,r_0})$. Using (5.3.2.4), we obtain $\nu(K_{z_0,r_0}) - 1/3[\nu(K_{z_0,r_0}) - \hat{\nu}(K_{z_0,r_0})] \leq \hat{\nu}(K_{z_0,r_0})$ and hence $\nu(K_{z_0,r_0}) \leq \hat{\nu}(K_{z_0,r_0})$, that contradicts (5.3.2.3). Since ν and $\hat{\nu}$ are symmetric in this reasoning the lemma is proved. □

Let us finish the proof of Theorem 5.3.1.4 for non-integer ρ.

We construct a distribution n with integer masses concentrated at points $z_{j,k,n}$. Set

$$n(z_{j,k,n}) := [\hat{\mu}(z_{j,k,n})]$$

and estimate the growth of the difference

$$\delta\mu := \hat{\mu} - n$$

that is also a mass distribution concentrated at the same points.

Since

$$\delta\mu(z_{j,k,n}) \leq 1$$

it is sufficient to count the number of points in the disc K_R.

The number of points in the annulus $\{R_j \leq |z| < R\}$ is found from (5.3.2.2),

$$\begin{aligned}\delta\mu(\{R_j \leq |z| < R\}) &\leq \left[\log\left(\frac{1+\delta_j}{1-\delta_j}\right)\right]^{-1} \frac{2\pi}{\delta_j} \log\frac{R}{R_j} \\ &\leq \text{const} \times \frac{\log(j+1)}{\delta_j^2} = \text{const} \times (j+1)^4 \log(j+1).\end{aligned}$$

The mass of the disc K_R is estimated by the inequality

$$\delta\mu(K_R) \leq \text{const} \times \sum_{k=0}^{n-1} (k+1)^4 \log(k+1) = o(n^6) = o(R^\epsilon) \tag{5.3.2.5}$$

for any $\epsilon > 0$ because $R > R_{n-1} = ((n-1)!)^{4/\kappa}$.

Exercise 5.3.2.4 Check this in detail.

The estimate (5.3.2.5) shows that

$$\delta\mu_t \to 0 \tag{5.3.2.6}$$

as $t \to \infty$.

Lemma 5.3.2.2 and (5.3.2.6) imply that

$$\mu_t - n_t \to 0. \tag{5.3.2.7}$$

Set

$$u_1(z) := \Pi(z, n, p)$$

(see (2.9.2.1)) where Π is a canonical potential. This is a subharmonic function in the plane with integral masses. Thus it is the logarithm of the modulus of the entire function

$$f(z) = \prod E(z/z_{j,k,n}).$$

(5.3.2.7) implies by Lemma 5.3.2.1 that $u_t - (u_1)_t \to 0$ and this is the assertion of Theorem 5.3.1.4 for non-integer ρ. □

5.4 Lower indicator of A.A. Gol′dberg. Description of lower indicator Description of the pair: indicator-lower indicator

5.4.1 Now we consider the *lower indicator*. For an entire function of finite order ρ and normal type it can be defined in one of the following ways:

$$\underline{h}_1(\phi, f) := \sup_{C\in\mathcal{C}}\{\liminf_{re^{i\phi}\to\infty, re^{i\phi}\notin C} \log|f(re^{i\phi})|r^{-\rho(r)}\}, \tag{5.4.1.1}$$

where $\mathcal{C}$ is the set of C_0-sets (see [Le, Ch. II, § 1]), i.e., the sets that can be covered by a union of discs $K_{\delta_j}(z_j) := \{z : |z - z_j| < \delta_j\}$ such that

$$\lim_{R\to\infty} \frac{1}{R} \sum_{|z_j|<R} \delta_j = 0.$$

The exclusion of C_0-sets is necessary because we must exclude from our consideration some neighborhoods of roots of $f(z)$ where $\log|f(z)|$ is near $-\infty$.

Similarly, define

$$\underline{h}_2(\phi, f) := \sup_{E(\phi)\in\mathcal{E}}\{\liminf_{r\to\infty, r\notin E(\phi)} \log|f(re^{i\phi})|r^{-\rho(r)}\}, \tag{5.4.1.2}$$

where $\mathcal{E}$ is the set of E_0-sets (see [Le, Ch. III]), i.e., sets $E \subset (0, \infty]$ satisfying the condition

$$\lim_{R\to\infty} \text{mes}\{E \cap (0, R)\}R^{-1} = 0.$$

The definition (5.4.1.1) was introduced by A.A. Gol′dberg (see [Go(1967)]). We will use the definition (3.2.1.2)

$$\underline{h}(\phi, f) = \inf\{v(e^{i\phi}) : v \in \mathbf{Fr}[f]\}. \tag{5.4.1.3}$$

It was proved in [AP, Thm. 1] that the definitions (5.4.1.1), (5.4.1.2) and (5.4.1.3) coincide.

Let us note that (5.4.1.3) uses the definition (3.2.1.2) only on the circle $\{|z| = 1\}$. However, it is easy to check, by using Theorem 3.2.1.2 that for $\underline{h}(z) = |z|^\rho\underline{h}(\arg z)$ properties h1) and h2), Theorem 3.2.1.1, are preserved.

Exercise 5.4.1.1 Check this.

We are going to prove

Theorem 5.4.1.1 *Let $g(\phi)$ be a 2π-periodic function that is either semicontinuous from above or $\equiv -\infty$ and $\rho(r) \to \rho$ be an arbitrary approximate order. Then there exists an entire function $f \in A(\rho, \rho(r))$ such that*

$$\underline{h}(\phi, f) = g(\phi) \tag{5.4.1.4}$$

for all $\phi \in [0, 2\pi)$.

5.4.2 We will use the following assertion that is a corollary of Theorem 4.3.5.1 and Corollary 5.3.1.5:

Theorem 5.4.2.1 *Let $\Lambda \subset U[\rho]$ be a compact, connected and $T_\bullet$-invariant subset. Then for any proximate order $\rho(r) \to \rho$ there exists $f \in A(\rho, \rho(r))$ such that*

$$h(\phi, f) = \sup\{v(e^{i\phi}) : v \in \Lambda\}, \tag{5.4.2.1}$$

$$\underline{h}(\phi, f) = \inf\{v(e^{i\phi}) : v \in \Lambda\}. \tag{5.4.2.2}$$

Exercise 5.4.2.1 Prove Theorem 5.4.2.1.

For the sake of clarity let us restrict ourselves to non-integer ρ. We will construct a set Λ such that

$$\inf\{v(e^{i\phi}) : v \in \Lambda\} = g(\phi).$$

Denote

$$H(z,p) := \log|1-z| + \Re\sum_{k=1}^{p}\frac{z^k}{k};\ p = [\rho],$$

$$\gamma(z,K,\lambda) := -\lambda + K|z-1|,\ \lambda, K \geq 0.$$

Note the following properties of these functions:

a) $\min\limits_{|z-1|\geq\delta} \delta H(z,p)|z|^{-\rho} \to 0$, as $\delta \to 0$;

b) $\delta H(z,p)|z|^{-\rho} \leq A\delta$, for all $z \in \mathbb{C}$, where A depends only on p;

c)
$$\max_{|z-1|\leq 0.5K} \gamma(z,K,\lambda) \leq -\frac{1}{2}. \tag{5.4.2.3}$$

Exercise 5.4.2.2 Prove properties a), b), c).

Let us note that $H(1,p) = -\infty$. Consider the family:

$$\Lambda_\infty = \{v_{\theta,\tau}(z) := H(ze^{-i\theta}\tau, p)\tau^{-\rho} : \theta \in [0,2\pi),\ \tau \in (0,\infty)\} \cup 0.$$

This family is contained in $U[\rho]$ because of b) and closed in $\mathcal{D}'$-topology. It is also $T_\bullet$-invariant, hence, satisfies the conditions of Theorem 5.4.2.1. For every $\phi \in [0,2\pi)$ there exists $\theta_0(=\phi)$, and $\tau_0(=1)$ such that $v_{\theta_0,\tau_0}(e^{i\phi}) = H(1,p) = -\infty$ Hence

$$\inf\{v(e^{i\phi}) : v \in \Lambda_\infty\} = -\infty. \tag{5.4.2.4}$$

For the general case this construction will be improved, cutting the "trunk" of the function $H(ze^{-i\theta}, p)$.

Take δ small enough so that the following conditions hold:

$$\delta H(z,p)|z|^{-\rho} \geq -\frac{1}{4}, \text{ for } |z-1| \geq \delta, \tag{5.4.2.5}$$

$$\delta H(z,p) \geq -\frac{1}{4}, \text{ for } |z-1| = \delta, \tag{5.4.2.6}$$

$$\delta \leq \frac{1}{2K}. \tag{5.4.2.7}$$

Then

$$\delta H(z,p) > \gamma(z,K,\lambda), \text{ for} |z-1| = \delta. \tag{5.4.2.8}$$

Denote

$$W(z,K,\delta,\lambda) := \begin{cases} \max\{\delta H(z,p), \gamma(z,K,\lambda)\}, & \text{for } |z-1| < \delta, \\ \delta H(z,p), & \text{for } |z-1| \geq \delta. \end{cases} \tag{5.4.2.9}$$

Lemma 5.4.2.2 *The following holds:*

aw) *The function $W(z,K,\delta,\lambda)$ is subharmonic in $\mathbb{C}$;*

bw) $\operatorname{supp} \mu_W \Subset \{|z-1| < \delta\}$;

cw)
$$\sup_{z\in\mathbb{C}} W(z,\bullet,\delta,\lambda)|z|^{-\rho} \leq A\delta, \tag{5.4.2.10}$$

where A depends only on p.

Proof. For $|z-1| < \delta$, W is subharmonic as the maximum of two subharmonic functions. For $|z-1| \geq \delta$ it is harmonic even in the neighborhood of the circle $|z-1| = \delta$, because of inequality (5.4.2.8). So aw) and bw) hold. The assertion cw) follows from b) and c) (5.4.2.3) above.

Now we get to the proof of (5.4.1.4). Let $g_n \downarrow g$ be a sequence of continuously differentiable functions that converges to g monotonically. This is possible, because g is semicontinuous from above.

Exercise 5.4.2.3 Prove that Theorem 2.1.2.9 and the Weierstrass theorem of approximation of every periodic function by trigonometrical polynomials imply the last assertion.

We write

$$M_n := \max_{\phi} g_n^+(\phi)$$

where as usual $a^+ = \max(a,0)$. Set

$$v_{\theta,n}(z) := W(ze^{-i\theta}, K_n, \delta_n, M_n + 1 - g_n(\theta)) + (M_n+1)|z|^\rho,$$

where δ_n is chosen small and K_n is chosen large. Set $z = \tau e^{i\phi}$. It is clear that

$$v_{\phi,n}(e^{i\phi}) = g_n(\phi) \tag{5.4.2.11}$$

for all K_n, δ_n.

We can choose K_n so large and δ_n so small that

$$\gamma(z, K_n, M_n + 1 - g(\theta))|z|^{-\rho} \geq g_n(\phi)$$

for $|z - 1| \leq \delta_n$, because g_n has bounded derivative.

After that we can make δ_n smaller so that for $|z - 1| \geq \delta_n$ the inequality (5.4.2.8) would hold.

Exercise 5.4.2.4 Estimate exactly K_n and δ_n via the derivative of g_n.

Then

$$v_{\theta,n}(z)|z|^{-\rho} \geq g_n(\phi)$$

for all $z = re^{i\phi}$. Thus

$$\min_{\theta,\tau} v_{\theta,n}(\tau e^{i\phi})\tau^{-\rho} = g_n(e^{i\phi}),$$

and the minimum is attained for $\tau = 1, \theta = \phi$.

Let us note that from (5.4.2.10) we have

$$\sup_{\theta} \sup_{z \in \mathbb{C}} v_{\theta,n}(z)|z|^{-\rho} \leq A\delta_n + M_n + 1 \leq A + M_1 + 1.$$

Consider now the family of functions

$$\Lambda_0 := \{v_{\theta,n}(z\tau)|\tau|^{-\rho} : \theta \in [0; 2\pi),\ n = 1, 2, \ldots, \tau \in (0; \infty)\}.$$

It is contained in $U[\rho, \sigma]$ for $\sigma = A + M_1 + 1$ and is $T_\bullet$-invariant. Let Λ be its closure in $\mathcal{D}'$. Let us show that

$$g(\phi) = \inf\{v(e^{i\phi}) : v \in \Lambda\}. \tag{5.4.2.12}$$

Indeed, for every sequence $v_j \in \Lambda_1$

$$v_j(e^{i\phi}) \geq \inf_n g_n(\phi) = g(\phi).$$

Let $v \in \Lambda$. By Theorem 2.7.4.1 ($\mathcal{D}'$ and Quasi-everywhere Convergence)

$$v(z) := (\mathcal{D}' - \lim_{j\to\infty} v_j)(z) = (\limsup_{j\to\infty} v_j)^*(z).$$

Hence

$$v(e^{i\phi}) \geq g(\phi).$$

However, the infimum is attained for every ϕ on the sequence $v_{\phi,n}(z)$ because of (5.4.2.11). Hence (5.4.2.12) holds and Theorem 5.4.1.4 is proved. □

5.4.3 Now we describe the pair: indicator-lower indicator. Let h be a 2π-periodic, ρ-trigonometrically convex function (ρ-t.c.f) and let g be a 2π-periodic upper semi-continuous function. Further they are indicator and lower indicator of an entire function, and hence must satisfy the condition

$$h(\phi) \geq g(\phi),\ \phi \in [0, 2\pi). \tag{5.4.3.1}$$

An interval $(a,b) \subset [0,2\pi)$ is called a *maximal interval of ρ-trigonometricity* of the function h if

$$h(\phi) = A\cos\rho\phi + B\sin\rho\phi,\ \ \phi \in (a,b) \tag{5.4.3.2}$$

for some constants A, B, and h has no such representation on any larger interval $(a',b') \supset (a,b)$.

A function h is said to be *strictly* ρ-t.c.f. if it is a ρ-t.c.f. and is not ρ-trigonometrical on any interval.

If the function h is a strictly ρ-t.c.f., then h and g (satisfying other previous bounds) could be an indicator and lower indicator of an entire function $f \in A(\rho(r))$. However this is not so if the function h has an interval of trigonometricity.

Recall, for example, the famous M. Cartwright Theorem [Le, Ch. IV, § 2, Thm. 6]: if an indicator of an entire function is trigonometrical on an interval (a,b) with $b-a > \pi/\rho$, then the function is a CRG -function on this interval, i.e.,

$$h(\phi) = g(\phi),\ \ \phi \in (a,b). \tag{5.4.3.3}$$

Let us formulate all the necessary conditions of such kind. Let (a,b) be a maximal interval of ρ-trigonometricity of the function h. The M. Cartwright theorem can be formulated as the implication

$$(b-a > \pi/\rho) \Rightarrow (5.4.3.3). \tag{5.4.3.4}$$

The following implications are also necessary:

$$(\exists\phi_0 \in (a,b) : h(\phi_0) = g(\phi_0)) \Rightarrow (5.4.3.3), \tag{5.4.3.5}$$

$$(h(a) = g(a) \wedge h'_+(a) = h'_-(a)) \Rightarrow (5.4.3.3), \tag{5.4.3.6a}$$

$$(h(b) = g(b) \wedge h'_+(b) = h'_-(b)) \Rightarrow (5.4.3.3), \tag{5.4.3.6b}$$

where $h'_\pm(a)$ and $h'_\pm(b)$ are the right and left derivatives of the function h at the points a and b.

$$(b-a = \pi/\rho \wedge h'_+(a) = h'_-(a)) \Rightarrow (5.4.3.3), \tag{5.4.3.7a}$$

$$(b-a = \pi/\rho \wedge h'_+(b) = h'_-(b) \Rightarrow (5.4.3.3), \tag{5.4.3.7b}$$

$$\left(\liminf_{\phi\to a+0} \frac{h(\phi)-g(\phi)}{\phi - a} = 0\right) \Rightarrow (5.4.3.3), \tag{5.4.3.8a}$$

$$\left(\liminf_{\phi\to b-0} \frac{h(\phi)-g(\phi)}{b-\phi} = 0\right) \Rightarrow (5.4.3.3). \tag{5.4.3.8b}$$

Now we shall give an exact formulation. The functions h and g are said to be *concordant* if at least one of the following conditions holds:

1. h is strictly ρ-t.c.;
2. for each (a,b) that is a maximal interval of ρ-trigonometricity of the function h the implications (5.4.3.4)–(5.4.3.8b) are satisfied.

Theorem 5.4.3.1 *Let $0 < \rho < \infty$, $h(\phi)$ be a 2π-periodic, ρ-t.c.f., $g(\phi)$ be an upper semicontinuous, 2π-periodic function, $h(\phi) \geq g(\phi)$ for all ϕ, and $h \not\equiv g$.*

A function $f \in A(\rho(r))$ which simultaneously satisfies the identity $h_f \equiv h, \underline{h}_f \equiv g$ with an arbitrary proximate order $\rho(r) \to \rho$ exists if and only if the functions h and g are concordant.

5.4.4

Proof of necessity. Note that implication (5.4.3.4) is a corollary of (5.4.3.6a) or (5.4.3.6b), because every ρ-trigonometrical function is continuous and has continuous derivative in (a, b). Recall that $(\bullet)_{[t]}$ was defined by (3.1.2.4a).

From properties of the limit set $\mathbf{Fr}[f]$ (Theorem 3.1.2.2, fr2), fr3)) and the definition of indicators ((3.1.2.1), (3.1.2.2)) we can obtain for every function $v \in \mathbf{Fr}[f]$ the inequality

$$v(\tau e^{i\phi}) \leq \tau^\rho h(\phi), \ \ \phi \in [0, 2\pi), \tau > 0. \tag{5.4.4.1}$$

Since $h(\phi)$ is ρ-trigonometrical for $\phi \in (a, b)$, the function

$$H(re^{i\phi}) := r^\rho h(\phi)$$

is harmonic in the angle

$$\Gamma(a, b) := \{re^{i\phi} : \phi \in (a, b), \ r \in (0, \infty)\},$$

whence the function $v - H$ is subharmonic and nonpositive in $\Gamma(a, b)$. By virtue of the maximum principle, either $v < H$ in $\Gamma(a, b)$ or $v \equiv H$ in $\Gamma(a, b)$ for each $v \in \mathbf{Fr}[f]$. Note that the condition $v \equiv H$ in $\Gamma(a, b)$ implies $v \equiv H$ in $\Gamma[a, b]$ for the closed interval because of the upper semicontinuity of v.

Let us prove (5.4.3.5). For every $v \in \mathbf{Fr}[f]$ we have $v(re^{i\phi_0}) - H(re^{i\phi_0}) = 0$ whence by the maximum principle $v = H$ in $\Gamma(a, b)$. Hence (5.4.3.3) holds.

Let us prove (5.4.3.6a). Assume the contrary:$h(a) = g(a) \wedge h'_+(a) = h'_-(a)$ holds, but there exists $\phi_0 \in (a, b)$ such that $h(\phi_0) > g(\phi_0)$. Then there exists $v \in \mathbf{Fr}[f]$ such that

$$g(\phi_0) \leq v(e^{i\phi_0}) < h(\phi_0)$$

whence

$$v(\tau e^{i\phi}) < \tau^\rho h(\phi) \in \Gamma(a, b). \tag{5.4.4.2}$$

Without loss of generality, we can assume that $v(z) > -\infty$, otherwise we can replace v with $\max(v, -C)$ for a large positive constant $C > 0$.

We choose $0 < \tau_1 < \tau_2$ and to every function

$$W_j(re^{i\phi}) := v_{[\tau_j]}(re^{i\phi + a}) - r^\rho h(\phi + a), \ j = 1, 2, \gamma = b - a, \ re^{i\phi} \in \Gamma(0, \gamma)$$

we apply the following lemma due to A.E. Eremenko and M.L. Sodin [So] (see also [PW, Ho]):

Lemma 5.4.4.1 (E.S.) *Let W be a subharmonic nonpositive function inside the angle $\Gamma(0,\gamma)$, $\gamma > 0$. Then the following implication is valid,*

$$\left(\limsup_{\phi\to 0}\frac{W(e^{i\phi})}{\phi}=0\right)\Rightarrow W\equiv 0.$$

If the condition of this theorem is not satisfied for

$$W^*(re^{i\phi})=\max_{\tau\in[\tau_1,\tau_2]} v_{[\tau]}(re^{i\phi})$$

it would be possible to insert a ρ-t.c.function between $h(\phi)-\epsilon(\phi-a)$ (for a small ϵ) and $v(e^{i\phi})$. However, such a function does not exist, because of the negative jump of the derivative. So it will be a contradiction. See further for details.

From Lemma 5.4.4.1 we get

$$\liminf_{\phi\to a+0}\frac{h(\phi)-v_{[\tau_1]}(e^{i\phi})}{\phi-a}:=\alpha_1>0$$

and likewise

$$\liminf_{\phi\to a+0}\frac{h(\phi)-v_{[\tau_2]}(e^{i\phi})}{\phi-a}:=\alpha_2>0.$$

So a $\Delta>0$ can be chosen such that $a+\Delta<b$ and the inequalities

$$H(\tau_j e^{i\phi})-v_{[\tau_j]}(e^{i\phi})>\alpha\tau_j^{\rho}(\phi-a),\ j=1,2, \tag{5.4.4.3}$$

where $\alpha := 1/2\min(\alpha_1,\alpha_2)$, hold for all $\phi\in[a,a+\Delta]$.

We write

$$\beta:=\min_{\tau\in[\tau_1,\tau_2]}\left(H(\tau e^{i(a+\Delta)})-v(\tau e^{i(a+\Delta)})\right)$$

which is positive because of (5.4.4.2).

Let us choose $\epsilon>0$ small enough to

$$\epsilon<\min(\alpha,\beta(\tau_2)^{-\rho}\Delta^{-1}) \tag{5.4.4.4}$$

and let us consider the ρ-trigonometrical function

$$h_\epsilon(\phi):=\rho^{-1}(h'(a)-\epsilon)\sin\rho(\phi-a)+h(a)\cos\rho(\phi-a),\ \phi\in(a,b)$$

that coincides with

$$h(\phi)=\rho^{-1}h'(a)\sin\rho(\phi-a)+h(a)\cos\rho(\phi-a),\ \phi\in(a,b)$$

in the point $\phi=a$ but has a tangent that is lower than the tangent of h.

Further

$$h(\phi)-h_\epsilon(\phi)=\rho^{-1}\epsilon\sin\rho(\phi-a)\le\epsilon(\phi-a),\ \phi\in[a,a+\Delta]. \tag{5.4.4.5}$$

Combining (5.4.4.3)–(5.4.4.5) we obtain

$$v_{[\tau_j]}(e^{i\phi}) < \tau_j{}^\rho h(\phi) - \alpha(\phi - a) < \tau_j{}^\rho h(\phi) - \epsilon(\phi - a) \qquad (5.4.4.6)$$
$$\leq \tau_j{}^\rho h_\epsilon(\phi),\ \phi \in [a, a+\Delta],\ j = 1, 2,$$
$$v(\tau e^{i\phi}) \leq \tau^\rho h_\epsilon(a+\Delta) + \tau^\rho \epsilon\Delta - \beta \qquad (5.4.4.7)$$
$$< \tau^\rho h_\epsilon(a+\Delta),\ \tau \in [\tau_1, \tau_2].$$

We write

$$G := \{re^{i\phi} : \phi \in [a, a+\Delta],\ \tau \in [\tau_1, \tau_2]\}. \qquad (5.4.4.8)$$

It follows from (5.4.4.6), (5.4.4.7) that

$$v(re^{i\phi}) < r^\rho h_\epsilon(\phi),\ re^{i\phi} \in \partial G,$$

where ∂G is the boundary of the domain G. Since the functions $v(re^{i\phi})$ and $r^\rho h_\epsilon(\phi)$ are subharmonic in G, by virtue of the maximum principle we have

$$v(re^{i\phi}) < r^\rho h_\epsilon(\phi),\ re^{i\phi} \in G. \qquad (5.4.4.9)$$

Let us consider the function

$$H_1(re^{i\phi}) := r^\rho h_1(\phi),\ re^{i\phi} \in \Gamma(a - \Delta, a + \Delta)$$

where

$$h_1(\phi) := \begin{cases} h(\phi), & \phi \in (a - \Delta, a], \\ h_\epsilon(\phi), & \phi \in [a, a+\Delta). \end{cases}$$

The function H_1 is continuous in $\Gamma(a - \Delta, a + \Delta)$ and subharmonic in the angles $\Gamma(a - \Delta, a)$ and $\Gamma(a, a + \Delta)$. Let us prove that it is subharmonic at the point $z = e^{ia}$. Let $\mathcal{M}(z, R, v)$ be the mean value of v over the circle $\{\zeta : |\zeta - z| = R\}$ (see (2.6.1.1)). Taking into consideration (5.4.4.9) and subharmonicity of v (see (2.6.1.1)), for all small R we have

$$\mathcal{M}(e^{ia}, R, H_1) \geq M(e^{ia}, R, v) \geq v(e^{ia}) = H_1(e^{ia}).$$

Hence H_1 is subharmonic for $z = e^{ia}$. Since H_1 is homogeneous, i.e., $H_1(kz) = k^\rho H_1(z)$,

$$\mathcal{M}(ke^{ia}, kR, H_1) = k^\rho \mathcal{M}(e^{ia}, R, H_1) \geq k^\rho H_1(e^{ia}) = H_1(ke^{ia}).$$

So H_1 is subharmonic on the ray $\{z = ke^{ia} : k \in (0, \infty)\}$ and hence in the angle $\Gamma(a - \Delta, a + \Delta)$. Thus $h_1(\phi)$ is a ρ-t.c.f. for $\phi \in (a - \Delta, a + \Delta)$. However, by construction

$$(h_1)'_-(a) = h'_-(a) = h'_+(a) = (h_\epsilon)'_+ + \epsilon = (h_1)'(a) + \epsilon$$

and this contradicts the fact that h_1 is ρ-t.c.f.

Concordance of the implication (5.4.3.6a) is proved.

5.4.5 Here we continue the proof of necessity. Pass to the proof of necessity of the condition (5.4.3.7a). Assume the contrary. Then there exists $v \in \mathbf{Fr}[f]$ and $\phi_0 \in [a,b]$ such that $g(\phi_0) \leq v(e^{i\phi_0}) < h(\phi_0)$, whence by virtue of the maximum principle, $v(\tau e^{i\phi}) < \tau^\rho h(\phi)$ for $\tau e^{i\phi} \in \Gamma(a,b)$. Actually $v(\tau e^{i\phi}) \leq \tau^\rho h(\phi)$ everywhere and on the circle we have strict inequality. If $v(\tau e^{ia}) = H(\tau e^{ia})$ for a $\tau > 0$, then $v_{[\tau]}(e^{ia}) = h(a)$, and it will suffice to repeat the arguments used in proving (5.4.3.6a) with $v_{[\tau]}$ instead of v.

Exercise 5.4.5.1 Do that.

So it is sufficient to examine the case $v(\tau e^{ia}) < H(\tau e^{ia})$, $\tau > 0$. Denote

$$T(\phi) := h'(a)\rho^{-1} \sin \rho(\phi - a) + h(a) \cos \rho(\phi - a).$$

This is a ρ-trigonometrical function, the graph of which is tangent to the graph of $h(\phi)$ at the point a.

There are two possibilities for $T(\phi)$ on some small interval $\phi \in (a-\gamma, a)$, $\gamma > 0$: either $T(\phi) < h(\phi)$ or $T(\phi) = h(\phi)$.

Inequality $T(\phi) > h(\phi)$ contradicts ρ-t.convexity at the point a. The equality on the sequence of points $\phi_j \to a - 0$ contradicts the maximum principle for ρ-t.c.functions.

Exercise 5.4.5.2 Why is it?

If $T(\phi) = h(\phi), \phi \in (a - \gamma, a)$, then h is ρ-trigonometrical on the interval $(a-\gamma, b) \supset (a,b)$ that was already considered in the case (5.4.3.4) (M. Cartwright's Theorem).

So we assume $T(\phi) < h(\phi), \phi \in (a - \gamma, a)$. We set

$$h_1(\phi) := h(\phi) - T(\phi), \phi \in (a - \gamma, a),$$
$$v_1(re^{i\phi}) := v(re^{i\phi}) - r^\rho T(\phi), re^{i\phi} \in \Gamma(a - \gamma, b).$$

Then $h_1(\phi) = 0$ for $\phi \in [a,b]$, $h_1(\phi) > 0$ for $\phi \in (a - \gamma, a)$ and $h'(a) = 0$.

The function $v_1(e^{i\phi}) < 0, \phi \in [a,b)$. Let us analyze the behavior of the function $v_1(e^{i\phi})$ at the point b. Either $v_1(e^{ib}) < 0$ or $v_1(e^{ib}) = 0$ but

$$\limsup_{\phi \to b-0} v_1(e^{i\phi})(b - \phi)^{-1} \leq -C$$

for some $C > 0$ by Lemma 5.4.4.1 (E.S.).

From the other side $v_1(e^{i\phi})$ is strictly negative also in some left (say, $(a - \Delta, a)$) neighborhood of a because of upper semicontinuity. In any case $v_1(e^{i\phi})$ can be majorated on the interval $(a - \Delta, b)$ by the function

$$h_\epsilon := -A \sin(\rho - \epsilon)(b - \phi)$$

with sufficiently small A.

A point of intersection of the graph of h_ϵ with the axis $0, \phi$ can be regulated by ϵ and can be chosen so close to the point a that the graph of h_ϵ also intersects the graph of $h_1(\phi)$, at some point $\theta_0 < a$ because $h_1(a) = h_1'(a) = 0$.

Exercise 5.4.5.3 Make the precise proof with all the estimates.

Let the parameters A, ϵ, θ_0 be fixed as above. Denote

$$S := \{re^{i\phi} : \phi \in (\theta_0, b), 0 < r < 1\}.$$

Then $H_\epsilon(re^{i\phi}) := r^{\rho-\epsilon}h_\epsilon(\phi)$ is harmonic in the sector S and satisfies the inequality $H_\epsilon(re^{i\phi}) \geq v_1(re^{i\phi})$ on ∂S. Hence $H_\epsilon(re^{i\phi}) \geq v_1(re^{i\phi})$ on S. Thus

$$v(re^{i\phi}) \leq H(re^{i\phi}) + H_\epsilon(re^{i\phi}), \ re^{i\phi} \in S. \tag{5.4.5.1}$$

Let $\mathcal{M}(r, v)$ be the mean value of the function on the circle $\{\zeta : |\zeta| = r\}$ (see 2.6.1.1). Using (5.4.5.1) we have

$$\begin{aligned}\mathcal{M}(r, v) &\leq \int_{\theta_0}^{b} [H(re^{i\phi}) + H_\epsilon(re^{i\phi})]d\phi + \int_{[0,2\pi)\setminus(\theta_0,b)} H(re^{i\phi})d\phi \\ &\leq d_1 r^\rho - d_2 r^{\rho-\epsilon}, \quad d_1, \ d_2 > 0.\end{aligned}$$

So we get $\mathcal{M}(r, v) < 0 = v(0)$ for sufficiently small $r > 0$ which contradicts the subharmonicity of the function v at zero.

5.4.6 Now we complete proof of necessity, proving (5.4.3.8a,b). Assume the contrary: suppose

$$\liminf_{\phi \to a+0} \frac{h(\phi) - g(\phi)}{\phi - a} = 0 \tag{5.4.6.1}$$

but there exists a $\phi_0 \in (a, b)$ such that $h(\phi_0) > g(\phi_0)$. Then there exists a function $v \in \mathbf{Fr}[f]$ such that

$$g(\phi_0) \leq v(e^{i\phi_0}) < h(\phi_0). \tag{5.4.6.2}$$

Then the function $v_1(re^{i\phi}) := v(re^{i\phi}) - H(re^{i\phi})$ is subharmonic and nonpositive in $\Gamma(a, b)$. By virtue of the maximum principle $v_1(re^{i\phi}) < 0, \ re^{i\phi} \in \Gamma(a, b)$.

From (5.4.6.1) we obtain

$$0 = \liminf_{\phi \to a+0} \frac{h(\phi) - g(\phi)}{\phi - a} \geq \liminf_{\phi \to a+0} \frac{h(\phi) - v(e^{i\phi})}{\phi - a} = -\liminf_{\phi \to a+0} \frac{v_1(e^{i\phi})}{\phi - a}$$

whence, recollecting that $v_1(e^{i\phi}) < 0$, we get

$$\limsup_{\phi \to a+0} \frac{v_1(e^{i\phi})}{\phi - a} = 0.$$

Applying Lemma 5.4.4.1 (E.S.) to the function

$$W(re^{i\phi}) = v_1(re^{i\phi + a}), \ re^{i\phi} \in \Gamma(0, \gamma), \ \gamma = b - a$$

we get $v_1 \equiv 0$ in $\Gamma(a, b)$ which leads to a contradiction. The implication (5.4.3.8b) is proved in the same way. So the proof of necessity in Theorem 5.4.3.1 is completed.

□

We do not include here the proof of sufficiency and refer the readers to the original paper [Po(1992)].

5.5 Asymptotic extremal problems. Semiadditive integral

5.5.1 Suppose some class of entire functions is determined by asymptotic behavior of their zeros, and we want to know what is the restriction on asymptotic behavior of functions: for example, to estimate the indicator of such a function. The first example of such a problem was considered by B.Ya. Levin in [Le, Ch. IV, § 1, Example]. A developed theory of such estimates was constructed in the papers of A.A. Gol′dberg [Go(1962)] and his pupils [Kon], [KF]. We consider this theory from the point of view of limit sets.

Let $\mathcal{M} \Subset \mathcal{M}(\rho)$ (see (3.1.3.4)) be a convex set of measures which is closed in $\mathcal{D}'$ and is invariant with respect to the transformation $(\bullet)_t$ (see (3.1.3.1), (3.1.3.2)) and let $A(\mathcal{M})$ be a class of entire functions f for which $\mathbf{Fr}[n_f] \subset \mathcal{M}$. We suppose ρ is non-integer. Recall that canonical potential $\Pi(z, \nu, p)$ is defined by: (see (2.9.2.1))

$$\Pi(z, \nu, p) := \int_{\mathbb{C}} G_p(z/\zeta)\nu(d\zeta),$$

where ν is a measure and

$$G_p(z) := \log|1 - z| + \Re \sum_{k=1}^{p} \frac{z^k}{k}.$$

Theorem 5.5.1.1 [AP] *The relation*

$$h(\phi, f) = \sup\{\Pi(e^{i\phi}, \nu, p) : \nu \in \mathcal{M}\} \tag{5.5.1.1}$$

is valid. There exists $f \in A(\mathcal{M})$ for which the equality holds in (5.5.1.1) *for all ϕ.*

Proof. We should only prove that there exists an entire function with such indicator. Consider the set

$$\Lambda := \{\Pi(e^{i\phi}, \nu, p) : \nu \in \mathcal{M}\}.$$

It is a convex set contained in $U[\rho]$. Thus there exists a subharmonic (see Corollary 4.1.4.2) and hence entire (see Corollary 5.3.1.5) function f such that $\mathbf{Fr}[f] = \Lambda$. By Theorem 5.4.2.1, (5.5.1.1) holds. □

For some $\mathcal{M}$ it is possible to compute the supremum in (5.5.1.1) and thus to obtain explicit precise estimates of indicators in the respective class $A(\mathcal{M})$. As an example, we shall present an estimate given by A.A. Gol′dberg.

We recall that the *upper density of zeros* of an entire function $f \in A(\rho)$ is defined by the equality

$$\overline{\Delta}[n_f] := \limsup_{r\to\infty} \frac{n_f(r)}{r^\rho}$$

where n_f is the distribution of zeros of the function f, and denote

$$K(t,\phi) := -\left[\frac{d}{dt}G_p^+(e^{i\phi}/t)\right]^- \tag{5.5.1.2}$$

where $a^+ := \max(a,0)$, $a^- := \min(a,0)$. This function is piecewise continuous.

Corollary 5.5.1.2 [Go(1962)] *Let the distribution of zeros n_f of a function f be concentrated on the positive ray, and let $\overline{\Delta}[n_f] \le \Delta < \infty$. Then*

$$h(\phi,f) \le \Delta \int_0^\infty t^\rho K(t,\phi)dt, \ \phi \in [0,2\pi) \tag{5.5.1.3}$$

and there exists a function from the same class for which equality is attained for all ϕ.

Proof of Corollary 5.5.1.2. We exploit Theorem 5.5.1.1. The class of functions f satisfying the assumption of the corollary coincides with the class of f for which

$$\mathbf{Fr}[n_f] \subset \mathcal{M} = \{\nu \in \mathcal{M}(\rho) : \operatorname{supp}\nu \subset [0,\infty] \wedge \nu(r) \le \Delta r^\rho\}. \tag{5.5.1.4}$$

Exercise 5.5.1.1 Show this by using Corollary 3.3.2.6.

Thus

$$\Pi(e^{i\phi},\nu,p) = \int_0^\infty G_p(e^{i\phi}/t)\nu(dt) \le \int_0^\infty G_p^+(e^{i\phi}/t)\nu(dt).$$

Integrating by parts we obtain

$$\Pi(e^{i\phi},\nu,p) \le -\int_0^\infty \nu(t)[\frac{d}{dt}G_p^+(e^{i\phi}/t)]^- dt.$$

By (5.5.1.4) we get (5.5.1.3). □

We write

$$M_p(r) := \max\{G_p(re^{i\phi}) : \phi \in [0,2\pi)\}.$$

In the same way one can prove

Corollary 5.5.1.3 [Go(1962), Thm. 4.1] *Let distribution of zeros of the function $f \in A(\rho)$ satisfy only the condition $\overline{\Delta}[n_f] \le \Delta < \infty$. Then*

$$h(\phi,f) \le \Delta\rho \int_0^\infty t^{\rho-1} M_p(1/t)dt, \ \phi \in [0,2\pi) \tag{5.5.1.5}$$

and there exists a function from the same class for which equality is attained for all ϕ.

Exercise 5.5.1.2 Prove this corollary exploiting

$$\mathcal{M} := \{\nu \in \mathcal{M}(\rho) : \nu(r) \le \Delta r^{\rho},\ \forall r > 0\}.$$

5.5.2 To be able to obtain explicit estimates for more diverse classes of entire functions defined by a restriction on the density of zeros, Gol′dberg introduced an integral with respect to a nonadditive measure and obtained estimates for indicators in terms of a one-dimensional integral (along a circumference) with respect to such a measure ([Go(1962)]. Gol′dberg initially constructed the integral sum of a special form. The construction presented here is based on the Levin–Matsaev–Ostrovskii theorem (see [Go(1962), Thm. 2.10]). Fainberg (1983) developed this approach using a two-dimensional integral. This made it possible to extend significantly the set of classes of entire functions for which the estimate expressed by a nonadditive integral is precise. We shall present these results after the necessary definitions.

Let $\delta(X)$ be a nonnegative monotonic function of $X \subset \mathbb{C}$, the function being finite on bounded sets and $\delta(\varnothing) = 0$. For a given family of sets $\mathcal{X} := \{X\}$ we denote by $N(\delta, \mathcal{X})$ the class of countable-additive measures μ defined by the relation

$$N(\delta, \mathcal{X}) := \{\mu : \mu(X) \le \delta(X),\ X \in \mathcal{X}\}.$$

For a Borel function $f \ge 0$ we define the quantity

$$(\mathcal{X}) \int f d\delta := \sup\left\{ \int f d\mu : \mu \in N(\delta, \mathcal{X}) \right\},$$

called an $(\mathcal{X})$-*integral* with respect to a *nonnegative measure* δ. For a Borel set $E \subset \mathbb{C}$ we set

$$(\mathcal{X}) \int_E f d\delta := (\mathcal{X}) \int f I_E d\delta,$$

where I_E is an indicator of the set E, i.e.,

$$I_E(z) := \begin{cases} 1, & \text{if } z \in E; \\ 0 & \text{if } z \notin E. \end{cases}$$

This integral possesses a number of natural properties: it is monotonic with respect to f and δ and the family $\mathcal{X}$, positively homogeneous and semi-additive with respect to the function f and δ. If δ is a measure, if $\mathcal{X}$ is a Borel ring, and if f is a measurable function, then $(\mathcal{X})$-integral coincides with the Lebesgue-Stieltjes integral.

Exercise 5.5.2.1 Check these properties.

Let $\delta(\Theta)$ be a nonadditive measure on the unit circle $\mathbb{T}$, defined initially on the family of all open sets $\Theta \subset \mathbb{T}$. It can be naturally extended to all closed sets Θ^F using the equality

$$\delta(\Theta^F) := \inf\{\delta(\Theta) : \Theta \supset \Theta^F\}.$$

Let χ_Θ be a set of open sets containing the set $\mathbb{T}$. We write

$$D_{r,\Theta} := \{z = te^{i\theta} : 0 < t < r, e^{i\theta} \in \Theta\}, \ \chi_z := \{D_{r,\Theta} : r > 0, \ \Theta \in \chi_\Theta\}.$$

The subscripts Θ and z at χ indicate that the families under consideration are located either on $\mathbb{T}$ or on the plane, respectively.

Let us define a nonadditive measure δ_z on χ_z by the equalities

$$\delta_z(D_{r,\Theta}) := r^\rho \delta(\Theta), \ D_{r,\Theta} \in \chi_z.$$

Now the integral $(\chi_z) \int G_p^+(e^{i\theta}/\zeta) d\delta_z$ is defined.

Recall that the classical angular upper density of zeros of an entire function $f \in A(\rho)$ is defined by the equality (compare (3.3.2.7))

$$\overline{\Delta}^{\mathrm{cl}}[n_f, \Theta] := \limsup_{r\to\infty} n_f(D_{r,\Theta}) r^{-\rho}.$$

Consider the class of entire functions $A^{\mathrm{cl}}(\delta, \chi_\Theta)$ defined by the equality

$$A^{\mathrm{cl}}(\delta, \chi_\Theta) := \{f : \overline{\Delta}^{\mathrm{cl}}[n_f, \Theta] \le \delta(\theta), \ \forall \Theta \in \chi_\Theta\} \tag{5.5.2.1}$$

for a given non-additive measure $\delta(\Theta)$ and a family χ_Θ.

Theorem 5.5.2.1 [Fa] *Let $\delta(\Theta)$ satisfy the condition*

$$\delta(\Theta) = \delta(\overline{\Theta}), \ \forall \Theta \in \chi_\Theta \tag{5.5.2.2}$$

(*the dash means the closure of a set*). *Then*

$$h(\phi, f) \le (\chi_z) \int G_p^+(e^{i\theta}/\zeta) d\delta_z. \tag{5.5.2.3}$$

There exists a function $f \in A^{\mathrm{cl}}(\delta, \chi_\Theta)$ such that equality in (5.5.2.7) *is attained for all $\phi \in [0, 2\pi)$ simultaneously.*

Proof. Let us note the following: If we replace in this theorem $\overline{\Delta}^{\mathrm{cl}}[n_f, \Theta]$ with its $\mathcal{D}'$ counterpart $\overline{\Delta}(Co_\Theta(I_1))$ (see Theorem 3.3.1.2) and consider the corresponding class of entire functions $A(\delta, \chi_\Theta)$, the assertion of the theorem holds without conditions (5.1.5.6). You should only apply Theorem 5.5.1.1 with the corresponding $\mathcal{M}$. The condition (5.5.2.7) is exploited only for replacing "$\mathcal{D}'$" quantities by the classic ones using results of Sections 3.3.2. □

Exercise 5.5.2.2 Prove this theorem in detail.

It is also worth noting that every family χ_Θ can be replaced by a family χ'_Θ that is dense in χ_Θ (see 3.2.2) and such that for χ'_Θ (5.5.2.6) already holds (see Theorem 3.3.2.3).

5.6 Entire functions of completely regular growth. Levin-Pfluger Theorem. Balashov's theory

5.6.1 The most famous definition of a function of *completely regular growth* (CRG-function) is the following:

A function $f \in A(\rho(r))$ is a function of completely regular growth, if the limit

$$\lim_{z\to\infty} r^{-\rho(r)} \log|f(z)|, \; r := |z|$$

exists when $z \to \infty$ uniformly outside some C_0^1-set (see Section 5.2.1.)

Actually, it is equivalent to all other definitions of the functions of completely regular growth in the plane (compare [Le, Ch. III], [Pf(1938)], [Pf(1939)]).

By A.A. Gol′dberg ([Go(1967)]) this definition was reduced to the following: A function $f \in A(\rho(r))$ is a function of completely regular growth, if

$$\underline{h}_f(\phi) = h_f(\phi), \; \forall \phi \in [0, 2\pi).$$

Because of the formulae (3.2.1.1), (3.2.1.2) (see also Section 3.2.7) we have the following

Theorem 5.6.1.1 *A function $f \in A(\rho(r))$ is a function of completely regular growth (CRG-function) iff* $\mathbf{Fr}[f]$ *consists of only one subharmonic function $h(z)$.*

Because of (3.2.1.11) the function $h(z)$ has the form

$$h(z) = r^{\rho} h(e^{i\phi}). \tag{5.6.1.1}$$

The function $h(\phi) := h(e^{i\phi})$ is ρ-trigonometrically convex and it was studied in Sections 3.2.3, 3.2.4, 3.2.5.

5.6.2 The initial definition of *regular zero distribution* [Le, Ch. II, § 1] is the following:

Let n be a zero distribution (divisor, or mass distribution) of convergence exponent $\rho_1 := \rho[n]$ (see Section 2.8.3), and let $\rho_1 > [\rho_1]$. Let $\rho_1(r) \to \rho_1$ be a proper proximate order of $n(r)$ (see Theorem 2.8.1.2). It means that $n \in \mathcal{M}(\rho(r))$, $\rho(r) \to \rho_1$ (see Section 3.1.3).

The initial definition of *regular zero distribution* for ρ_1 being non-integer is: A zero distribution n is regular if the limit

$$\lim_{r\to\infty} \frac{n(Co_{(\alpha,\beta)}(I_t))}{t^{\rho_1(t)}} := \Delta((\alpha,\beta)) \quad {}^{1}$$

exists for all $\alpha > \beta$ except perhaps for a countable set on the circle.

[1] For the definition of $Co_{(\alpha,\beta)}(I_t)$, see Exercise 3.3.1.5.

By using results of Section 3.3, one can prove

Theorem 5.6.2.1 *The zero distribution n is regular iff* $\mathbf{Fr}[n]$ *consists of only one measure* ν_{reg}.

Exercise 5.6.2.1 Prove this exploiting Theorems 3.3.3.1 and 3.3.2.4.

Recall that for $f \in A(\rho(r)), \rho(r) \to \rho$ we have $n_f \in \mathcal{M}(\rho(r)), \rho(r) \to \rho$ (see Theorem 2.9.3.2). Now we can formulate

Theorem 5.6.2.2 (Levin-Pfluger) [Le, Ch. II, Ch. III] *An entire function* $f \in A(\rho(r))$, $\rho(r) \to \rho$ *of non-integer order* ρ *is of completely regular growth function iff its zero distribution is regular.*

After Theorems 5.6.1.1, 5.6.2.1 this theorem is a direct corollary of Theorem 3.1.5.1.

5.6.3 Consider now the case of integer ρ. In general, this case differs from the case of non-integer ρ. For example, Theorem 2.9.4.2 (Brelot-Lindelöf) implies that

$$(f \in A(\rho(r)), \rho(r) \to \rho) \Longleftrightarrow n_f \in \mathcal{M}(\rho(r)), \rho(r) \to \rho$$

iff the family of polynomials (2.9.4.4a) is compact.

To describe the regularity of zero distribution for the case of integer ρ we assume that the limit

$$\lim_{R\to\infty} \delta_R(z,\nu,\rho) := \Re[\delta_\infty z^\rho] \tag{5.6.3.1}$$

exists, where

$$\delta_\infty := \lim_{R\to\infty}\left[\int\limits_{|\zeta|<R} |\zeta|^{-\rho}\cos\arg\zeta n(d\zeta) + i\int\limits_{|\zeta|<R} |\zeta|^{-\rho}\sin\arg\zeta n(d\zeta)\right].$$

Now a zero distribution $n \in \mathcal{M}(\rho(r))$, $\rho(r) \to \rho$ with *integer* ρ is called *regular* if $\mathbf{Fr}[n]$ consists of only one measure ν_{reg} as in Theorem 5.6.2.1 and the limit (5.6.3.1) exists.

Under this definition Theorem 5.6.2.2 still holds, because the set $(\mathcal{H}, \mathbf{Fr})[\log|f|]$ from Theorem 3.1.5.2 consists of only one element $(\Re[\delta_\infty z^\rho], \nu_{\text{reg}})$.

Note also

Proposition 5.6.3.1 *The measure* ν_{reg} *has the form*

$$\nu_{\text{reg}}(drd\phi) = \rho r^\rho dr \otimes \Delta(d\phi)$$

where Δ *is a measure of bounded variation on the unit circle.*

This assertion is a corollary of invariance of $\mathbf{Fr}[n]$, Theorem 3.1.3.3, frm3).

5.6.4 In the papers [Bal(1973), Bal(1976)] functions of *completely regular growth along curves of regular rotation* were considered. A *curve of regular rotation* is a curve that is described by the equation

$$z = te^{i(\gamma(t)\log t+\phi)}, 0 < t < \infty$$

for a fixed ϕ.

If $\gamma(t) \equiv \gamma$, then this curve is a logarithmic spiral. In the general case $\gamma(t)$ is a differentiable function such that

$$\gamma(t) \to \gamma, t\gamma'(t) \to 0,\ t \to \infty.$$

To describe this theory in terms of limit sets we consider the transformation

$$P_t = te^{i\gamma(t)\log t},$$

$$u_t(z) = u(P_t z)t^{-\rho(t)}.$$

The following theorem is similar to Theorem 3.1.2.1.

Proposition 5.6.4.1 (Existence of spiral Limit Set) *The following holds:*

esls 1) $u_t \in SH(\rho(r))$ *for all* $t \in (0, \infty)$;

esls 2) *the family* $\{u_t\}$ *is precompact at infinity.*

The set of all limits $\mathcal{D}' - \lim_{j\to\infty} u_{t_j}$ does not depend on $\gamma(t)$ but only on the constant γ since

$$\lim_{t\to\infty} (\gamma(t) - \gamma) \log t = \lim_{t\to\infty} t\gamma'(t) = 0.$$

So it is the same as that for

$$P_t = te^{i\gamma\log t},$$

i.e., the case that was already considered in the general theory.

In particular (3.2.1.8) for this case has the form

$$z^0(z) = e^{i(-\gamma\log r+\phi)}. \tag{5.6.4.1}$$

Hence, from Theorem 3.2.1.2 the indicator (see (3.2.1.1)) has the form

$$h(re^{i\phi}) = r^\rho h(-\gamma\log r + \phi),\ z = re^{i\phi},$$

where $h(\phi)$ is a ρ-trigonometrically convex 2π-periodic function (see Section 3.2.3).

All other assertions of Levin-Pfluger theory can be obtained analogously from other general assertions as it was done in the previous sections.

Theorem 3.1.6.1 connects limit sets for every γ.

Exercise 5.6.4.1 Formulate and prove Balashov's analogy of the Levin-Pluger Theorem 5.6.2.2 and Theorem 3.1.6.1 for $m = 2$.

For other generalizations of the Levin-Pfluger theory see [AD] and [Az(2007)].

5.7 General characteristics of growth of entire functions

5.7.1 A functional $\mathcal{F}(u)$ acting in the unit circle and defined on subharmonic functions $u \in SH(\rho(r))$ is called a *growth characteristic* if the following conditions are fulfilled:

1. *continuity*:

$$\mathcal{F}(u_j) \to \mathcal{F}(u), \tag{5.7.1.1}$$

if $u_j \to u$ uniformly on compacts (of course, for continuous functions u) or if $u_j \downarrow u$;

2. *positive homogeneity*:

$$\mathcal{F}(cu) = c\mathcal{F}(r,u); \tag{5.7.1.2}$$

for every constant $c > 0$.

Here we shall list some widely used functionals that satisfy these conditions:

$$H_\phi(u) := u(e^{i\phi}); \tag{5.7.1.3}$$

$$T(u) = \frac{1}{2\pi}\int_0^{2\pi} u^+(e^{i\phi})d\phi; \tag{5.7.1.4}$$

$$M_\alpha(u) := \max\{u(e^{i\phi}) : |\phi| \le \alpha\}; \tag{5.7.1.5}$$

$$M(u) := M_\pi(u); \tag{5.7.1.6}$$

$$I_{\alpha\beta}(u) := \int_\alpha^\beta u(e^{i\phi})d\phi; \tag{5.7.1.7}$$

$$I(u,g) := \int_0^{2\pi} u(e^{i\phi})g(\phi)d\phi, \ g \in L^1[0,2\pi]. \tag{5.7.1.8}$$

Exercise 5.7.1.1 Check properties 1 and 2 for these functionals.

Let $\alpha(t)$ and $\alpha_\epsilon(\zeta)$ be the "hats" defined by the equalities (2.3.1.1)–(2.3.1.3) and let $R_\epsilon u$ be defined by (2.3.1.4).

This averaging has the following properties.

Proposition 5.7.1.1

1. *if u is subharmonic, then $R_\epsilon u$ is subharmonic;*
2. *$R_\epsilon u \downarrow u$ as $\epsilon \downarrow 0$ for every subharmonic function;*
3. *if $u_j \to u$ in $\mathcal{D}'$ and u_j, u are locally summable functions, $R_\epsilon u_j \to R_\epsilon u$ uniformly on every compact set.*

Exercise 5.7.1.2 Prove this using Theorem 2.3.4.5, 2.6.2.3.

Now we can define the *asymptotic characteristics of growth* of entire function $f \in A(\rho(r))$:

$$\overline{\mathcal{F}}[f] := \lim_{\epsilon \to 0} \limsup_{t \to \infty} \mathcal{F}(R_\epsilon u_t(\bullet)), \tag{5.7.1.9}$$

$$\underline{\mathcal{F}}[f] := \lim_{\epsilon \to 0} \liminf_{t \to \infty} \mathcal{F}(R_\epsilon u_t(\bullet)), \tag{5.7.1.10}$$

where $u = \log |f|$ and $(\bullet)_t$ is defined by (3.1.2.1).

Proposition 5.7.1.2 *For $\mathcal{F}(u)$ defined by* (5.7.1.3)

$$\overline{\mathcal{F}}[f] = h_f(\phi); \quad \underline{\mathcal{F}}[f] = \underline{h}_f(\phi).$$

For other functionals from the list (5.7.1.4)–(5.7.1.8) *one may replace $R_\epsilon u$ by u and omit* $\lim\limits_{\epsilon \to 0}$.

Exercise 5.7.1.3 Prove this.

The following assertion connects the asymptotic growth characteristics with limit sets.

Theorem 5.7.1.3 *The relations*

$$\overline{\mathcal{F}}[f] = \sup\{\mathcal{F}(v) : v \in \mathbf{Fr}[f]\},$$
$$\underline{\mathcal{F}}[f] = \inf\{\mathcal{F}(v) : v \in \mathbf{Fr}[f]\}$$

are true.

Proof. Let $v \in \mathbf{Fr}[f]$ and $u_{t_j} \to v$ in $\mathcal{D}'$. Then $R_\epsilon u_{t_j} \to R_\epsilon v$ uniformly on every compact set. Hence

$$\lim_{t_j \to \infty} \mathcal{F}(R_\epsilon u_{t_j}) = \mathcal{F}(R_\epsilon v).$$

Passing to the limit as $\epsilon \to 0$ we obtain

$$\lim_{\epsilon \to 0} \lim_{t_j \to \infty} \mathcal{F}(R_\epsilon u_{t_j}) = \mathcal{F}(v).$$

Choosing a sequence that corresponds to lim sup or lim inf we obtain the assertion of the theorem. □

Applying this theorem to the functional (5.7.1.3) we obtain the RHS's of (3.2.1.1), (3.2.1.2) and hence another definition for the indicator and lower indicator.

5.7.2 A *family of growth characteristics* $\chi_A := \{\mathcal{F}_\alpha(r, \bullet) : \alpha \in A\}$ is called *total* if the equation

$$\mathcal{F}_\alpha(v_1) = \mathcal{F}_\alpha(v_2), \ \forall r > 0, \ \alpha \in A \tag{5.7.2.1}$$

implies $v_1 \equiv v_2$ for $v_1, v_2 \in U[\rho]$ (see 3.1.2.4).

Here are some examples of the total families:

$$\chi_H := \{H_\phi(u(e^{i\phi}) : \phi \in [0, 2\pi)\}; \tag{5.7.2.2}$$
$$\chi_I := \{I_{\alpha,\beta}(u) : \alpha, \beta \in [0, 2\pi)\}; \tag{5.7.2.3}$$
$$\chi_{Fo} := \{c_k(u) = I(u, g_k) : k \in \mathbb{Z}\}; \tag{5.7.2.4}$$

where

$$g_0 := 1, \ g_k := \cos k\phi; \ g_{-k} = \sin k\phi, \ k \in \mathbb{N}. \tag{5.7.2.5}$$

It is easy to deduce from Theorem 5.6.1.1

Theorem 5.7.2.1 *Let a family $\{\mathcal{F}_\alpha(\bullet) : \alpha \in A\}$ be a total family of characteristics. An entire function f is a CRG-function iff*

$$\overline{\mathcal{F}}_\alpha[f] = \underline{\mathcal{F}}_\alpha[f]. \tag{5.7.2.6}$$

Exercise 5.7.2.1 Check this.

5.7.3 Let us consider a total family of characteristics of the form

$$\chi_\Psi := \{I(u, \psi) : \psi \in \Psi\}, \tag{5.7.3.1}$$

where Ψ is a set which is complete in $L^1[0, 2\pi]$. For instance, such are the families χ_I and χ_{Fo}.

Theorem 5.7.3.1 [Po(1985)] *Let $f \in A(\rho(r))$. The following assertions are equivalent:*

a) $\overline{\mathcal{F}}[fg] = \overline{\mathcal{F}}[f] + \overline{\mathcal{F}}[g], \forall \mathcal{F} \in \chi_\Psi$,
b) $\underline{\mathcal{F}}[fg] = \underline{\mathcal{F}}[f] + \underline{\mathcal{F}}[g], \forall \mathcal{F} \in \chi_\Psi$, *for all entire functions $g \in A(\rho(r))$.*
c) *f is a GRG-function.*

Proof of sufficiency of assertion c). Let us prove c) $\Longrightarrow$ a) and c) $\Longrightarrow$ b). Using Theorem 5.7.1.3 we obtain for every characteristic $\mathcal{F}$

$$\overline{\mathcal{F}}[fg] = \sup\{\mathcal{F}(w) : w \in \mathbf{Fr}[fg]\}. \tag{5.7.3.2}$$

Because of Theorem 3.1.2.4 fru1),

$$\mathbf{Fr}[fg] \subset \mathbf{Fr}[f] + \mathbf{Fr}[g].$$

Since f is a CRG-function, $\mathbf{Fr}[f]$ consists of only one subharmonic function v_{reg} (see Theorem 5.6.1.1) and it is easy to check that in this case we have equality

$$\mathbf{Fr}[fg] = v_{\text{reg}} + \mathbf{Fr}[g].$$

Exercise 5.7.3.1 Check this.

Since $\mathcal{F}(v_{\text{reg}} + v_g) = \mathcal{F}(v_{\text{reg}}) + \mathcal{F}(v_g)$, we obtain

$$\overline{\mathcal{F}}[fg] = \mathcal{F}(v_{\text{reg}}) + \sup\{\mathcal{F}(v_g) : v_g \in \mathbf{Fr}[g]\} = \overline{\mathcal{F}}[f] + \overline{\mathcal{F}}[g].$$

So c) $\Longrightarrow$ a) was proved. In the same way one can prove c) $\Longrightarrow$ b). □

Exercise 5.7.3.2 Prove this.

In the proof of sufficiency for c) of a) and b) we can suppose that ψ belong to the space $\mathcal{D}(\mathbb{T})$ of infinitely differentiable functions on the unit circle $\mathbb{T}$ because $\mathcal{D}(\mathbb{T})$ is complete in $L^1[0, 2\pi]$. We prove now sufficiency of b) in Theorem 5.7.3.1.

We recall that (see (3.1.2.4a))

$$v_{[t]}(z) = v(tz)t^{-\rho}, v \in U[\rho]$$

to distinguish it from $(\bullet)_t$ that we define as

$$u_t(z) = u(tz)t^{-\rho(t)}, u \in SH(\rho(r)).$$

The main constructive element for the proof sufficiency of b) in Theorem 5.7.3.1 is

Lemma 5.7.3.2 *Let $\psi^0 \in \mathcal{D}(S)$. There exists $v \in U[\rho]$ with the following properties:*

$$\mathcal{D}' - \lim_{t \to 0} v_{[t]} = \mathcal{D}' - \lim_{t \to \infty} v_{[t]} = \tilde{v}, \tag{5.7.3.3}$$

$$\langle v_{[t]}(e^{i\bullet}), \psi^0 \rangle > \langle v(e^{i\bullet}), \psi^0 \rangle \text{ for } t \in (0, \infty), t \neq 1, \tag{5.7.3.4}$$

$$\langle \tilde{v}(e^{i\bullet}), \psi^0 \rangle > \langle v(e^{i\bullet}), \psi^0 \rangle. \tag{5.7.3.5}$$

Proof of Lemma 5.7.3.2. Let ψ^0 be represented by Fourier series

$$\psi^0(\phi) = \frac{a_0}{2} + \sum_{n=1}^{\infty} (a_n \cos n\theta + b_n \sin n\theta)$$

Since $\psi^0 \not\equiv 0$ there exists $a_k \neq 0$ or $b_k \neq 0$. Suppose there exists $a_k \neq 0$. In the proof we will consider three cases:

1. $k = 0$;
2. $k \neq 0 \wedge k \leq p$;
3. $k \geq p + 1$.

The number ρ is supposed non-integer and $p = [\rho]$.

Consider the case $a_0 \neq 0$, $a_0 > 0$. Set

$$\psi(x) := \log(-e^{-\alpha|x|} + C), \alpha > 0, C > 1,$$

$$v(z) := |z|^{\rho} e^{\psi(\log|z|)} = \exp(\rho \log r + \psi(\log r)). \tag{5.7.3.6}$$

Applying the Laplace operator, we obtain:

$$\begin{aligned}\Delta v &= \frac{1}{r^2} r \frac{\partial}{\partial r} r \frac{\partial}{\partial r} v(r) = e^{-2x} \frac{\partial^2}{\partial x^2} e^{\rho x + \psi(x)} \\ &= \exp((\rho - 2)x + \psi(x)) \left[(\rho + \psi'(x))^2 + \psi''(x)\right], \ x = \log r.\end{aligned} \tag{5.7.3.7}$$

Since

$$\psi'(x) = \alpha \operatorname{sgn} x \exp(-\alpha|x|) \to 0, \ \psi''(x) = -\alpha^2 \operatorname{sgn} x \exp(-\alpha|x|) \to 0$$

as $x \to \pm\infty$, it is possible to choose α such that the expression (5.7.3.7) is positive. So $v(z)$ is subharmonic.

It is easy to check that all the assertions of the lemma are satisfied and $\tilde{v}(z) = b|z|^\rho$ where $b(> 0)$ is a constant.

Exercise 5.7.3.3 Check this.

If $a_0 = -|a_0| < 0$, consider the function

$$v^0(z) := \begin{cases} \log|z|, & |z| \geq 1, \\ 0, & |z| < 1; \end{cases}$$

it is subharmonic and

$$\langle v^0_{[t]}(e^{i\bullet}), \psi^0 \rangle = a_0 t^{-\rho} \log^+ t. \tag{5.7.3.8}$$

Since the RHS of (5.7.3.8) is minimized for $t_0 = e^{\rho^{-1}}$, the function

$$v(z) := v^0_{[t_0^{-1}]}(z)$$

satisfies the assertions of the lemma with $\tilde{v} = \lim\limits_{t\to 0,\infty} v_{[t]} = 0$

Now let $a_0 = 0, a_k \neq 0, 0 < k < p$. We will search for a function v of the form

$$v(re^{i\phi}) := \int_0^{2\pi} G_p(re^{i(\phi-\theta)})(1 - \operatorname{sgn} a_k \cos k\theta)d\theta. \tag{5.7.3.9}$$

This is the convolution $G_p(re^{i\bullet}) * g$ of the primary kernel (see Section 2.9.1)

$$G_p(z) = \log|1 - z| + \Re \sum_{n=1}^{p} z^n/n$$

with a positive function $g(\theta) := (1 - \operatorname{sgn} a_k \cos k\theta)$ on the circle. So it is subharmonic. Recall that the cos-Fourier coefficients of the function $G_p(re^{i\theta})$ are (see Exercise 2.3.7.2).

$$\hat{G}_p(m, r) = \begin{cases} 0, & m = 0, 1, \dots, p \\ (1/m)r^m, & m = p + 1, \dots \end{cases} \quad \text{if } r \leq 1, \tag{5.7.3.10}$$

and

$$\hat{G}_p(m,r) = \begin{cases} \log r, & m = 0 \\ \frac{1}{m}(r^m - r^{-m}), & m = 1,\dots,p \quad \text{if } r \geq 1. \\ (1/m)r^m, & m = p+1,\dots \end{cases} \tag{5.7.3.11}$$

All the sin-Fourier coefficients are equal to zero. The Fourier coefficients of the function g are 1 and $-\operatorname{sgn} a_k$.

Using well-known properties of Fourier coefficients, we obtain for $0 < k \leq p$,

$$\hat{v}_{[t]}(0) = \begin{cases} 0, & t \leq 1, \\ \frac{\log t}{t^\rho}, & t \geq 1, \end{cases}$$

$$\hat{v}_{[t]}(k) = \begin{cases} 0, & t \leq 1, \\ -\frac{1}{k}\frac{t^k - t^{-k}}{t^\rho} \operatorname{sgn} a_k & t \geq 1, \end{cases}$$

$$\langle v_{[t]}(e^{i\bullet}), \psi^0 \rangle = \begin{cases} 0, & t \leq 1, \\ -1/k(t^{k-\rho} - t^{-k-\rho})|a_k| & t \geq 1. \end{cases}$$

The function $t \mapsto \langle v_t(e^{i\bullet}), \psi^0 \rangle$ tends to zero when $t \to 0, \infty$ and has its only minimum at the point

$$t_0 = \left(\frac{\rho + k}{\rho - k} \right)^{1/k}.$$

Thus $v_{[(t_0)^{-1}]}$ satisfies the conditions of the lemma with $\tilde{v} = 0$.

For $k \geq p+1$ we should take the same g and then

$$\langle v_t(e^{i\bullet}), \psi^0 \rangle = \begin{cases} -(1/k)t^{k-\rho}|a_k|, & t \leq 1, \\ -(1/k)t^{-k-\rho}|a_k| & t \geq 1. \end{cases}$$

So the corresponding function $t \mapsto \langle v_t(e^{i\bullet}), \psi^0 \rangle$ obtains its minimum at the point $t_0 = 1$ and the function v satisfies the assertions of the lemma with $\tilde{v} = 0$. □

Exercise 5.7.3.4 Prove the lemma for the case $b_k \neq 0$.

Lemma 5.7.3.3 *Let $v \in U[\rho]$ with the condition*

$$\mathcal{D}' - \lim_{t \to 0} v_{[t]} = \mathcal{D}' - \lim_{t \to \infty} v_{[t]} = \tilde{v}$$

*fulfilled, and let $u \in SH(\rho(r))$ with some $v^0 \in$ **Fr**$[u]$. Then there exists $w^0 \in SH(\rho(r))$ such that*

$$\mathbf{Fr}[w^0] = \{v_{[t]} : t \in (0,\infty)\} \cup \tilde{v} \tag{5.7.3.12}$$

and the following condition holds:

1. *If the sequence $\lim_{t_n \to \infty} w^0_{t_n} = v_{[t]}$ for some $t \in (0,\infty)$ and the sequence u_{t_n} converges in $\mathcal{D}'$ as $t_n \to \infty$, then $\lim_{n \to \infty} u_{t_n} = v^0_{[t]}$.*

For the proof see Corollary 4.4.1.4.

Lemma 5.7.3.4 *Let* $w \in SH(\rho(r))$, $\psi \in \mathcal{D}(S)$. *Then the following holds:*

$$\liminf_{t\to\infty}\langle w,\psi\rangle = \min_{v\in\mathbf{Fr}\ w}\langle v,\psi\rangle,$$
$$\limsup_{t\to\infty}\langle w,\psi\rangle = \max_{v\in\mathbf{Fr}\ w}\langle v,\psi\rangle.$$

Exercise 5.7.3.5 Prove this exploiting completeness of **Fr**.

Proof of sufficiency of b) *in Theorem* 5.7.3.1. In assumption b) we should prove that f is a CRG-function, i.e., by Theorem 5.6.1.1 its $\mathbf{Fr}[f]$ consists of only one function. Since $\log|f| \in SH(\rho(r))$ and because of Theorem 5.3.1.4 (Approximation) it is enough to prove the corresponding theorem for subharmonic functions. Suppose

$$\underline{\mathcal{F}}[u+w] = \underline{\mathcal{F}}[u] + \underline{\mathcal{F}}[w], \forall\mathcal{F} \in \chi_\Psi \tag{5.7.3.13}$$

for all $w \in SH(\rho(r)$. We exploit Lemma 5.7.3.4 and write (5.7.3.13) in the form:

$$\min_{v\in\mathbf{Fr}[u+v]}\langle v,\psi\rangle = \min_{v\in\mathbf{Fr}u}\langle v,\psi\rangle + \min_{v\in\mathbf{Fr}w}\langle v,\psi\rangle, \forall\psi\in\Psi.$$

Suppose the contrary, i.e., u is not a CRG-function and $\mathbf{Fr}u$ does not consist of only one $v_{\min} \in U[\rho]$. Then there exists $v^0 \neq v_{\min}$. The family χ_Ψ is total; therefore there exists $\psi^0 \in \Psi$ such that $\langle v^0,\psi^0\rangle \neq \langle v_{\min},\psi^0\rangle$ and hence

$$\langle v^0,\psi^0\rangle > \langle v_{\min},\psi^0\rangle. \tag{5.7.3.14}$$

Using Lemma 5.7.3.2, construct for the function ψ^0 a function $v \in U[\rho]$ satisfying the conditions (5.7.3.3), (5.7.3.4) and (5.7.3.5). Apply Lemma 5.7.3.3 to construct a function w^0 satisfying (5.7.3.12) and the condition 1. Under conditions of the theorem,

$$\min_{\omega\in\mathbf{Fr}(u+w^0)}\langle\omega,\psi^0\rangle = \min_{\omega\in\mathbf{Fr}(u)}\langle\omega,\psi^0\rangle + \min_{\omega\in\mathbf{Fr}(w^0)}\langle\omega,\psi^0\rangle. \tag{5.7.3.15}$$

Let $\gamma \in \mathbf{Fr}(u+w^0)$ be the function on which the minimum of LRH in (5.7.3.15) is attained. Using (5.7.3.4), (5.7.3.5) and (5.7.3.12), we can rewrite (5.7.3.15) in the form

$$\langle\gamma,\psi^0\rangle = \min_{\omega\in\mathbf{Fr}u}\langle\omega,\psi^0\rangle + \langle v,\psi^0\rangle. \tag{5.7.3.16}$$

Since $\gamma \in \mathbf{Fr}(u+w^0), \gamma = \mathcal{D}' - \lim_{n\to\infty}(u+w^0)_{t_n}$. Passing to subsequences, we can suppose that the sequences $\{u_{t_n}\}$ and $\{w^0_{t_n}\}$ have limits. Since $\mathbf{Fr}w^0$ has the form (5.7.3.12), there are two possible cases : $w^0_{t_n} \to v_{[t]}$, $t \in (0,\infty)$ and $w^0_{t_n} \to \tilde{v}$.

Consider the first case. Because of condition 1 from Lemma 5.7.3.3, $u_{t_n} \to v^0_{[t]}$ and $\gamma = v^0_{[t]} + v_{[t]}$. Substituting this in (5.7.3.15), we obtain

$$\langle v^0_{[t]},\psi^0\rangle - \min_{\omega\in\mathbf{Fr}u}\langle\omega,\psi^0\rangle = \langle v,\psi^0\rangle - \langle v_{[t]},\psi^0\rangle.$$

This equality leads to a contradiction because for $t = 1$ it contradicts (5.7.3.14) and for $t \neq 1$ it contradicts (5.7.3.4).

Consider the second case, when $w^0_{t_n} \to \tilde{v}$. Denote $v^2 = \lim_{n\to\infty} u_{t_n}$ and rewrite (5.7.3.15) in the form

$$\langle v^2, \psi^0\rangle - \min_{\omega\in \mathbf{Fr}u} \langle \omega, \psi^0\rangle = \langle v, \psi^0\rangle - \langle \tilde{v}, \psi^0\rangle.$$

The last equality contradicts (5.7.3.5). □

Sufficiency of condition a) of Theorem 5.7.3.1 can be proved using the Lemmas 5.7.3.3, 5.7.3.4 and a variation of Lemma 5.7.3.2.

Lemma 5.7.3.2′ *Let $\psi^0 \in \mathcal{D}(S)$. There exists $v \in U[\rho]$ with the following properties:*

$$\mathcal{D}' - \lim_{t\to 0} v_{[t]} = \mathcal{D}' - \lim_{t\to\infty} v_{[t]} = \tilde{v}, \tag{5.7.3.3′}$$

$$\langle v_{[t]}(e^{i\bullet}), \psi^0\rangle < \langle v(e^{i\bullet}), \psi^0\rangle \ \textit{for}\ t \in (0,\infty), t \neq 1, \tag{5.7.3.4′}$$

$$\langle \tilde{v}(e^{i\bullet}), \psi^0\rangle < \langle v(e^{i\bullet}), \psi^0\rangle. \tag{5.7.3.5′}$$

Exercise 5.7.3.6 Prove this lemma and sufficiency of a) in Theorem 5.7.3.1.

5.7.4 In this section we consider the question of summing the asymptotic characteristics connected with the functional (5.7.1.3), i.e., indicator and lower indicator. Recall that $f \in A(\rho(r))$ is completely regular on the ray $\{\arg z = \phi\}$ ($f \in A_{\mathrm{reg},\phi}$) if

$$h_f(\phi) = \underline{h}_f(\phi). \tag{5.7.4.0}$$

We are going to prove the following assertions:

Theorem 5.7.4.1 *Let $f \in A_{\mathrm{reg},\phi}$. Then for every $g \in A(\rho(r))$,*

$$h_{fg}(\phi) = h_f(\phi) + h_g(\phi), \tag{5.7.4.1}$$

$$\underline{h}_{fg}(\phi) = \underline{h}_f(\phi) + \underline{h}_g(\phi). \tag{5.7.4.2}$$

Theorem 5.7.4.2 *Suppose the equality* (5.7.4.1) *holds for every $g \in A(\rho(r))$. Then $f \in A_{\mathrm{reg},\phi}$.*

Let us note that the assertion of Theorem 5.7.4.2 holds also if the equality (5.7.4.1) fulfilled for some sequence $\phi_n \to \phi$, because the indicator is a continuous function (see Section 3.2.5). So if the equality (5.7.4.1) holds for the set Φ of ϕ that is dense in $[0, 2\pi)$ (or the set

$$e^{i\Phi} := \{e^{i\phi} : \phi \in \Phi\} \tag{5.7.4.3}$$

is dense on the unit circle), then $f \in A_{\mathrm{reg},\phi}$ for all ϕ, i.e., f is a CRG-function.

On the other hand, the following assertion holds

Theorem 5.7.4.3 *If the set* Θ *of* θ *is not dense in* $[0, 2\pi)$*, there exists* $f \in A_{\mathrm{reg},\theta}, \theta \in \Theta$ *that is not a CRG-function.*

The situation with a lower indicator is analogous, but in another topology.

A set E is called *non-rarefied* at a point z_0 if for every function v subharmonic in a neighborhood of z_0 the following holds:

$$v(z_0) = \limsup_{z \in E, z \to z_0, z \neq z_0} v(z) = \limsup_{z \in E, z \to z_0} v(z).$$

A set is *rarefied* if it is not non-rarefied.

Note that if $\underline{h}_f(\phi) = -\infty$, then $\underline{h}_{fg}(\phi) = -\infty$ for every $g \in A(\rho(r))$. It is obvious that $f \notin A_{\mathrm{reg},\phi}$.

The next theorems were proved in [GPS].

Theorem 5.7.4.4 *Let* (5.7.4.2) *be fulfilled for* $\psi \in E$ *for all* $g \in A(\rho(r))$ *and* e^{iE} *be non-rarefied at the point* $e^{i\phi}$*. Then* $f \in A_{\mathrm{reg},\phi}$*.*

Theorem 5.7.4.5 *Let* E_0 *be a set such that* e^{iE_0} *is rarefied at all points of the unit circle. Then there exists* $f \in A(\rho(r))$ *for which* (5.7.4.2) *is fulfilled for all* $\phi \in E_0$ *and all* $g \in A(\rho(r))$*, but* $f \notin A_{\mathrm{reg},\phi}$ *for all* ϕ *and* $\underline{h}_f(\phi) > -\infty$, $\forall \phi$*.*

Let us note that E_0 can be dense in $[0, 2\pi)$ and E from Theorem 5.7.4.4 can even be of zero measure.

The proof of Theorems 5.7.4.4 and 5.7.4.5 is based on the following assertion that gives a criterion for (5.7.4.2) in terms of limit sets $\mathbf{Fr}[f]$.

Theorem 5.7.4.6 *Let* $f \in A(\rho(r))$ *and* $\underline{h}_f(\phi) > -\infty$*. The condition* (5.7.4.2) *holds for every* $g \in A(\rho(r))$*, such that* $h_g(\phi) > -\infty$ *iff*

$$\liminf_{t \to 1} v(te^{i\phi}) = \underline{h}_f(\phi) \tag{5.7.4.4}$$

for all $v \in \mathbf{Fr}[f]$*.*

An analogous criterion holds for (5.7.4.1).

Theorem 5.7.4.7 *Let* $f \in A(\rho(r))$*. Then* (5.7.4.1) *holds for every* $g \in A(\rho(r))$*, iff*

$$\limsup_{t \to 1} v(te^{i\phi}) = h_f(\phi), \tag{5.7.4.5}$$

for all $v \in \mathbf{Fr}[f]$*.*

Corollary 5.7.4.8 *The equality* (5.7.4.5) *implies* $f \in A_{\mathrm{reg},\phi}$*.*

Actually, for every $v \in \mathbf{Fr}[f]$ we have, using semicontinuity of subharmonic functions and the definition (3.2.1.1) of the indicator,

$$h_f(\phi) = \limsup_{t\to 1} v(te^{i\phi}) \le v(e^{i\phi}) \le h_f(\phi)$$

for all $v \in \mathbf{Fr}[f]$. So $\mathbf{Fr}[f]$ consists of functions v that coincide at the point $e^{i\phi}$ and hence on the ray $\{re^{i\phi} : r \in (0,\infty)\}$.

Note also that the set e^{iE} for which (5.7.4.1) holds is closed and Theorem 5.7.4.4 means that the set where (5.7.4.2) holds is *thinly closed*, i.e., closed in *thin topology* (see [Br, § 6]).

Therefore if $e^{i\phi_0}$ is a limit point of e^{iE} in the euclidian (respectively, thin) topology, then (5.7.4.1) ((5.7.4.2), respectively) is also a sufficient condition for completely regular growth at ϕ_0.

5.7.5 The main constructive element for proving Theorem 5.7.4.6 is

Lemma 5.7.5.1 *Let $\epsilon > 0, t_0 > 0$ and $\phi_0 \in [0, 2\pi)$ be fixed. Then there exists $v \in U[\rho]$ with the following properties:*

$$\mathcal{D}' - \lim_{t\to 0} v_{[t]} = \mathcal{D}' - \lim_{t\to\infty} v_{[t]} = 0, \tag{5.7.5.1}$$

$$v(e^{i\phi_0}) < v_{[t]}(e^{i\phi_0}),\ t \in (0,1) \cup (1,\infty), \tag{5.7.5.2}$$

$$-\infty < v(e^{i\phi_0}) < -\epsilon, \tag{5.7.5.3}$$

and the inequality

$$v_{[t]}(e^{i\phi_0}) - v(e^{i\phi_0}) \le \epsilon/2 \tag{5.7.5.4}$$

implies

$$t \in [1/t_0, t_0]. \tag{5.7.5.5}$$

The last condition means that the function $\psi(t) := v_{[t]}(e^{i\phi_0})$ can be less than $\psi(1) + \epsilon/2$ only in a neighborhood of $t = 1$.

Proof. Set

$$w(z) := \max(\log|1 - ze^{-i\phi_0}|, -N) + \Re\sum_{n=1}^{p} \frac{1}{n}(ze^{-i\phi_0})^n,$$

$$N > 0, p = [\rho]. \tag{5.7.5.6}$$

It is obvious that w is subharmonic, with masses ν_w concentrated in a neighborhood of the point $e^{i\phi_0}$. Thus $\nu_w \in \mathcal{M}[\rho]$ (see (3.1.3.4)) and

$$\mathcal{D}' - \lim_{t\to 0}(\nu_w)_{[t]} = \mathcal{D}' - \lim_{t\to\infty}(\nu_w)_{[t]} = 0.$$

Hence (see Theorem 3.1.4.2) $w \in U[\rho]$, and (see (3.1.5.0))

$$\mathcal{D}' - \lim_{t\to 0} w_{[t]} = \mathcal{D}' - \lim_{t\to\infty} w_{[t]} = 0. \tag{5.7.5.7}$$

Let us capitalize on the behavior of $w_{[t]}$ on the ray $\{\arg z = \phi_0\}$:

$$w_{[t]}(e^{i\phi_0}) := \psi(t) = \left(\max(\log|1-t|, -N) + \Re\sum_{n=1}^{p}\frac{1}{n}\right)t^{-\rho}t^n. \tag{5.7.5.8}$$

It is possible to prove directly the following properties of $\psi(t)$.

i) outside interval $(1-e^{-N}, 1+e^{-N})$, $\psi(t) = G_p(t)t^{-\rho}$; where G_p is the Primary Kernel (see Section 2.9.1) and inside this interval the first summand is $-N$;

ii) $\psi(t) > 0$ for $t > t_1$ where t_1 is a zero of the equation $G_p(t) = 0$, $\psi(t)$ decreases monotonically on the interval $(0, 1-e^{-N})$ and increases monotonically on the interval $(1-e^{-N}, t_1)$.

Exercise 5.7.5.1 Prove this.

Now set $t_2 := 1 - e^{-N}$ and $v(z) := Dw_{t_2}(z)$, where D is a constant. This function satisfies the conditions (5.7.5.1) and (5.7.5.2) of the lemma and $v_{[t]}(e^{i\phi_0})$ has only one negative minimum for $t = 1$. Thus it is possible to take D sufficiently large to satisfy the conditions (5.7.5.3) and (5.7.5.4) for fixed ϵ and t_0. □

Exercise 5.7.5.2 Prove this in detail.

In the proof of Theorem 5.7.4.6 we also use Lemma 5.7.3.3. We can prove all the assertions for subharmonic functions from $SH(\rho(r))$.

Proof of Theorem 5.7.4.6. Necessity. We should prove that if the equality

$$\underline{h}(e^{i\phi_0}, u+w) = \underline{h}(e^{i\phi_0}, u) + \underline{h}(e^{i\phi_0}, w) \tag{5.7.5.9}$$

holds for a fixed $u \in SH(\rho(r))$, ϕ_0 and every $w \in SH(\rho(r))$, then

$$\liminf_{t\to 1} v(te^{i\phi}) = \underline{h}(e^{i\phi}, u) \tag{5.7.5.10}$$

for all $v \in \mathbf{Fr}u$. Assume that $\underline{h}(e^{i\phi}, u) > -\infty$ and $\underline{h}(e^{i\phi_0}, w) > -\infty$. Suppose the contrary, i.e., there exists $v^0 \in \mathbf{Fr}u$ such that

$$\liminf_{t\to 1} v^0(te^{i\phi_0}) > \underline{h}(e^{i\phi_0}, u). \tag{5.7.5.11}$$

The inequality (5.7.5.11) implies that there exists $\epsilon > 0$ and $t_0 > 0$ such that for every $t \in [1/t_0, t_0]$ the inequality

$$v^0(te^{i\phi_0}) > \underline{h}(e^{i\phi_0}, u) + \epsilon \tag{5.7.5.11a}$$

holds. Let us construct by Lemma 5.7.5.1 for these ϵ, t_0, ϕ_0 a function v and by Lemma 5.7.3.3 for the functions u, v^0 and the already found v a function w^0. Let us show that for w^0 the equality (5.7.5.9) does not hold.

Compute $\underline{h}(e^{i\phi}, w^0)$. From (3.2.1.2)

$$\underline{h}(e^{i\phi_0}, w^0) = \min\{0, \inf\{v_{[t]}(e^{i\phi_0} : t \in (0, \infty)\}\}.$$

The inequalities (5.7.5.3) imply that 0 can be omitted and (5.7.5.2) implies that the infimum is attained at $t = 1$, i.e.,

$$\underline{h}(e^{i\phi_0}, w^0) = v(e^{i\phi_0}). \tag{5.7.5.12}$$

Find $v^\epsilon \in \mathbf{Fr}(u + w^0)$ such that $\underline{h}(e^{i\phi_0}, u + w^0) > v^\epsilon(e^{i\phi_0}) - \epsilon/3$. Let $t_n \to \infty$ and $(u + w^0)_{t_n} \to v^\epsilon$ in $\mathcal{D}'$. Passing to subsequences we can assume that u_{t_n} and $w^0_{t_n}$ also converge. Consider two cases. The first, when

$$\mathcal{D}' - \lim w^0_{t_n} = v_{[t]},\ t \in (0, \infty). \tag{5.7.5.13}$$

By Lemma 5.7.3.3 $\lim u_{t_n} = v^0_{[t]}$ and hence $v^\epsilon = \lim(w^0 + u)_{t_n} = v_{[t]} + v^0_{[t]}$. If $t \notin [1/t_0, t_0]$, then by (5.7.5.4)

$$v_{[t]}(e^{i\phi_0}) > v(e^{i\phi_0}) + \epsilon/2 = \underline{h}(e^{i\phi_0}, w^0) + \epsilon/2. \tag{5.7.5.14}$$

In this case we have

$$\underline{h}(e^{i\phi_0}, u + w^0) \geq v^\epsilon(e^{i\phi_0}) - \epsilon/3 \geq v_{[t]} + v^0_{[t]} - \epsilon/3. \tag{5.7.5.14a}$$

Using (5.7.5.14a), we obtain

$$\underline{h}(e^{i\phi_0}, u + w^0) \geq \underline{h}(e^{i\phi_0}, w^0) + \underline{h}(e^{i\phi_0}, u) + \epsilon/6. \tag{5.7.5.15}$$

If $t \in [1/t_0, t_0]$, then from (5.7.5.11) we have

$$\underline{h}(e^{i\phi_0}, u + w^0) \geq \underline{h}(e^{i\phi_0}, w^0) + \underline{h}(e^{i\phi_0}, u) + 2\epsilon/3. \tag{5.7.5.16}$$

So the case (5.7.5.13) is settled.

Let $\mathcal{D}' - \lim w^0_{t_n} = 0$. In this case we have

$$\underline{h}(e^{i\phi_0}, u + w^0) \geq v^\epsilon(e^{i\phi_0}) - \epsilon/3 \geq \underline{h}(e^{i\phi_0}, u) - \epsilon + \epsilon - \epsilon/3 = \underline{h}(e^{i\phi_0}, u) - \epsilon + 2\epsilon/3.$$

Using (5.7.5.12) and (5.7.5.3) we obtain

$$\underline{h}(e^{i\phi_0}, u + w^0) \geq \underline{h}(e^{i\phi_0}, u) + \underline{h}(e^{i\phi_0}, w^0) + 2\epsilon/3.$$

So we proved in any case that (5.7.5.9) does not hold if (5.7.5.10) does not hold.

Let us prove sufficiency in Theorem 5.7.4.6. We prove it for subharmonic functions. Let $u \in SH(\rho(r))$ and for every $v \in \mathbf{Fr}u$ (5.7.5.10) holds. Let us show that for all $w \in SH(\rho(r))$ (5.7.5.9) holds. It is sufficient to prove that

$$\underline{h}(e^{i\phi_0}, u + w) \leq \underline{h}(e^{i\phi_0}, u) + \underline{h}(e^{i\phi_0}, w) \tag{5.7.5.17}$$

holds since the inverse inequality holds for every $w \in SH(\rho(r))$ (see (3.2.1.5)). Let us begin by noting that for every $v^2 \in \mathbf{Fr}w$ there exist $v \in \mathbf{Fr}(u+w)$ and $v^1 \in \mathbf{Fr}u$ such that

$$v = v^1 + v^2. \tag{5.7.5.18}$$

Indeed, let $t_n \to \infty$ be a sequence such that $w_{t_n} \to v^2$. We can suppose, in choosing a subsequence, that $u_{t_n} \to v^1$ and $(u+w)_{t_n} \to v$. Then (5.7.5.18) holds.

Let ϵ be arbitrarily small. Choose $v^2 \in \mathbf{Fr}w$ such that $v^2(e^{i\phi}) < h(e^{i\phi}, w) + \epsilon$ holds. From upper semicontinuity of v^2 we have

$$\limsup_{t\to 1} v^2(e^{i\phi}) \leq \underline{h}(e^{i\phi_0}, w) + \epsilon. \tag{5.7.5.19}$$

Let $v^1 \in \mathbf{Fr}u$ and $v \in \mathbf{Fr}(u+w)$ satisfy (5.7.5.18). Then we have

$$\underline{h}(e^{i\phi_0}, u+w) \leq (v^1 + v^2)_{[t]}(e^{i\phi_0}) = v^1_{[t]}(e^{i\phi_0}) + v^2_{[t]}(e^{i\phi_0}), \ \forall t.$$

Hence

$$\underline{h}(e^{i\phi_0}, u+w) \leq \liminf_{t\to 1} v^1_{[t]}(e^{i\phi_0}) + \limsup_{t\to 1} v^2_{[t]}(e^{i\phi_0}).$$

Using (5.7.5.10) and (5.7.5.19) we obtain

$$\underline{h}(e^{i\phi_0}, u+w) \leq \underline{h}(e^{i\phi_0}, u) + \underline{h}(e^{i\phi_0}, w) + \epsilon.$$

This proves the inverse inequality and hence the equality (5.7.5.9), because ϵ is arbitrarily small. □

5.7.6 Now we are going to prove Theorem 5.7.4.4. We need the following assertion from Potential Theory.

Lemma 5.7.6.1 *Let E be a set that is non-rarefied at the point $e^{i\phi_0}$. Let E' be a set in $\mathbb{C}$, such that $\forall e^{i\phi} \in E$ and $\forall \delta > 0$ there exists a point $z' \in E'$ on the ray $\{\arg z = \phi\}$ such that $|z' - e^{i\phi}| < \delta$. Then E' is also non-rarefied at the point $e^{i\phi_0}$.*

Proof. We can suppose without loss of generality that E' has no intersection with some neighborhood of zero. Denote by $P(z)$ the map $z \mapsto e^{i \arg z}$. It is easy to see that for all pairs $z_1', z_2' \in E'$ the inequality $|P(z_1') - P(z_2')| < A|z_1' - z_2'|$ holds for some constant A. Thus the logarithmic capacity (2.5.2.5) satisfies ([La, Ch. II, § 4, it. 11, 15]).

$$\mathbf{cap}_l(M) < A\mathbf{cap}_l(M') \tag{5.7.6.1}$$

where $M' \subset E', M = P(M')$. Now we exploit the following properties of non-rarefied sets. First, if E is non-rarefied at a point z_0, then there exists a compact set that is non-rarefied at z_0 ([La, Ch. V, § 1, it. 5, § 3, it. 9]). Second, for a compact set K that is non-rarefied at z_0,

$$\sum_{n=1}^{\infty} \frac{n}{\log(\mathbf{cap}_l K_n)^{-1}} = \infty \tag{5.7.6.2}$$

where $K_n := K \cap \{z : q^{n+1} \leq |z - z_0| \leq q^n\},\ 0 < q < 1.$

Using the inequality (5.7.6.1), we obtain that divergence of the series (5.7.6.2) for a compact $K \subset E$ implies divergence for $K' \subset E'$ where $K = P(K')$, i.e., E' is non-rarefied at the point $P(e^{i\phi_0}) = e^{i\phi_0}$. □

Proof of Theorem 5.7.4.4. Let $\epsilon(\phi) \to 0$ as $\phi \to \phi_0$ and let $v \in \mathbf{Fr}u$. Suppose (5.7.5.9) holds for $e^{i\phi} \in E$. By Theorem 5.7.4.6 the equality (5.7.5.10) holds. Thus $\forall \Delta > 0,\ \exists z' = z'(e^{i\phi}, \Delta)$ such that

$$|z' - e^{i\phi}| < \Delta, \arg z' = \phi, v(z') < \underline{h}(e^{i\phi}) + \epsilon(\phi). \tag{5.7.6.3}$$

Set

$$E' := \bigcup_{\phi \in E} \bigcup_{n=1}^{\infty} z'(e^{i\phi}, 1/n).$$

By (5.7.6.3) and upper semicontinuity of $h(e^{i\phi})$ we obtain

$$\limsup_{z' \to e^{i\phi_0},\ z' \in E'} v(z') \le \underline{h}(e^{i\phi_0}). \tag{5.7.6.4}$$

Since E' is non-rarefied, by Lemma 5.7.6.1 the upper limit of v coincides with $v(e^{i\phi_0})$ and hence $v(e^{i\phi_0}) \le \underline{h}(e^{i\phi_0})$. The inverse inequality holds always. Thus $v(e^{i\phi_0}) = \underline{h}(e^{i\phi_0}), \forall v \in \mathbf{Fr}u$. Hence $h(e^{i\phi_0}) = \underline{h}(e^{i\phi_0})$. □

5.7.7 Now we are going to prove Theorem 5.7.4.5. Before this we need to describe a construction and prove some auxiliary assertions.

Let $B_j := \{z : T^j < |z| < T^{j+1}\},\ j = 0, \pm 1, \pm 2, \ldots$ where $T > 1$ is a fixed number. Denote $L_{E_0} := \{z : e^{i \arg z} \in e^{iE_0}\}$. Recall that e^{iE_0} is a set rarefied at every point of the unit circle. Let Q be the set of rational numbers on the interval $(1, T)$. Set

$$\begin{aligned} S_Q &:= \{z : |z| \in Q\}, \\ T^j S_Q &:= \{zT^j : z \in S_Q\}, \\ A_j &:= L_{E_0} \cap T^j S_Q, \quad j = 0, \pm 1, \pm 2, \ldots. \end{aligned}$$

Lemma 5.7.7.1 *There exists $v \in U[\rho]$ such that*

$$v(z) = -\infty \tag{5.7.7.1}$$

for $z \in A_0$ and

$$\mu_v(e) = 0,\ \forall e \subset \mathbb{C} \setminus B_0. \tag{5.7.7.2}$$

Proof. The set E is rarefied at every point, hence it is polar ([Br, Ch. 7, § 4]). Thus the set $\{z : |z| = r\} \cap L_{E_0}$ is polar (see [Br, Ch. 3, § 2]). A countable union of polar sets is polar ([Br, Ch. 3, § 2]). Thus A_0 is polar.Hence there exists a positive measure μ concentrated on B_0 for which the potential $v(z) := \int G_p(z/\zeta)\mu(d\zeta)$ is equal to $-\infty$ on A_0 (see [Br, Ch. 4, § 6, Applications]). It is easy to see that $\mu \in \mathcal{M}(\rho)$ and hence $v \in U[\rho]$ (see Theorem 3.1.4.2). □

Lemma 5.7.7.2 *There exists $\omega \in U[\rho]$ such that the following conditions are fulfilled:*

$$\omega(z) = -\infty, \; z \in A := \cup_{j=-\infty}^{+\infty} A_j; \; \omega(Tz) = T^{\rho}\omega(z). \tag{5.7.7.3}$$

Proof. Set for every $E \Subset \mathbb{C} \setminus 0$,

$$\nu(E) := \sum_{j=-\infty}^{j=+\infty} T^{j\rho}\mu_v(T^{-j}E \cup B_0) \tag{5.7.7.4}$$

(compare Theorem 4.1.7.1). We have $\nu \in \mathcal{M}(\rho)$. Set

$$\omega(z) := \int G_p(z/\zeta)\nu(d\xi d\eta), \; \zeta = \xi + i\eta.$$

This ω satisfies (5.7.7.3). □

Exercise 5.7.7.1 Prove this using Theorem 4.1.7.1.

Lemma 5.7.7.3 *Let ω be a subharmonic function in $\mathbb{C}$. Denote*

$$m(\phi) := \max\{\omega(re^{i\phi}) : r \in [1, T]\}.$$

Then there exists a constant $C > -\infty$ such that $m(\phi) > C \; \forall \phi$.

Proof. If not, there exists a sequence ϕ_n that we can assume to converge to ϕ_∞ such that $m(\phi_n) \to -\infty$. By upper semicontinuity of ω we have $\omega(z) = -\infty$, $ze^{-i\phi_\infty} \in [1, T]$. Thus $\omega(z) \equiv -\infty$ because the capacity of the segment in the plane is positive and hence it is not polar for some subharmonic function. □

Recall that for $v \in U[\rho]$ (see (4.1.3.1))

$$\mathbb{C}(v) := \mathcal{D}' - \text{clos}\{v_{[t]} : 0 < t < \infty\}, \tag{5.7.7.5}$$

$$\Omega(v) := \{v' \in U[\rho] : (\exists t_k \to \infty)(v' = \lim_{k\to\infty} v_{[t_k]}\}, \tag{5.7.7.6}$$

$$A(v) := \{v' \in U[\rho] : (\exists \tau_k \to 0)(v' = \lim_{k\to\infty} v_{[t_k]}\}. \tag{5.7.7.7}$$

By Theorems 4.1.3.3 and 4.2.1.2, if

$$A(v) \cap \Omega(v) \neq \varnothing, \tag{5.7.7.8}$$

there exists $u \in SH(\rho(r))$ such that

$$\mathbf{Fr}u = \mathbb{C}(v). \tag{5.7.7.9}$$

Lemma 5.7.7.4 *There exists* $v^1 \in U[\rho]$ *such that the following holds:*

$$A(v^1) = \Omega(v^1), \tag{5.7.7.10}$$

$$\inf\{v(e^{i\phi}) : v \in \mathbb{C}(v^1)\} = \liminf_{t\to 1} v(te^{i\phi}) = 0, \ \forall v \in \mathbb{C}(v^1), \ \forall e^{i\phi} \in e^{iE_0}, \tag{5.7.7.11}$$

$$\sup\{v(e^{i\phi}) : v \in \mathbb{C}(v^1)\} \neq \inf\{v(e^{i\phi}) : v \in \mathbb{C}(v^1)\}. \qquad (5.7.7.12.)$$

Proof. Let $\omega(z)$ be constructed by Lemma 5.7.7.2. Set

$$v(z) := \omega(z) + D\log^+ 2|z|.$$

The condition (5.7.7.3) implies

$$A(\omega) = \Omega(\omega) = \{\omega_{[t]} : t \in [1, T]\}$$

because it is a Periodic Limit Set (see Theorem 4.1.7.1).

Since $(\log^+ 2|z|)_{[t]} \to 0, \ t \to 0, \ t \to \infty$, the function v satisfies the condition

$$A(v) = \Omega(v) = \{\omega_{[t]} : t \in [1, T]\}.$$

By Theorem 2.1.7.4 for the function $v^1 := v^+$ we have

$$A(v^1) = \Omega(v^1) = \{\omega^+_{[t]} : t \in [1, T]\}.$$

Note that $v^1(z) = 0$ for $z \in A$ and since A is dense in L_{E_0} (5.7.7.11) holds. Choosing D sufficiently large it is possible (using Lemma 5.7.7.3) to find on every ray $\{\arg z = \phi\}$ a point z_ϕ where $v^1(z_\phi) > 0$. Hence $\sup\{v(e^{i\phi}) : v \in \mathbb{C}(v^1)\} > 0$. Because of (5.7.7.11) and upper semicontinuity of $\inf\{v(e^{i\phi}) : v \in \mathbb{C}(v^1)\}$ it is zero for every $e^{i\phi}$. Thus (5.7.7.12) holds. □

Proof of Theorem 5.7.4.5. Let us construct by Theorems 4.1.3.3 and 4.2.1.2 a function $u \in SH(\rho(r)$ such that $\mathbf{Fr}u = \mathbb{C}(v^1)$ where v^1 is taken from Lemma 5.7.7.4. It does not belong to $A_{\mathrm{reg},\phi}$ for any ϕ. The equality (5.7.5.9) holds for every $\phi \in E_0$ because of (5.7.7.11) by Theorem 5.7.4.6. □

5.7.8 The proof of Theorem 5.7.4.1 is a copy of the proof of sufficiency of assertion c) in Theorem 5.7.3.1.

Exercise 5.7.8.1 Prove Theorem 5.7.4.1.

Now we are going to prove Theorem 5.7.4.7 which implies (as it was shown in Corollary 5.7.4.8) Theorem 5.7.4.2.

The main constructive element of the proof of necessity is

Lemma 5.7.8.1 *Let $\epsilon > 0, t_0 > 0$ and $\phi_0 \in [0, 2\pi)$ be fixed. Then there exists $v \in U[\rho]$ with the following properties:*

$$\mathcal{D}' - \lim_{t\to 0} v_{[t]} = \mathcal{D}' - \lim_{t\to\infty} v_{[t]} = 0, \tag{5.7.8.1}$$

$$v(e^{i\phi_0}) > v_{[t]}(e^{i\phi_0}),\ t \in (0,1) \cap (1,\infty), \tag{5.7.8.2}$$

and the inequality

$$v_{[t]}(e^{i\phi_0}) - v(e^{i\phi_0}) \geq -\epsilon/2 \tag{5.7.8.3}$$

implies

$$t \in [t_0, 1/t_0]. \tag{5.7.8.4}$$

The last condition means that the function $\psi(t) := v_{[t]}(e^{i\phi_0})$ can be more than $\psi(1) - \epsilon/2$ only in a neighborhood of $t = 1$.

Proof. Consider the function

$$w(z) := \log^+ |z|. \tag{5.7.8.5}$$

It is subharmonic and satisfies (5.7.8.1). Since the function

$$\psi(t) := w_{[t]}(e^{i\phi_0}) = t^{-\rho} \log^+ t$$

has its only strict maximum in the point $t_{\max} > 1$, the function

$$v(z) := w(z/t_{\max})$$

has all the properties (5.7.8.1)–(5.7.8.4). □

After this lemma all the proof of Theorem 5.7.4.6 can be repeated with minimal changes.

Exercise 5.7.8.2 Prove Theorem 5.7.4.7.

5.7.9 Now we are going to prove Theorem 5.7.4.3. Let us prove the following

Lemma 5.7.9.1 *Let $\overline{\Theta}$ be a closed subset of $[0, 2\pi)$. Then for every $\sigma > 0$ there exists a 2π-periodic ρ-trigonometrically convex function $h(\phi)$ such that*

$$h(\phi) = \sigma \tag{5.7.9.1}$$

for $\phi \in \overline{\Theta}$ and

$$h(\phi) > \sigma \tag{5.7.9.2}$$

for $\phi \notin \overline{\Theta}$.

Proof. We can suppose that $0 \in \overline{\Theta}$, otherwise we can shift it a little. The set $[0, 2\pi) \setminus \overline{\Theta}$ is open and it can be represented as the union of non-intersecting open intervals. If length of an interval is $\leq \pi/\rho$ we can construct a ρ-trigonometrical function that is equal to σ on the ends of the interval. It is greater than σ in all inner points of the interval because $f(\phi) \equiv \sigma$ is a strictly ρ-trigonometrical function. If the length of the interval is greater than π/ρ, for example $(-l/2, l/2)$ with $l > \pi/2\rho$, we cover it by intersecting intervals of length less then π/ρ, construct $h_I(\phi)$ as before for every interval I and set $h(\phi) = \max_I h_I(\phi)$. It is obvious that $h(\phi)$ is greater than σ and it is ρ-trigonometrically convex. □

Theorem 5.7.4.3 is a corollary of Lemma 5.7.9.1 and the following

Theorem 5.7.9.2 *Let h_1 and h_2 be two ρ-trigonometrically convex functions. Then there exists a function $f \in A(\rho(r))$ such that*

$$h_f(\phi) = \max(h_1(\phi), h_2(\phi)), \ \underline{h}_f(\phi) = \min(h_1(\phi), h_2(\phi)).$$

Proof. Consider the set

$$U := \{v(z) = cr^\rho h_1(\phi) + (1-c)r^\rho h_2(\phi) : 0 \leq c \leq 1\}. \tag{5.7.9.3}$$

It consists of invariant subharmonic functions and is contained in $U[\rho]$ and satisfies the condition of Theorem 4.1.4.1. Hence (Theorems 4.2.1.2, Corollary 5.3.1.5) there exists a function $f \in A(\rho(r))$ such that

$$\mathbf{Fr} f = U. \tag{5.7.9.4}$$

By formulae (3.2.1.1), (3.2.1.2) we obtain the assertion of the theorem, using (5.7.9.3). □

Exercise 5.7.9.1 Prove Theorem 5.7.4.3.

5.7.10 The family of characteristics $\{\mathcal{F}_\alpha,\ \alpha \in A\}$ is called *independent* if for every subset $A' \subset A$ (or subset in some class of subsets, for example, measurable or closed) there exists a function $f = f_{A'} \in A(\rho(r))$ such that

$$\underline{\mathcal{F}}_\alpha[f] = \overline{\mathcal{F}}_\alpha[f], \ \alpha \in A',$$
$$\underline{\mathcal{F}}_\alpha[f] \neq \overline{\mathcal{F}}_\alpha[f], \ \alpha \in A \setminus A'.$$

It means that for every pointed subset of characteristics there exists a function that has regular growth with respect to this subset of characteristics and is not of regular growth with respect to all other characteristics.

Theorem 5.7.4.3 can be considered as an assertion of independence of the family (5.7.2.2).

Theorem 5.7.10.1 *The family* χ_{Fo} (5.7.2.4) *is independent.*

I.e., for every $A \subset \mathbb{Z}$ there exists $f \in A(\rho(r)$ such that

$$\lim_{r\to\infty} r^{-\rho(r)} \int_0^{2\pi} \log|f(re^{i\phi})g_k(\phi)d\phi$$

exists for all $k \in A$ and does not exist for $k \in \mathbb{Z} \setminus A$. To begin we prove

Lemma 5.7.10.2 *There exist two* ρ*-trigonometrically convex functions* h_1 *and* h_2 *for which*

$$\int_0^{2\pi} h_1(\phi)g_k(\phi)d\phi = \int_0^{2\pi} h_2(\phi)g_k(\phi)d\phi,\ k \in A, \tag{5.7.10.1}$$

$$\int_0^{2\pi} h_1(\phi)g_k(\phi)d\phi \neq \int_0^{2\pi} h_2(\phi)g_k(\phi)d\phi,\ k \in \mathbb{Z}\setminus A. \tag{5.7.10.2}$$

Proof. Let $g(\phi) \in C^2$ be a function, the Fourier coefficients of which with indices $k \in A$ are equal to zero. We can represent it as a difference of ρ-trigonometrically convex functions in the following way. Suppose for simplicity that ρ is non-integer. Then take $T_\rho g = g'' + \rho^2 g$ and consider

$$h_1(\phi) := \frac{1}{2\rho\sin\pi\rho}\int_0^{2\pi} \widetilde{\cos\rho}(\phi-\psi-\pi)(T_\rho g)^+(\phi)d\phi;$$

$$h_2(\phi) := \frac{1}{2\rho\sin\pi\rho}\int_0^{2\pi} \widetilde{\cos\rho}(\phi-\psi-\pi)(T_\rho g)^-(\phi)d\phi.$$

By Theorem 3.2.3.3 these functions are ρ trigonometrically convex and $h_1 - h_2 = g$. Hence (5.7.10.11), (5.7.10.12) holds. □

Proof of Theorem 5.7.10.1. We consider a function $f \in A(\rho(r))$ with the limit set $U := \{v(z) = cr^\rho h_1(\phi) + (1-c)r^\rho h_2(\phi) : 0 \le c \le 1\}$ with h_1, h_2 from the conditions of lemma, and we exploit Theorem 5.7.1.3. □

Exercise 5.7.10.1 Do this in detail.

5.8 A generalization of the Valiron-Titchmarsh theorem

5.8.1 The point of departure on this topic is the following

Theorem VT [Va, Ti] *Let $f \in A(\rho)$, $\rho < 1$ have its zeros on the negative ray. If the limit*

$$\lim_{r\to\infty} r^{-\rho}\log|f(r)|$$

exists, then the limit

$$\lim_{r\to\infty} r^{-\rho} n(r)$$

exists.

The latter means that f is a CRG-function.

The general problem is the following. Let ρ be any non-integer number, $f \in A(\rho(r))$, and suppose all zeros of f lie on a finite system of rays

$$K_{S_1} := \{z = re^{i\phi} : 0 < r < \infty, \phi \in S_1\} \tag{5.8.1.1}$$

where

$$S_1 := \{e^{i\theta_j} : j = 1, 2, \dots, m\}. \tag{5.8.1.2}$$

We write $n_f \in \mathcal{M}_{S_1}$.

Let n_j be a zero distribution on the ray $\{\arg z = \theta_j\}$ and all the limits

$$\lim_{r\to\infty} r^{-\rho} n_j(r) := \Delta_j \tag{5.8.1.3}$$

exist. In such a case we write $n_f \in \mathcal{M}_{\mathrm{reg},S_1}$.

Let K_S be one more system of rays

$$S = \{e^{i\psi_k} : k = 1, 2, \dots, n\}. \tag{5.8.1.4}$$

Some ψ_k can coincide with some θ_j. Suppose that f has regular growth on this system, i.e.,

$$h_f(\phi) = \underline{h}_f(\phi), \; e^{i\phi} \in S. \tag{5.8.1.5}$$

In such a case we write $f \in A_{\mathrm{reg},S}$.

The problem is, what is the connection between S and S_1 so that the implication $(f \in A_{\mathrm{reg},S}) \Longrightarrow (n_f \in \mathcal{M}_{\mathrm{reg},S_1})$ holds.

This problem can be reformulated in another way. For $n_f \in \mathcal{M}_{\mathrm{reg},S_1}$ if $n_f \in \mathcal{M}_{S_1}$ it is necessary and sufficient that f is a CRG-function, because existence of an angle density is equivalent to existence of all the limits. So the problem can be reformulated in the form: what is the connection between S and S_1, so that the implication $(f \in A_{\mathrm{reg},S}) \Longrightarrow (f$ is CRG-function) holds.

We write

$$G(t,\gamma,\rho) := G_p(e^{t-i\gamma})e^{-\rho t},\ p = [\rho] \tag{5.8.1.6}$$

where G_p is the Primary Kernel:

$$G_p(z) = \log|1-z| + \Re\sum_{k=1}^{p} z^k/k.$$

Set

$$\hat{G}(s,\gamma,\rho) := \int_{-\infty}^{\infty} G(t,\gamma,\rho)e^{-ist}dt.$$

This is the Fourier transformation of $G(t,\gamma,\rho)$. It can be computed (see, e.g,[Oz, Lem. 3]);

$$\hat{G}(s,\gamma,\rho) = \frac{\pi\cos(\pi+\gamma)(\rho+is)}{(\rho+is)\sin\pi(\rho+is)}.$$

Consider the matrix

$$\hat{\mathbb{G}}(s, S_1 - S) := \|\hat{G}(s,\theta_j - \psi_k,\rho)\|. \tag{5.8.1.7}$$

We are going to prove (see [Az(1998)])

Theorem 5.8.1.1 *The implication*

$$\{f \in A_{\text{reg},S}\} \wedge \{n_f \in \mathcal{M}_{S_1}\} \Longrightarrow \{f \textit{ is a CRG-function}\}$$

holds iff

$$\text{rank } \hat{\mathbb{G}}(s, S_1 - S) = m,\ \forall s \in (-\infty,\infty). \tag{5.8.1.8}$$

As a corollary we obtain the following ([De])

Theorem 5.8.1.2 (Delange) *Suppose that S_1 and S consist of one ray, i.e.,*

$$S_1 = \{e^{i\theta_1}\},\ S = \{e^{i\psi_1}\}.$$

The implication (5.8.1.5) *holds iff*

$$\theta_1 - \psi_1 \neq (1-(2k+1)/2\rho)\pi,\ k = 1,2,\dots. \tag{5.8.1.9}$$

5.8.2 A Fourier transformation for distribution ν on the real axes is a distribution in the standard space $\mathcal{S}'$ (see [Hö, vol. 1, Ch. 7, § 7.1]). For a locally bounded measure whose variation is "not very quickly" growing, it can be defined by

$$(\mathcal{F}\nu)(s) := \lim_{\epsilon\to 0}\int_{-\infty}^{\infty} e^{its}e^{-\frac{\epsilon t^2}{2}}\nu(dt)$$

where the right side is understood in the sense of distributions.

For example, if $\nu(dt) := e^{is_0t}dt$, we have $\mathcal{F}\nu(s) = \delta(s-s_0)$ where δ is the Dirac function.

Exercise 5.8.2.1 Check this.

For distribution and a summable function one can define a convolution for which the property $\mathcal{F}(f * \nu)(s) = \mathcal{F}f(s)\mathcal{F}\nu(s)$ holds.

Proof of Theorem 5.8.1.1. Since $f \in \mathcal{M}_{S_1}$ the limit set $\mathbf{Fr}n_f$ is concentrated on K_{S_1}. So every $v \in \mathbf{Fr}[f]$ can be represented in the form (see Theorem 3.1.5.1):

$$v(z) = \sum_{j=1}^{j=m} \int_0^\infty G_p(z/re^{i\theta_j})\mu_j(dr) \tag{5.8.2.1}$$

where μ_j is concentrated on the ray $\{\arg\zeta = \theta_j\}$ and belongs to $U[\rho]$. After changing variables,

$$r = e^\tau, |z| = e^t,$$

we obtain from (5.8.2.1)

$$v^1(te^{i\phi}) = \sum_{j=1}^{j=m} \int_{-\infty}^\infty G(t-\tau, \phi-\theta_j, \rho)\mu_j^1(d\tau) \tag{5.8.2.1a}$$

where

$$e^{\rho\tau}\mu_j^1(d\tau) := \mu_j(dr), \ v^1(te^{i\phi}) := v(|z|e^{i\phi})e^{-\rho|z|}. \tag{5.8.2.2}$$

The equality (5.8.2.1a) can be written as

$$v^1(te^{i\phi}) = \sum_{j=1}^{j=m} [G(\bullet, \phi-\theta_j, \rho) * \mu_j^1](t) \tag{5.8.2.3}$$

where $*$ stands for convolution. Then $f \in A_{reg,S}$ with $n_f \in \mathcal{M}_{S_1}$, iff every pair $v_1, v_2 \in \mathbf{Fr}[f]$ satisfies the condition

$$v_1(z) = v_2(z), z \in K_S. \tag{5.8.2.4}$$

Denote by $\mu_{1,j}, \mu_{2,j}$ the restriction of $\mu_{v_1}\mu_{v_1}$ to the ray $\{\arg z = \theta_j\}$. Set $\nu_j := \mu_{1,j} - \mu_{2,j}$ Using (5.8.2.3) we can rewrite (5.8.2.4) in the form

$$\sum_{j=1}^{j=m} [G(\bullet, \phi_k - \theta_j, \rho) * \nu_j^1](t) \equiv 0, k = 1, 2, \ldots, n. \tag{5.8.2.5}$$

Applying Fourier transforms we obtain a system of linear equations:

$$\sum_{j=1}^{j=m} [\hat{G}(\bullet, \psi_k - \theta_j, \rho) \cdot \hat{\nu}_j^1](t) \equiv 0, k = 1, 2, \ldots, n. \tag{5.8.2.6}$$

Suppose now that rank $\hat{\mathbb{G}}(s, S-S_1) = m$ for every $s \in \mathbb{R}$. The system (5.8.2.6) has only the trivial solution for every s. Thus $\hat{\nu}_j^1(s) \equiv 0$, for $j = 1, 2, \ldots, m$. This implies $\nu_j^1(t) \equiv 0$ for $j = 1, 2, \ldots, m$ and $\nu_j \equiv 0$ for $j = 1, 2, \ldots, m$. Thus $\mu_{v_1} = \mu_{v_2}$, i.e., (by (5.8.2.3)) $\mathbf{Fr}[f]$ consists of one function $v \in U[\rho]$. Thus f is a CRG-function.

Conversely, suppose that rank $\hat{\mathbb{G}}(s, S-S_1) < m$ for some s_0.

Then there exists a nontrivial solution $(b_1, \ldots, b_m)$ that satisfies the corresponding system. We obtain that $\{\hat{\nu}_j^1 b_j \delta(s-s_0),\ j = 1, 2 \ldots, m\}$ is a solution of (5.8.2.6) for all $s \in \mathbb{R}$ and hence

$$\nu_j^1(dt) = b_j e^{its_0} dt,\ j = 1, 2, \ldots, m.$$

Since ν_j^1 have bounded densities $d\nu_j^1/dt$, we can find a constant C such that $\sup\{|d\nu_j^1/dt| : 0 < t < \infty,\ j = 1, 2 \ldots, m\} \leq C$.

Set

$$\mu_{1,j}^1(dt) = Cdt + \nu_j^1(dt);\ \mu_{2,j}^1 = Cdt. \tag{5.8.2.7}$$

Both of these are measures. Now we pass to $m_{1,j},\ m_{2,j}$ via (5.8.2.2). It is easy to check that $m_{1,j},\ m_{2,j} \in \mathcal{M}(\rho)$.

Exercise 5.8.2.2 Check this.

Consider $\mu_1, \mu_2 \in \mathcal{M}(\rho)$ which are defined uniquely by their restrictions $\mu_{1,j}, \mu_{2,j}$ respectively on K_{S_1}. Set

$$v_1(z) := \int_{\mathbb{C}} G_p(z/\zeta)\mu_1(d\xi d\eta); v_2(z) := \int_{\mathbb{C}} G_p(z/\zeta)\mu_2(d\xi d\eta);\ \zeta = \xi + i\eta.$$

It is easy to check that the equality

$$v_1(z) = v_2(z),\ z \in K_S \tag{5.8.2.8}$$

holds.

Exercise 5.8.2.3 Check this.

Since μ_1 and μ_2 are finite sums of trigonometrical functions, for v_1 and v_2 the condition (4.1.3.3) is satisfied. Thus by Theorem 4.3.6.1 there exists a function $f \in A(\rho(r))$ for which

$$\mathbf{Fr}[f] = \bigcup_{0 \leq c \leq 1} \mathbb{C}(cv_1 + (1-c)v_2).$$

Since for $v \in \mathbb{C}(cv_1 + (1-c)v_2)$ (5.8.2.8) also holds, the same holds for $v \in \mathbf{Fr}[f]$ and this function is not a CRG-function. $\square$

Chapter 6

Application to the Completeness of Exponential Systems in Convex Domains and the Multiplicator Problem

The completeness of exponential systems in convex domains is intimately connected to the multiplicator problem. Considering a special form of exponent system is related to the study of special subharmonic functions that determine the periodic limit set, the so-called automorphic subharmonic functions. The next Sections 6.1, 6.2 are devoted to these problems.

6.1 The multiplicator problem

6.1.1 Let $\Phi \in A(\rho(r))$ and let $H(\phi)$ be a ρ-trigonometrically convex function. A function $g \in A(\rho(r))$ is called an H-multiplicator of Φ if the indicator $h_{g\Phi}$ of the product $g\Phi$ satisfies the inequality

$$h_{g\Phi}(\phi) \leq H(\phi), \forall \phi.$$

In some questions (see Section 6.3) we need to determine whether a given function Φ has a multiplicator. We shall study this problem in terms of the limit set of Φ. Define $H(z) := r^{\rho} H(\phi), \ z = re^{i\phi}$. Let $v \in U[\rho]$ (see (3.1.2.4)). Consider the function

$$m(z, v, H) := H(z) - v(z).$$

As will be proved in Corollary 6.1.9.3, the maximal subharmonic minorant of $m(z, v, H)$ exists and is continuous. The maximal subharmonic minorant of m

(m.s.m.) belonging to $U[\rho]$ will be denoted by $\mathcal{G}_H v$, while the domain of definition of the operator $\mathcal{G}_H$ will be denoted by D_H. Though $m(0,\bullet,\bullet)=0$, the m.s.m. of m can differ from zero (as was remarked by A.E. Eremenko and M.L. Sodin), but if the m.s.m. of m equals zero at zero, then it belongs to $U[\rho]$.

Exercise 6.1.1.0 Prove this.

Exercise 6.1.1.1 Consider the function

$$w(z) = \begin{cases} |z|\log|z|, & \text{if } |z| \le 1; \\ |z|-1, & \text{if } |z| \ge 1. \end{cases}$$

It is subharmonic and belongs to $U[1]$. Show that the maximal subharmonic minorant of $K|z| - w(z)$ is different from zero in 0 for every $K > 0$.

Theorem 6.1.1.1 ([AG(1992)]) *$\Phi \in A(\rho(r))$ has an H-multiplicator iff*

$$\mathbf{Fr}[\Phi] \subset D_H. \tag{6.1.1.1}$$

Proof of necessity. Let g be a multiplicator of Φ, i.e.,

$$h_{g\Phi}(\phi) \le H(\phi) \tag{6.1.1.2}$$

and let $v \in \mathbf{Fr}[\Phi]$. We can choose $v_{g\Phi} \in \mathbf{Fr}[g\Phi]$ and $v_g \in \mathbf{Fr}[g]$ such that $v_{g\Phi} = v + v_g$ (see Theorem 3.1.2.4, fru1)).

Exercise 6.1.1.2 Prove this directly.

By definition of indicator (3.2.1.1) and (6.1.1.2) we have $v_{g\Phi}(z) \le H(z)$ or $v_g(z) \le m(z, v, H)$. Since $v_g \in U[\rho]$, $v \in D_H$. □

For proving sufficiency we need the following

Theorem 6.1.1.2 *The operator $\mathcal{G}_H$ is*

1. *upper semicontinuous in the $\mathcal{D}'$-topology,* 6.1.1.5, *i.e.,*

$$(v_j \to v) \wedge (\mathcal{G}_H v_j \to w) \Longrightarrow (w \in U[\rho]) \wedge (w(z) \le \mathcal{G}_H(z), z \in \mathbb{C});$$

2. *invariant:* $(\mathcal{G}_H v)_{[t]} = \mathcal{G}_H v_{[t]}$; *(see* (3.1.2.4a) *for* $P_t \equiv tI$*);*
3. *concave:*

$$(\forall v_1, v_2 \in D_H,\ c \in [0;1]) \Longrightarrow (v_c := cv_1 + (1-c)v_2 \in D_H)$$

and

$$\mathcal{G}_H(v_c) \ge c\mathcal{G}_H(v_1) + (1-c)\mathcal{G}_H(v_2).$$

Proof. Let us prove 1). Suppose $v_j \in U[\rho] \to v$ and $\mathcal{G}_H v_j \to w$. Then

$$\mathcal{G}_H v_j \leq H(z) - v_j(z), \ z \in \mathbb{C}. \tag{6.1.1.3}$$

Applying $(\bullet)_\epsilon$ from (2.6.2.2) and Theorem 2.3.4.5, reg 3), we obtain

$$w_\epsilon \leq (H)_\epsilon(z) - (v)_\epsilon(z), \ z \in \mathbb{C} \ w_\epsilon(0) \geq 0.$$

Passing to the limit as $\epsilon \downarrow 0$ we obtain by Theorem 2.6.2.3, ap2)

$$w(z) \leq H(z) - v(z) = m(z, H, v), z \in \mathbb{C}.$$

Since $0 \leq w(0) \leq m(0, H, v) = 0$ we have $w(0) = 0$ and, hence, $w \in U[\rho]$. Thus $v \in D_H$ and $w(z) \leq \mathcal{G}_H v(z)$.

Let us prove 2). Since $H(z)$ is invariant with respect to $(\bullet)_{[t]}$,

$$(\mathcal{G}v)_{[t]} \leq H(z) - v_{[t]}.$$

Hence,

$$(\mathcal{G}v)_{[t]}(z) \leq (\mathcal{G}(v_{[t]}))(z), \tag{6.1.1.4}$$

because $\mathcal{G}(v_{[t]})$ is the maximal subharmonic minorant. We can replace v with $v_{[1/t]}$ and obtain $(\mathcal{G}v_{[1/t]})_{[t]}(z) \leq \mathcal{G}v(z)$. Applying $(\bullet)_{[t]}$ to the two sides of the inequality, we obtain $\mathcal{G}v_{[1/t]}(z) \leq (\mathcal{G}v(z))_{[1/t]}$. Now we can replace $1/t$ with t and obtain the reverse inequality to (6.1.1.4), which, together with (6.1.1.4), proves 2).

3). Let $v_1, v_2 \in D_H$ and $c \in [0; 1]$. One has

$$\mathcal{G}v_i(z) \leq H(z) - v_i(z), \ i = 1, 2, \ \forall z.$$

Then

$$[c\mathcal{G}v_1 + (1-c)\mathcal{G}v_2](z) \leq H(z) - [cv_1 + (1-c)v_2](z).$$

Thus

$$[c\mathcal{G}v_1 + (1-c)\mathcal{G}v_2](z) \leq \mathcal{G}[cv_1 + (1-c)v_2](z).$$

□

Proof of sufficiency in Theorem 6.1.1.1. Assume that $\mathbf{Fr}[\Phi] \subset D_H$ and consider the set

$$\boldsymbol{U} := \{(v', v'') : v'' \leq \mathcal{G}v', v' \in \mathbf{Fr}[\Phi]\}. \tag{6.1.1.5}$$

Then $\boldsymbol{U}$ is nonempty, because of (6.1.1.1), closed, because of Theorem 6.1.1.2, 1), and invariant, because of Theorem 6.1.1.2, 2).

Every fiber $\boldsymbol{U}'' = \{v'' : v'' \leq \mathcal{G}v'\}$ is convex because of Theorem 6.1.1.2, 3). By Theorem 4.4.1.2 there exists $u'' \in U(\rho(r))$ such that for the curve $\boldsymbol{u} := (u', u'')$,

$$\mathbf{Fr}[\boldsymbol{u}] = \boldsymbol{U}. \tag{6.1.1.6}$$

By Theorem 5.3.1.4 (Approximation Theorem) the function u'' can be replaced with $\log|g|$, where $g \in A(\rho(r))$, retaining the property (6.1.1.6).

Let us prove that g is an H-multiplicator of Φ. Indeed, set $\Pi := g\Phi$. It is enough to prove that for every $v_\Pi \in \mathbf{Fr}[\Pi]$,

$$v_\Pi(z) \leq H(z). \tag{6.1.1.7}$$

Note that every v_Π has the form $v_\Pi = v_g + v$, where $(v, v_g) \in \boldsymbol{U}$. Thus, because of definitions (6.1.1.5) and (6.1.1.6), v_Π satisfies (6.1.1.7). □

Let us note that the pair $(v, \mathcal{G}_H v) \in \boldsymbol{U}$ because of closeness of $\boldsymbol{U}$. Hence the following assertion holds.

Proposition 6.1.1.3 *Every $\Phi \in A(\rho)$ that satisfies* (6.1.1.1) *has a multiplicator $g \in A(\rho)$ such that*

$$v + \mathcal{G}_H v \in \mathbf{Fr}[g\Phi]. \tag{6.1.1.8}$$

Exercise 6.1.1.3 Check this in detail.

Although $v \in U[\rho]$ is in general an upper semicontinuous function, we need

Theorem 6.1.1.4 *The function $\mathcal{G}_H v(z)$, $v \in U[\rho]$, is a continuous function that is harmonic outside the set $E = \{z : \mathcal{G}_H v(z) = m(z, v, H)\}$.*

Proof. $\mathcal{G}_G v(z)$ is continuous because of Corollary 6.1.9.3. If $\mathcal{G}_H v(z_0) < v(z_0)$ and if $\mathcal{G}_H v(z)$ is not harmonic in a neighborhood of z_0, we can make sweeping of masses from a small disc $\{|z - z_0| < \epsilon\}$ (see Theorem 2.7.2.1). The obtained subharmonic function will be greater than $\mathcal{G}_H v(z)$, contradicting maximality. □

6.1.2 Suppose that some H-multiplicator $g = g(z, \Phi, H)$ of the function Φ is found. We examine the function $\Pi = g\Phi$. The structure of its limit set is described by the following statement:

Proposition 6.1.2.1 *Every $v_\Pi \in \mathbf{Fr}[g\Phi]$ can be written as $v_\Pi = v + w_1$, where $v \in \mathbf{Fr}[\Phi]$ and $w_1 \in U[\rho]$ with the condition*

$$w_1(z) \leq \mathcal{G}_H(z), \forall z \in \mathbb{C}, \tag{6.1.2.1}$$

and, conversely, for every $v \in \mathbf{Fr}[\Phi]$ there exists a v_g, $v_g(z) \leq \mathcal{G}_H v(z)$, such that

$$v + v_g \in \mathbf{Fr}[g\Phi].$$

Exercise 6.1.2.1 Prove this the same way as in Exercise 6.1.1.2.

An H-multiplicator G of the function Φ will be called *ideally complementing* if it satisfies the condition

$$\mathbf{Fr}[G\Phi] = \{v_\Pi = v + \mathcal{G}_H v : v \in \mathbf{Fr}[\Phi]\}.$$

If a multiplicator is ideally complementing, then equality is achieved in (6.1.2.1) for all $v \in \mathbf{Fr}[\Phi]$. This make the multiplicator optimal in another respect. Recall that an entire function f is of *minimal type* with respect to a proximate order $\rho(r), \rho(r) \to \rho$ if (see (2.8.1.6))

$$\sigma_f := \limsup_{r\to\infty} \log M(r,f) r^{-\rho(r)} = 0.$$

Proposition 6.1.2.2 *Let $G = G(\bullet, \Phi, H)$ be an ideally complementing H-multiplicator of a function Φ. Then each H-multiplicator of the function $\Pi = G\Phi$ is of minimal type.*

This proposition is proved in Section 6.1.3.

A function Φ is said to be ideally complementable if for each H the condition (6.1.1.1) implies that Φ has an ideally complementing multiplicator. For instance, if Φ is a function of completely regular growth (see Section 5.6) then it is ideally complementable.

Exercise 6.1.2.2 Prove this.

Theorem 6.1.2.3 *Every function with periodic limit set is ideally complementable.*

This theorem is proved in Section 6.1.6.

Let $C \subset \mathbb{R}^l$ be an l-dimensional connected compact and let $\{h(\phi, c) : c \in C\}$ be a set of ρ-t.c. functions that is continuous with respect to $c \in C$. For example, $c \in [0,1]$ and $h(\phi, c) = ch_1(\phi) + (1-c)h_2(\phi)$. The set

$$U_{\text{ind}} := \{v(re^{i\phi}s) = r^\rho h(\phi, c) : c \in C\} \tag{6.1.2.2}$$

is the limit set of an entire function.

Exercise 6.1.2.3 Prove this using Theorem 4.3.6.1.

Such a set is called a *set of indicators*. Entire functions with such limit sets can be also considered as a generalization of CRG-functions.

Theorem 6.1.2.4 *Every function Φ whose limit set is a set of indicators is ideally complementable.*

This theorem is proved in Section 6.1.7.

The existence of an ideally complementing H-multiplicator depends, of course, both on $\Phi \in A(\rho(r))$ (or, more precisely, on its limit set $\mathbf{Fr}[\Phi]$) and on H.

Theorem 6.1.2.5 *Let Φ and H be such that the condition* (6.1.1.1) *is satisfied. The function Φ has an ideally complementing H-multiplicator if and only if the operator $\mathcal{G}_H$ is continuous on $\mathbf{Fr}\Phi$.*

This theorem is proved in Section 6.1.6.

Now we formulate a sufficient condition for continuity of the operator $\mathcal{G}_H$. We shall say that *the maximum principle for* $U[\rho]$ *is valid in the domain* G, (which is, generally speaking, unbounded), if the conditions $w \in U[\rho]$, $w(z) = 0$ for $z \notin G$ imply $w(z) \equiv 0$.

Let us denote by $\mathcal{H}_w$ a region of harmonicity of $w \in U[\rho]$, i.e., a region where the conditions "w is harmonic in G" and "$G \supset \mathcal{H}_w$" imply $G = \mathcal{H}_w$.

We remark that $\mathcal{H}_w$ is a connected component of the open set on which w is harmonic. Generally it is not unique.

The image of $U \in U[\rho]$ will be denoted by $\mathcal{G}_H U$, while its closure in the $\mathcal{D}'$-topology will be denoted by $\operatorname{clos} \mathcal{G}_H U$.

Theorem 6.1.2.6 *Suppose for every* $w \in \operatorname{clos} \mathcal{G}_H U$ *and every* $\mathcal{H}_w$ *the maximum principle for* $U[\rho]$ *holds. Then* $\mathcal{G}_H$ *is continuous on* U.

This theorem is proved in Section 6.1.5.

In Section 6.1.8 we will construct an example of Φ and H such that the operator $\mathcal{G}_H$ is not continuous on $\mathbf{Fr}[\Phi]$. This is also an example of the function that has no ideally complementing multiplicator.

6.1.3

Proof of Proposition 6.1.2.2. Let g be an ideally complementing multiplicator of the function $\Pi = G\Phi$. We write

$$\theta(z) := (gG\Phi)(z). \tag{6.1.3.1}$$

Let $v_g \in \mathbf{Fr}[g]$. Let us choose $t_j \to \infty$ such that:

$$(\log |g|)_{t_j} \to v_g; \ (\log |\Pi|)_{t_j} \to v_\Pi \in \mathbf{Fr}[\Pi]; \ (\log |\theta|)_{t_j} \to v_\theta \in \mathbf{Fr}[\theta].$$

It follows from (6.1.3.1) that $v_\theta = v_g + v_\Pi$. Since g is a multiplicator of Π, we have

$$v_\theta(z) = v_g(z) + v_\Pi(z) \le H(z). \tag{6.1.3.2}$$

Since G is an ideally complementing multiplicator, $v_\Pi = v + \mathcal{G}_H v$. So for all $z \in \mathbb{C}$ (6.1.3.2) implies

$$(v_g + \mathcal{G}_H v)(z) \le (H - v)(z).$$

Since $\mathcal{G}_H v$ is the maximal subharmonic minorant, $v_g(z) \le 0$ and hence $v_g(z) \equiv 0$. Thus (see (3.2.1.1)) we have $h_g(\phi) \equiv 0$ and therefore

$$\sigma_g = \max_{0 \le \phi \le 2\pi} h_g(\phi) = 0. \qquad \square$$

6.1.4 In order to prove Theorem 6.1.2.6 we need a number of auxiliary statements.

Lemma 6.1.4.1 *Let the maximum principle be valid in* G *for* $U[\rho]$ *and for some continuous functions* $w_1, w \in U[\rho]$ *satisfy:*

a) w *is harmonic in* G;

b) $w_1(z) = w(z)$ *outside of* G.

Then

$$w_1(z) \leq w(z), \;\; z \in G. \tag{6.1.4.1}$$

Proof. We set

$$w_0(z) := \begin{cases} (w_1 - w)^+(z), & z \in G, \\ 0, & z \notin G. \end{cases}$$

This function is continuous in $\mathbb{C}$ and, evidently, subharmonic both in G and in $\mathbb{C} \setminus \overline{G}$. Since $w_0(z) \geq 0$, the inequality for the mean values

$$0 = w_0(z) \leq \frac{1}{2\pi} \int_0^{2\pi} w_0(z + \epsilon r e^{i\phi}) d\phi, \;\; z \in \partial G,$$

implies the subharmonicity on ∂G. Since G satisfies the maximum principle for $U[\rho]$, we have $w_0 \equiv 0$, which is equivalent to (6.1.4.1). □

Now we shall dwell on some properties of maximal subharmonic minorants and, in particular, of $w = \mathcal{G}_H v$. Let

$$E_v := \{z \in \mathbb{C} : \mathcal{G}_H v(z) = m(z, v, H)\}. \tag{6.1.4.2}$$

We remark that $m(z, v, H)$ is a δ-subharmonic function in $\mathbb{C}$ whose charge will be denoted by $\nu(\bullet, v)$, its positive and negative parts will be denoted by ν^+ and ν^-.

Let us denote by μ_H the measure of $H(z)$. It is decomposed into the product of measures (see Section 3.2 and Proposition 5.6.3.1)

$$\mu_H = \Delta_H \otimes \rho r^{\rho - 1} dr, \tag{6.1.4.3}$$

where Δ_H is the measure on the unit circle and $\rho r^{\rho-1} dr$ is the measure on the ray. It is obvious that

$$\nu^+(\bullet, v) \leq \mu_H(\bullet). \tag{6.1.4.4}$$

We shall denote the mass distribution of $w \in U[\rho]$ by μ_w.

The modulus of continuity of w (if w is continuous) will be denoted by $\omega_w(z, h)$, $z \in \mathbb{C}$, $h > 0$.

The following lemma lists various properties of $w \in \mathcal{G}_H U$, $U \subset U[\rho]$ which will be useful in the sequel:

Lemma 6.1.4.2 *Let $w \in \mathcal{G}_H U$. Then*

1. *$w \in U[\rho, \sigma]$ where*

$$\sigma = 4 \cdot 2^\rho [\max\{H(e^{i\phi}) : \phi \in [0, 2\pi]\} + 2\sigma_1],$$
$$\sigma_1 = \max\{v(z)|z|^{-\rho} : z \in \mathbb{C}, v \in U\};$$

2. *the charge restriction* $\nu(\bullet, v)|_{E_v}$ *to* E_v *is nonnegative, i.e.,*

$$\nu(\bullet, v)|_{E_v} = \nu^+(\bullet, v)|_{E_v};$$

3. *outside* E_v *the function* w *is harmonic, i.e.,*

$$\mu_w|_{\mathbb{C}\setminus E_v} = 0;$$

4. *the measure* μ_w *is bounded from above by* $\nu^+(\bullet, v)$*, i.e.,*

$$\mu_w \le \nu^+(\bullet, v);$$

5. $\mathcal{G}_H U$ *is equicontinuous on each compact set, i.e.,*

$$\omega_w(z, h) \le C(R, \sigma, \rho)\sqrt{h}\log(1/h), \ |z| \le R,$$

where $C(R, \sigma, \rho)$ *is independent of* $w \in \mathcal{G}_H U$*.*

Proof. Let us prove property 1. We have

$$T(r, w) := \frac{1}{2\pi}\int_0^{2\pi} w^+(re^{i\phi})d\phi$$
$$\le \frac{1}{2\pi}\left[r^\rho \int_0^{2\pi} H^+(e^{i\phi})d\phi + \int_0^{2\pi} v^+(re^{i\phi})d\phi + \int_0^{2\pi} v^-(re^{i\phi})d\phi\right].$$

Since $v(0) = 0$, we have

$$\int_0^{2\pi} v^-(re^{i\phi})d\phi \le \int_0^{2\pi} v^+(re^{i\phi})d\phi.$$

Therefore

$$T(r, w) \le [\max\{H(e^{i\phi}) : \phi \in [0, 2\phi]\} + 2\sigma_1]r^\rho.$$

It is known (see Theorem 2.8.2.3, (2.8.2.5)) that $M(r) \le 4T(2r)$. So we conclude that

$$w(z) \le 4 \cdot 2^\rho[\max\{H(e^{i\phi}; \phi \in [0, 2\pi]\} + 2\sigma_1]|z|^\rho = \sigma|z|^\rho.$$

Let us prove property 2. To this end we shall use the following theorem (Grishin's Lemma) [Gr].

Theorem A.F.G *Let* g *be a nonnegative* δ*-subharmonic function, and let* ν_g *be its charge. Then the restriction* $\nu_g|_E$ *to the set* $E = \{z : g(z) = 0\}$ *is a measure.*

Applying this theorem to the function $g := m(z, v, H) - \mathcal{G}_H v(z)$, we get

$$\nu(\bullet, v)|_{E_v} \ge \mu_w|_{E_v}, \tag{6.1.4.5}$$

hence, we obtain property 2.

Let us prove property 3. Since w and v are upper semicontinuous, and H is continuous (see Theorem 3.2.5.5), the set $\{z : (w+v)(z) - H(z) < 0\}$ is open.

Let us take a neighborhood of an arbitrary point of this set and replace the function w within it with the Poisson integral constructed using this function, i.e., let us sweep out the mass from this neighborhood. The subharmonic function obtained would be strictly greater than the initial one, if the latter were not harmonic. This means that the initial w was not the maximal minorant. We have arrived at a contradiction, which proves property 3.

Property 4 immediately follows from property 3 and (6.1.4.5).

In order to prove property 5 we shall need an auxiliary statement which will be stated as a number of lemmas. Let

$$P(z, \phi, R) := \frac{1}{2\pi} \frac{R^2 - |z|^2}{|z - Re^{i\phi}|}$$

be the Poisson kernel in the disc $K_R = \{z : |z| < R\}$.

Below, C's with indices will denote constants.

Lemma 6.1.4.3 *In the disc $K_{R/2}$, we have*

$$|\operatorname{grad}_z P(z, \phi, R)| \leq C_1(R),$$

where $C_1(R)$ depends only on R.

Exercise 6.1.4.1 Prove this.

We shall introduce the notation for the Green function for the Laplace operator in the disc K_R:

$$G(z, \zeta, R) := \log \left| \frac{R^2 - \zeta\overline{z}}{R(z - \zeta)} \right|.$$

The disc $\{\zeta : |\zeta - z| < t\}$ will be denoted by $K_{z,t}$.

Lemma 6.1.4.4 *Let $z \in K_{R/2} \setminus K_{\zeta,\sqrt{h}}$. Then for a small h,*

$$|\operatorname{grad}_z G(z, \zeta, r)| \leq C_2(R)/\sqrt{h}.$$

Exercise 6.1.4.2 Prove this.

Let us write $\mu(z, t) := \mu(K_{z,t})$.

Lemma 6.1.4.5 *For $z \in K_{R/2}, 0 < t < R/10$, we have*

$$\mu_H(z, t) \leq C_3(\sigma, R)t.$$

Proof. Applying Theorem 2.6.5.1 (Jensen-Privalov) to the function $H(z)$, we obtain

$$M_H = \max\{H(e^{i\phi} : \phi \in [0; 2\pi]\} = \Delta_H(\mathbb{T})/\rho$$

where $\mathbb{T}$ is the unit circle.

Now

$$\mu(z,t) \leq \Delta_H(\mathbb{T}) \int_{|z|-t}^{|z|+t} r^{\rho-1} dr \leq \rho^2 M_H R^{\rho-1} t \leq \sigma C(\rho) R^{\rho-1} t$$

where $C(\rho)$ is a constant depending only on ρ. This proves the lemma. □

Lemma 6.1.4.6 *Let $h < 1$ and suppose that a monotonic function $\mu(t)$ satisfies the condition*

$$\mu(t) < ct. \tag{6.1.4.6}$$

Then

$$\int_0^{\sqrt{h}} \log(1/t)\mu(dt) \leq (3/2)c\sqrt{h}\log h.$$

Exercise 6.1.4.3 Prove this by integrating by parts and using (6.1.4.6).

Lemma 6.1.4.7 *Let $z \in K_{R/2}$ and $\zeta \in K_R$. Then*

$$|\log|(R^2 - \zeta\bar{z}/R|| \leq C_4(R).$$

Exercise 6.1.4.4 Prove this.

Now we pass to the proof of assertion 5 from Lemma 6.1.4.2. According to the F. Riesz theorem (Theorem 2.6.4.3) we represent w in the circle as

$$w(z) = H(z,w) - \int_{K_R} G(z,\zeta,R)\mu_w(d\xi d\eta), \;\; \zeta = \xi + i\eta, \tag{6.1.4.7}$$

where

$$H(z,w) = \frac{1}{2\pi} \int_0^{2\pi} P(z,\phi,R) w(Re^{i\phi}) d\phi.$$

It follows from Lemma 6.1.4.3 and 1 of Lemma 6.1.4.2 that

$$|\operatorname{grad}_z H(z,w)| \leq C_1(R) \frac{1}{2\pi} \int_0^{2\pi} |w|(Re^{i\phi}) d\phi \leq C_1(R) 2\sigma R^\rho. \tag{6.1.4.8}$$

We split the integral in (6.1.4.7) into three terms:

$$\psi_1(z,h) := \int\limits_{K_r\setminus K_{z_0,\sqrt{h}}} G(z,\zeta,R)\mu_w(d\xi d\eta),$$

$$\psi_2(z,h) = \int\limits_{K_{z_0,\sqrt{h}}} \log|(R^2-\overline{z}\zeta)/R|\mu(d\xi d\eta),$$

$$\psi_3(z,h) = \int\limits_{K_{z_0,\sqrt{h}}} \log|\zeta - z|\mu(d\xi d\eta),$$

where z_0 is an arbitrary fixed point in $K_{R/2}$.

Combining property 4 and inequality (6.1.4.4) we have

$$\mu_w(E) \le \mu_H(E),\ \forall E \subset K_R. \tag{6.1.4.9}$$

For all $z \in K_{z_0,\sqrt{h}/2}$ Lemma 6.1.4.4 yields

$$|\operatorname{grad}\psi_1(z,h)| \le C_2(r)\sigma R^\rho/\sqrt{h}. \tag{6.1.4.10}$$

Combining Lemmas 6.1.4.5 and 6.1.4.7 with inequality (6.1.4.9), we get

$$|\psi_2(z,h)| \le C_4(R)C_3(\sigma,R)\sqrt{h}. \tag{6.1.4.11}$$

Further, from Lemmas 6.1.4.5 and 6.1.4.6, taking into account the fact that $\log|\zeta - z| < 0$, we obtain

$$|\psi_3(z,h)| \le (3/2)C_3(\sigma,R)\sqrt{h}\log h. \tag{6.1.4.12}$$

Now consider the difference

$$\Delta w := w(z_0+\Delta z) - w(z_0),\ |\Delta z| < h < \sqrt{h}/2.$$

It can be represented as

$$\Delta w = \Delta\psi_1 + \Delta\psi_2 + \Delta\psi_3 + \Delta H(z,w). \tag{6.1.4.13}$$

Choosing h small enough, one may assume that $z_0 + \Delta z \in K_{\sqrt{h}/2,z_0}$. Thus, according to (6.1.4.11),

$$|\Delta\psi_2(z_0,h)| \le |\psi_2(z_0,h)| + |\psi_2(z_0+h,h)| \le C_6(\sigma,R)\sqrt{h}. \tag{6.1.4.14}$$

Likewise (6.1.4.12) yields

$$|\Delta\psi_3(z_0,h)| \le |\psi_3(z_0,h)| + |\psi_3(z_0+h,h)| \le C_7(\sigma,R)\sqrt{h}\log h. \tag{6.1.4.15}$$

Finally, from (6.1.4.10) and (6.1.4.8), respectively, we obtain

$$|\Delta\psi_1| \le C_3(\sigma,R)\sqrt{h},\ |\Delta H(z_0,w)| \le C_8(\sigma,R)h. \tag{6.1.4.16}$$

Substituting (6.1.4.14)–(6.1.4.16) into (6.1.4.13), we obtain relation 5 of Lemma 6.1.4.2. □

Thus we have completed the proof of Lemma 6.1.4.2.

6.1.5 In this item we are going to prove Theorem 6.1.2.6. However, before that, we prove

Lemma 6.1.5.1 *Let* $w_n = \mathcal{G}_H v_n$, $v_n \overset{\mathcal{D}'}{\to} v$ *and* $w_n \overset{\mathcal{D}'}{\to} w_\infty$. *Set*

$$E_\infty := \{z : w_\infty(z) = H(z) - v(z)\}.$$

Then w_∞ *is harmonic in* $\mathbb{C} \setminus E_\infty$.

Let us note that w_∞, in general, is not the maximal subharmonic minorant because the operator $\mathcal{G}_H$ can be only upper semicontinuous, as will be demonstrated by example in Section 6.1.8. However, w_∞ is a minorant of $H - v$ because of Theorem 6.1.1.2, 1.

Proof. Let $z_0 \notin E_\infty$. Then there exists a $\delta > 0$ such that

$$w_\infty(z_0) + v(z_0) \leq H(z_0) - 2\delta.$$

Since the function $b(z) := w_\infty + v(z) - H(z)$ is upper semicontinuous, there exists an $\epsilon = \epsilon(\delta)$ such that $b(z) < -\delta$ for all $z \in \{|z - z_0| < 2\epsilon\}$.

Let $(\bullet)_\epsilon$ be a smoothing operator from (2.6.2.3). If $w_n \overset{\mathcal{D}'}{\to} w$ then $(w_n)_\epsilon \to w_\epsilon$ uniformly on every compact set (Theorem 2.3.4.5, reg3) and for every subharmonic function v the sequence $v_\epsilon(z) \downarrow v(z)$,when $\epsilon \downarrow 0$ (Theorem 2.6.2.3, ap2).

Then $(b)_\epsilon(z) < -\delta$, for $|z - z_0| < \epsilon$ or $(w_\infty)_\epsilon(z) + (v)_\epsilon(z) \leq (H)_\epsilon(z) - \delta$. The function H is continuous, hence uniformly continuous on the circle $\{z : |z - z_0| \leq \epsilon\}$. Thus we can replace $(H)_\epsilon$ in the last inequality with H and δ with $\delta/2$. So, we have

$$(w_\infty)_\epsilon(z) + v_\epsilon(z) \leq H(z) - \delta/2, \ |z - z_0| < \epsilon. \tag{6.1.5.1}$$

Since $(\bullet)_\epsilon$ is monotonic on subharmonic functions, we can replace ϵ in (6.1.5.1) with any $\epsilon_1 < \epsilon$. So we obtain

$$(w_\infty)_{\epsilon_1}(z) + v_{\epsilon_1}(z) \leq H(z) - \delta/2, \ |z - z_0| < \epsilon. \tag{6.1.5.2}$$

Since $(w_n)_{\epsilon_1} \to (w_\infty)_{\epsilon_1}$ uniformly in the disc $|z - z_0| \leq \epsilon$ we can replace in (6.1.5.2) w_∞ with w_n and respectively v with v_n, changing $\delta/2$ with $\delta/4$. After that we can pass to the limit as $\epsilon_1 \downarrow 0$ for every sufficiently large n. So we obtain

$$w_n(z) + v_n(z) \leq H(z) - \delta/4, \ |z - z_0| < \epsilon.$$

It means that the disc $\{|z - z_0| < \epsilon\} \subset \mathbb{C} \setminus E_{v_n}$. Because of Lemma 6.1.4.2, 3, w_n is harmonic in this disc for all large n. Thus w_∞ is also harmonic, as the $\mathcal{D}'$-limit of w_n. □

Proof of Theorem 6.1.2.6. Let $v_n \overset{\mathcal{D}'}{\to} v$. Then the set $w_n = \mathcal{G}_H v_n$ is equicontinuous by Lemma 6.1.4.2, 5, and we can choose from it a subsequence uniformly converging to a continuous function w_∞. Let $w = \mathcal{G}_H v$, $E = E_w$, E_∞ being defined in Lemma 6.1.5.1.

Since

$$(w_\infty + v)(z) \le H(z),\ (w + v)(z) \le H(z),\ \forall z \in \mathbb{C}$$

and v is upper semicontinuous, whereas w and H are continuous, the sets E and E_∞ are closed.

Since $w_\infty(z) \le H(z) - v(z)$ we have

$$w_\infty(z) \le w(z),\ \forall z \in \mathbb{C}, \tag{6.1.5.3}$$

and therefore $E_\infty \subset E$.

The function w is subharmonic in $\mathbb{C} \setminus E_\infty$, and w_∞ is harmonic in $\mathbb{C} \setminus E_\infty$ by Lemma 6.1.5.1. They take the same values on E_∞. As the maximum principle holds in $\mathcal{H}_{w_\infty}$ by assumption we have, according to Lemma 6.1.4.1 the inequality

$$w(z) \le w_\infty(z),\ \forall z \in \mathbb{C}. \tag{6.1.5.4}$$

The inequalities (6.1.5.4) and (6.1.5.3) imply that $w(z) = w_\infty(z)$, i.e., $\mathcal{G}_H$ is continuous. □

6.1.6

Proof of Theorem 6.1.2.5. Sufficiency. We exploit the following criterion for existence of a limit set that follows from Theorems 4.2.1.1, 4.2.1.2, 4.3.1.2 and Corollary 5.3.1.5:

Proposition 6.1.6.1 *In order that $U \subset U[\rho]$ be a limit set of an entire function $f \in A(\rho(r))$ it is necessary and sufficient that there exists a piecewise continuous, ω-dense in U asymptotically dynamical pseudo-trajectory (a.d.p.t) $v(\bullet|t)$.*

Exercise 6.1.6.1 Check this.

Let $v_\Phi(\bullet|t)$ be an a.d.p.t. corresponding to $\mathbf{Fr}\Phi$. Consider the pseudo-trajectory $v_g(\bullet|t) := \mathcal{G}_H v_\Phi(\bullet|t)$. It exists because of (6.1.1.1). Prove that this pseudo-trajectory is asymptotically dynamical, i.e., (4.3.1.1) is fulfilled. Recall that $T_\tau \bullet = (\bullet)_{[e^\tau]}$.

Using the property of invariance of $\mathcal{G}_H$ (Theorem 6.1.1.2, 2) we have

$$T_\tau v_g(\bullet|e^t) - v_g(\bullet|e^{t+\tau}) = \mathcal{G}_H[T_\tau v_\Phi(\bullet|e^t) - v_\Phi(\bullet|e^{t+\tau})].$$

Thus (4.3.1.1) is fulfilled because of continuity of $\mathcal{G}_H$. Also the condition of ω-denseness (4.3.1.4) is fulfilled and

$$\{w \in U[\rho] : (\exists t_j \to \infty)\ w = \mathcal{D}' - \lim v_g(\bullet|e^{t_j})\} = \mathcal{G}_H(\mathbf{Fr}\Phi).$$

The corresponding entire function $g \in A(\rho(r))$ with the limit set $U_g = \mathcal{G}_H(\mathbf{Fr}\Phi)$ is an ideally complementing multiplicator, because

$$\mathbf{Fr}[g\Phi] = \{v + \mathcal{G}_H v : v \in \mathbf{Fr}[\Phi]\}.$$

Exercise 6.1.6.2 Check this.

Necessity. Let G be an ideally complementing multiplicator of Φ. Let us show that $\mathcal{G}_H$ is continuous on $\mathbf{Fr}[\Phi]$. Assume this is not the case, i.e., there exists a sequence $v_j \to v$ such that $\mathcal{G}_H v_j \to W$ and $W \neq \mathcal{G}_H v$. Since the limit set $\mathbf{Fr}[G\Phi]$ is closed, we have $v_j + \mathcal{G}_H v_j \to v + \mathcal{G}_H v$, $v_j \in \mathbf{Fr}[\Phi]$. On the other hand, $v_j + \mathcal{G}_H v_j \to v + W$. Thus, $W = \mathcal{G}_H v$, which is a contradiction. □

Proof of Theorem 6.1.2.3. Let $\mathbf{Fr}[\Phi]$ be a periodic limit set, that is

$$\mathbf{Fr}[\Phi] = \mathbb{C}(v) = \{v_{[t]} : 1 \leq t \leq e^P\},$$

where $v \in U[\rho]$. We shall show that $\mathcal{G}_H$ is continuous on $U[\rho]$. By Theorem 6.1.1.2, 2) the equality $(\mathcal{G}_H v)_{[t]} = \mathcal{G}_H v_{[t]}$ holds. Since the operation $(\bullet)_{[t]}$ is continuous for all t, $\mathcal{G}_H$ is continuous on $\mathbb{C}(v)$. □

6.1.7 Now we are going to prove Theorem 6.1.2.4. However, we need some preparation.

Let $h(\phi)$, $\phi \in [0, 2\pi)$ be a 2π-periodic ρ-t.c.function, satisfying the condition

$$\max_{\phi \in [0,2\pi]} h(\phi) = \sigma.$$

We denote this class as $TC[\rho, \sigma]$ and write

$$TC[\rho] := \bigcup_{\sigma > 0} TC[\rho, \sigma].$$

The class of functions $w = h_1 - h_2$ where $h_1, h_2 \in TC[\rho, \sigma]$ will be denoted as $\delta TC[\rho, \sigma]$ and we will also write

$$\delta TC[\rho] := \bigcup_{\sigma > 0} \delta TC[\rho, \sigma].$$

From properties of a ρ-t.c.function (see Sections 3.2.3–3.2.5) we can obtain the following properties of $\delta - \rho$-t.c.functions:

Proposition 6.1.7.1 *For $w \in \delta TC[\rho]$ the following holds:*

1. *$w'(\phi - 0)$ and $w'(\phi + 0)$ exist at each point and are bounded in $[0; 2\pi]$;*
2. *$w'(\phi - 0) = w'(\phi + 0)$ for all $\phi \in [0; 2\pi]$, except, perhaps, a countable set;*
3. *the charge Δ_w generated by the function*

$$\Delta_w := w'(\phi) + \rho^2 \int^{\phi} w(\theta) d\theta$$

has bounded variation $|\Delta_w|$; the variation $|\Delta_w|(\alpha,\beta)$ of the charge on the interval $(\alpha;\beta)$ and the variation of the charge generated by derivative $|w'|(\alpha;\beta)$ on the same interval satisfy the relation

$$|\Delta_w|(\alpha,\beta) \geq |w'|(\alpha;\beta) + \rho^2(\beta-\alpha).$$

4. *For all $w \in \delta TC[\rho,\sigma]$,*

$$\max(|w'(\phi-0)|, |w'(\phi+0)|) \leq C(\rho,\sigma),\ \ \phi \in [0;2\pi];$$

5. *if $r^\rho w_n \xrightarrow{\mathcal{D}'} r^\rho w$ and $w_n \in \delta TC[\rho,\sigma]$, then $w_n \to w$ uniformly on $[0;2\pi]$.*

Exercise 6.1.7.1 Prove this using properties of ρ-t.c.functions.

We also need a technical

Lemma 6.1.7.2 *Let $M_n(\phi)$ be a sequence of functions that satisfy the conditions:*

1. $M_n \geq 0$; $M_n(0) = 0$;
2. *M_n converges uniformly to $M_\infty(\phi) \geq A\sin\rho\phi$, $A > 0$;*
3. *$M_n'(\phi-0)$, $M_n'(\phi+0)$ exist at every point, and they coincide almost everywhere;*
4. *there exists a sequence $\phi_n \downarrow 0$ such that for each arbitrarily small $\epsilon > 0$ and arbitrarily large $n_0 \in \mathbb{N}$ there exists $n > n_0$ for which the inequality $M_n(\phi_n) < \epsilon\phi_n$ holds.*

Then there exists a sequence (ζ_n, η_n) of disjoint intervals and a subsequence M_{k_n} such that

$$M_{k_n}'(\zeta_n) - M_{k_n}'(\eta_n) \geq A\rho/2. \tag{6.1.7.1}$$

Proof. Set $\epsilon_0 = 1/2$, $\eta_0 = \pi/4$ and choose the required sequence recurrently. Let $\epsilon_n, \eta_n, \zeta_n$ be already chosen. Set $\epsilon_{n+1} = \epsilon_n/2$, find $\phi_{n+1} < \eta_n$ and choose $k_0 = k_0(n)$ so that for $k > k_0$,

$$M_k(\phi_{n+1}) - A\rho\phi_{n+1} > -\epsilon_{n+1}\phi_{n+1}.$$

This is possible because of condition 2 and $\sin\rho\phi \sim \rho\phi$, $\phi \to 0$. So we have

$$\frac{M_k(\phi_{n+1})}{\phi_{n+1}} > A\rho - \epsilon_{n+1}. \tag{6.1.7.2}$$

Now, choose $\psi_{n+1} < \phi_{n+1}$ and $k_{n+1} > k_0$ so that

$$M_{k_{n+1}}(\psi_{n+1}) < \epsilon_{n+1}\psi_{n+1}. \tag{6.1.7.3}$$

This is possible by condition 4. Thus for small ϵ_{n+1} from (6.1.7.2) and (6.1.7.3) we obtain

$$\frac{M_{k_{n+1}}(\phi_{n+1}) - M_{k_{n+1}}(\psi_{n+1})}{\phi_{n+1} - \psi_{n+1}} > (2/3)A\rho. \tag{6.1.7.4}$$

On the other hand

$$\frac{M_{k_{n+1}}(\psi_{n+1}) - M_{k_{n+1}}(0)}{\psi_{n+1} - 0} < \epsilon_{n+1}. \tag{6.1.7.5}$$

On the interval (ψ_{n+1}, ϕ_{n+1}) there is a point η_{n+1} where the derivative exists and the inequality

$$M'_{k_{n+1}}(\eta_{n+1}) \geq \frac{M_{k_{n+1}}(\phi_{n+1}) - M_{k_{n+1}}(\psi_{n+1})}{\phi_{n+1} - \psi_{n+1}} \tag{6.1.7.6}$$

is valid. Also there is a point $\zeta_{n+1} \in (0, \psi_{n+1})$ where the derivative exists and the inequality

$$M'_{k_{n+1}}(\zeta_{n+1}) \leq \frac{M_{k_{n+1}}(\psi_{n+1}) - M_{k_{n+1}}(0)}{\psi_{n+1} - 0} \tag{6.1.7.7}$$

is valid.

From the inequalities (6.1.7.4)–(6.1.7.7) we obtain (6.1.7.1). □

Proof of Theorem 6.1.2.4. Denote by $\hat{\mathcal{G}}_H h$ the maximal ρ-t.c.minorant of $H(e^{i\phi}) - h(\phi)$. It follows from Theorem 6.1.1.2, 2 that

$$\mathcal{G}_H(r^\rho h(\phi))(re^{i\phi}) = r^\rho \hat{\mathcal{G}}_H h(\phi).$$

Exercise 6.1.7.2 Prove this.

So taking in consideration Proposition 6.1.7.1, 5, one must prove

Proposition 6.1.7.3 *The operator $\hat{\mathcal{G}}_H$ is continuous on the set*

$$\hat{U}_{\text{ind}} := \{h(\phi, c) : c \in C\}$$

in the uniform topology.

Proof. Let $h_n \to h,\ h_n, h \in \hat{U}_{\text{ind}}$. Set $\hat{w}_n = \hat{\mathcal{G}}_H h_n,\ \hat{w} = \hat{\mathcal{G}}_H h,\ \hat{w}_\infty = \lim_{n\to\infty} \hat{w}_n$. We set also $\hat{M}_n = H - h_n - \hat{w}_n,\ \hat{M}\infty = H - h - \hat{w}_\infty,\ \hat{M} = H - h - \hat{w}$. Let $(\alpha_n; \beta_n)$ be a maximum interval where $\hat{M}_n(\phi) > 0$. We shall show that $\beta_n - \alpha_n \leq \pi/\rho$. Indeed, for a fixed n let us consider the function

$$\hat{W}_n := \hat{w}_n + \epsilon_n L(\phi - (\alpha_n + \beta_n)/2)$$

where

$$L(\phi) = \begin{cases} \cos|\phi|, & \phi \in (-\pi/2\rho; \pi/2\rho); \\ 0, & \phi \in [-\pi; \pi] \setminus (-\pi/2\rho; \pi/2\rho), \end{cases}$$

and ϵ_n is small enough. If $\beta_n - \alpha_n > \pi/\rho$, then $\hat{W}_n$ is also a ρ-t.c.minorant of $H - h_n$, i.e., $\hat{w}_n$ is not maximal.

If $\beta_n - \alpha_n = \pi/\rho$, then, to ensure that $\hat{w}_n$ is a maximal minorant, at least one of the conditions

$$\liminf_{\phi \to \alpha_n + 0} \frac{\hat{M}_n(\phi)}{\phi - \alpha_n} = 0, \ \liminf_{\phi \to \beta_n - 0} \frac{\hat{M}_n(\phi)}{\beta_n - \phi} = 0 \tag{6.1.7.8}$$

must be satisfied.

Let us choose (and preserve the previous notation) a subsequence $\hat{M}_n(\phi)$ for which $\alpha_n \to \alpha$, and $\beta_n \to \beta$.

If $\beta - \alpha < \pi/\rho$, then the maximum principle for ρ-t.c.functions is valid. Repeating arguments of proof of Theorem 6.1.2.6, we obtain $\hat{w}_\infty = \hat{w}$ for all ϕ, which proves Proposition 6.1.7.3 for the case considered.

Exercise 6.1.7.3 Repeat them.

Consider the case when $\beta - \alpha = \pi/\rho$. Set $q(\phi) = (w - w_\infty)(\phi)$. The function q is ρ-trigonometric on the interval $(\alpha;\beta)$ since $\hat{w}$ and $\hat{w}_\infty$ are ρ-trigonometric, i.e., have the form $A\sin\rho\phi + B\cos\rho\phi$.

Exercise 6.1.7.4 Explain this.

Besides, we have $q \geq 0$ and $q(\alpha) = q(\beta) = 0$. It is easy to see that q has the form

$$q(\phi) = A\sin\rho(\phi - \alpha), \ A > 0. \tag{6.1.7.9}$$

Exercise 6.1.7.5 Prove this.

Since $\hat{M}(\phi) \geq 0$, we have $\hat{M}_\infty(\phi) = \hat{M}(\phi) + (\hat{w} - \hat{w}_\infty)(\phi) \geq (\hat{w} - \hat{w}_\infty)(\phi)$, $\forall\phi$, whence

$$\hat{M}_\infty(\phi) \geq A\sin\rho(\phi - \alpha), \ A > 0. \tag{6.1.7.10}$$

Since $\beta_n - \alpha_n \leq \pi/\rho$, the segment $[\alpha, \beta]$ contains the infinite sequence α_n or β_n. Let us single out a subsequence, let it be, for example, $\alpha_n \to \alpha + 0$, $\alpha_n \in [\alpha;\beta]$.

Consider the sequence $M_n(\phi) = \hat{M}_n(\phi - a)$. From the definition of M_n and from relation (6.1.7.10) it follows that conditions 1 and 2 of Lemma 6.1.7.2 are fulfilled. Condition 3 is fulfilled because of property 1 of Lemma 6.1.7.2. Further, if $\alpha_n \not\equiv \alpha$, then condition 4 of Lemma 6.1.7.2 is trivially true,since $M_n(\alpha_n - \alpha) = 0$; otherwise, if $\alpha_n \equiv \alpha$, condition 4 follows from (6.1.7.8).

Applying Lemma 6.1.7.2, we obtain the union of intervals satisfying (6.1.7.7). The equality $H(\phi) = \hat{M}_n - h_n - \hat{w}_n$ yields the following inequality for the measure Δ_H:$\Delta_H((\eta_n;\zeta_n)) \geq A\rho/2$. Summing this inequality and taking into account the fact that the intervals do not intersect, we obtain $\Delta_H(\cup_n(\eta_n,\zeta_n)) = \infty$, which is impossible. So, Proposition 6.1.7.3 is proved. □

Hence, Theorem 6.1.2.4 is proved. □

6.1.8 In this item we show an example of H and an entire function without an ideally complementing H-multiplicator.

According to Theorem 6.1.2.5, to construct such an example it is sufficient to construct a limit set on which $\mathcal{G}_H$ is not continuous.

We set

$$L(\eta) = \begin{cases} \cos|\eta|, & \eta \in (-\pi/2\rho; \pi/2\rho); \\ 0, & \eta \in [-\pi; \pi] \setminus (-\pi/2\rho; \pi/2\rho). \end{cases}$$

Let us define $X \in C^\infty$ so that $X(\xi) = 1$ for $\xi < 0$ and $X = 0$ for $\xi > \alpha$.

We set

$$\kappa := (1/\rho^2) \max_{(-\infty;+\infty)} [2\rho X' + X''](\xi), \quad H_0(\eta) := L(\eta) + \kappa. \tag{6.1.8.1}$$

We also set

$$v(\zeta, c) := [H_0 - X(\xi - c)L(\eta)]e^{\rho\xi}, \quad \zeta = \xi + i\eta,$$

where H_0 and L have been periodically extended from the interval $[-\pi; \pi]$ to $(-\infty, +\infty)$.

As $H(z)$ we take

$$H(z) := H_0(\phi)r^\rho.$$

Lemma 6.1.8.1 *We have*

$$v(\log z, c) \in U[\rho, \sigma], \quad \sigma = 1 + \kappa, \tag{6.1.8.2}$$

$$\mathcal{G}_H v(\bullet, c) \equiv 0, \tag{6.1.8.3}$$

$$\lim_{c \to \infty} v(\log z, c) = \kappa r^\rho \tag{6.1.8.4}$$

uniformly with respect to $z \in K \Subset \mathbb{C}$, *and*

$$\mathcal{G}_H(\kappa r^\rho) = L(\phi)r^\rho. \tag{6.1.8.5}$$

Proof. For the Laplace operators in ζ and z it is true that $\Delta_\zeta = \Delta_z/|\zeta|^2$. Let us check that $v(\zeta, c)$ is subharmonic in ζ. We have

$$\Delta_\zeta v(\zeta, c) = \{[1 - X(\xi - c)](L'' + \rho^2 L)(\eta) + [\rho^2\kappa - L(\xi)[X''(x - c) + 2\rho X'(x - c)]\}e^{\rho\xi}.$$

Exercise 6.1.8.1 Check this computation.

Since $X(\xi) \leq 1$ and $L(\eta)$ is ρ-t.c.,

$$[1 - X(\xi - c)](L'' + \rho^2 L)(\eta) \geq 0.$$

Since $L(\xi) \leq 1$ and $[X''(x - c) + 2\rho X'(x - c)] \leq \kappa\rho^2$ we have

$$[\rho^2\kappa - L(\xi)[X''(x - c) + 2\rho X'(x - c)] \geq 0.$$

Thus $v(\log z, c)$ is subharmonic.

Exercise 6.1.8.2 Prove that $v(\log z, c) \in U[\rho, \sigma]$ for $\sigma = 1 + \kappa$.

Let us prove (6.1.8.3). We have

$$H(z) - v(\log z, c) = X(\log r - c)L(\phi)r^{\rho}.$$

Since $X = 0$ for $r > e^{c+\alpha}$, the maximal subharmonic minorant of $H - v$ is zero by the maximum principle.

Relation (6.1.8.4) is obvious, since $X(\log r - c)$ converges to 1 uniformly on every disc $\{|z| \leq R\}$. Relation 6.1.8.5 follows from the equality $H(z) - \kappa r^{\rho} = L(\phi)r^{\rho}$, since $L(\phi)r^{\rho} \in U[\rho]$. □

Now we pass to the construction of the example. Examine the set

$$U_1 := \text{clos}\{v(\log z, c) : c \in [0; \infty)\}.$$

It contains the function

$$\mathcal{D}' - \lim_{c \to \infty} v(\log z, c) = \kappa r^{\rho}.$$

Let us consider the minimal convex $(\bullet)_{[t]}$-invariant set U containing U_1. The set is contained in $U[\rho, 1 + \kappa]$. It is a limit set for a certain entire function Φ. Let us show that $\mathcal{G}_H$ is not continuous on $\mathbf{Fr}[\Phi]$. We take an arbitrary sequence $c_j \to \infty$ and set $v_j(z) := v(\log z, c_j) \in U$. Now $\mathcal{D}' - \lim_{j\to\infty} v_j = \kappa r^{\rho}$ by (6.1.8.4) and $\mathcal{G}_H v_j(z) = 0$, so $\mathcal{D}' - \lim_{j\to\infty} \mathcal{G}_H v_j = 0$ but

$$\mathcal{G}_H(\lim v_j) = \mathcal{G}_H(\kappa r^{\rho}) = L(\phi)r^{\rho} \not\equiv 0$$

which shows the lack of continuity.

By virtue of Theorem 6.1.2.5, Φ is not ideally complementable.

6.1.9 Here we prove existence and continuity of the maximal subharmonic minorant for some classes of functions $m(z)$.

Theorem 6.1.9.1 *Let $m(z)$ be a continuous function such that the set of subharmonic minorants is nonempty. Then the maximal subharmonic minorant of m exists and is continuous.*

Proof. The set of subharmonic minorants is nonempty and partially ordered. Indeed, for every subset $\{u_\alpha, \alpha \in A\}$ of subharmonic minorants there exists $u_A = (\sup\{u_\alpha : \alpha \in A\})^*$ which is subharmonic and is a minorant of m, because m is continuous.

Exercise 6.1.9.1 Explain this in detail.

Thus there exists a uniquely maximal element m.s.m.(z, m), which is a subharmonic minorant of m.

Let us prove that it is continuous at every point z_0. Since m.s.m.(z, m) is upper semicontinuous,

$$\text{m.s.m.}(z, m) > \text{m.s.m.}(z_0, m) - \epsilon$$

for $|z - z_0| < \delta$ for arbitrarily small ϵ and corresponding $\delta = \delta(\epsilon)$. So we need to prove the inequality

$$\text{m.s.m.}(z, m) < \text{m.s.m.}(z_0, m) + \epsilon$$

for arbitrarily small ϵ and corresponding $\delta = \delta(\epsilon)$.

Perform sweeping m.s.m.(z, m) from the disc $|z - z_0| < \delta$ such that the result $u(z, \delta)$ satisfies the inequality

$$\text{m.s.m.}(z, m) < u(z, \delta) < \text{m.s.m.}(z, m) + \epsilon < m(z) + \epsilon.$$

Thus $u(z, \delta) - \epsilon < m(z)$. Hence m.s.m.$(z, m) > u(z, \delta) - \epsilon$ for all z. Since $u(z, \delta)$ is continuous, $u(z, \delta) > u(z_0, \delta) - \epsilon$ in the disc $\{|z - z_0| < \delta_1\}$. So m.s.m.$(z, m) > u(z_0, \delta) - \epsilon >$ m.s.m.$(z_0, m) - \epsilon$. □

Theorem 6.1.9.2 *Let $m = m_1 - m_2$, where m_1, m_2 are subharmonic functions. Then the maximal subharmonic minorant of m exists. If m_1 is continuous, then the maximal subharmonic minorant is continuous.*

Proof. Set $\mathcal{M}_\epsilon(z, m) := \mathcal{M}_\epsilon(z, m_1) - \mathcal{M}_\epsilon(z, m_2)$, where $\mathcal{M}_\epsilon(z, m_i), i = 1, 2$ is defined by (2.6.1.1). Since $\mathcal{M}_\epsilon(z, m)$ is continuous (see Theorem 2.6.2.3 (Smooth approximation)), there exists m.s.m.$(z, \mathcal{M}_\epsilon(z, m))$. We have

$$u(z, m) := \limsup_{\epsilon \longrightarrow 0} \text{m.s.m.}(z, \mathcal{M}_\epsilon(\bullet, m)) \leq \lim_{\epsilon \longrightarrow 0} \mathcal{M}_\epsilon(z, m) = m_1 - m_2(z) = m(z).$$

Now we prove that the upper semicontinuous regularization $u^*(z, m)$ also satisfies the inequality $u^*(z, m) \leq m(z)$. Indeed, $m_2 + u(z, m) \leq m_1(z)$. Hence,

$$\mathcal{M}_\epsilon(z, m_2) + \mathcal{M}_\epsilon(z, u(\bullet, m)) \leq \mathcal{M}_\epsilon(z, m_1).$$

Passing to the limit we obtain three subharmonic functions and inequality

$$m_2(z) + u^*(z, m) \leq m_1(z).$$

We prove that $u^*(z, m)$ is the m.s.m.(z, m). If not, there would exist a subharmonic function u_1 which exceeds $u^*(z, m)$ on a set of positive measure (otherwise they coincide); thus we would have for some z and ϵ,

$$u^*(z, m) < \mathcal{M}_\epsilon(z, u_1) \leq \text{m.s.m.}(z, \mathcal{M}_\epsilon(z, m)).$$

This contradicts the definition of $u^*(z, m)$.

Now suppose that m_1 is continuous at a point z_0. From Theorem 2.6.5.1 (Jensen-Privalov) we obtain that it is equivalent to

$$\int_0^{\epsilon} \frac{\mu_{m_1}(\{z : |z - z_0| < t\})}{t} dt = o(1), \ \epsilon \to 0.$$

Similarly to the proof of Theorem 6.1.4.2, 5, we obtain

$$\mu_{\text{m.s.m.}(z,m)} \leq \mu_{m_1}.$$

Hence m.s.m.(z, m) is also continuous. □

Exercise 6.1.9.2 Prove continuity in detail.

Corollary 6.1.9.3 *For $m = m(z, v, H)$ the function $\mathcal{G}_H v(z) :=$ m.s.m.(z, m) exists and is continuous.*

Exercise 6.1.9.3 Prove Corollary 6.1.9.3.

6.2 A generalization of ρ-trigonometric convexity

6.2.1 One of the important and useful kinds of limit sets is periodic limit sets. They are determined by one subharmonic function $v \in U[\rho]$ that satisfies the condition

$$v(Tz) = T^{\rho} v(z), \ z \in \mathbb{C}. \tag{6.2.1.1}$$

Such a function is called *automorphic.* They generate the class of so-called L_ρ-subfunctions, that is a generalization of ρ-trigonometrically convex functions. In this part we are going to review properties of such functions from different points of view that will be useful for applications (see [ADP]).

In connection with property (6.2.1.1) it is natural to consider so-called T-*homogeneous domains* in $\mathbb{C}$, i.e., such domains G that satisfy the condition $\{Tz : z \in G\} = G$ or shortly $TG = G$. As we can see they are invariant with respect to dilation by T. For example, every component of an open set of harmonicity of an automorphic function is a T-homogeneous domain.

Let v satisfy (6.2.1.1). Then the function

$$q(z) := v(e^z) e^{-\rho x} \tag{6.2.1.2}$$

is a 2π periodic function in y and P-periodic in x, where $P = \log T$.

The function q can be considered as a function on a torus $\mathbb{T}_P^2$, obtained by identifying the opposite sides of the rectangle $\Pi = (0, T) \times (-\pi, \pi)$.

The homology group of $\mathbb{T}_P^2$ is nontrivial, and generated by the cycles γ_x, γ_y, where $\gamma_x = \mathbb{T}_P^2 \cap \{y = 0\}$, $\gamma_y = \mathbb{T}_P^2 \cap \{x = 0\}$.

Let π be the covering map of $\mathbb{C}$ onto $\mathbb{T}^2_P$, then $\phi = \pi \circ \log$ is a well-defined covering map of $\mathbb{C} \setminus \{0\}$ onto $\mathbb{T}^2_P$, where the group of deck transformations is given by the dilations by T^m for $m \in \mathbb{Z}$. So if G is a given T-homogeneous domain, then

$$D = \pi \circ \log G = \phi(G) \tag{6.2.1.3}$$

is a domain in $\mathbb{T}^2_P$. On the other hand, not every domain in $\mathbb{T}^2_P$ has a T-homogeneous domain as its preimage under ϕ. The preimage $\phi^{-1}(D)$ under ϕ is a possibly disconnected set which is invariant under dilations by T^m for $m \in \mathbb{Z}$. An intrinsic description is given by the next proposition.

Proposition 6.2.1.1 *Let γ be a closed curve in a domain $D \subset \mathbb{T}^2_P$ that is homologous in $\mathbb{T}^2_P$ to a cycle $\gamma = n_x\gamma_x + n_y\gamma_y$, $n_x, n_y \in \mathbb{Z}$.*

1. *If $n_x = 0$ for every such γ in D, then*
$$\phi^{-1}(D) = \cup_{j=-\infty}^{\infty} G_j,$$
where $G_j = T^j G_0$, G_0 is an arbitrary connected component of $\phi^{-1}(D)$, and $G_j \cap G_l = \varnothing$ for $j \neq l$.
2. *If there exists a curve γ as above with $n_x \neq 0$, then*
$$\phi^{-1}(D) = \cup_{q=0}^{k-1} G_q,$$
where $k = \min |n_x|$ with the minimum taken over all such curves γ; G_0 is an arbitrary component of $\phi^{-1}(D)$; G_j, $j = 0, 1, \ldots, k-1$, are disjoint T^k-homogeneous domains, and for every $m \in \mathbb{Z}$, $T^m G_0 = G_q$, provided $m = lk + q$, for some $q \in \mathbb{Z}$, $0 \le q \le k-1$, $l \in \mathbb{Z}$.

We call domains as in part 2 of Proposition 6.2.1.1 *connected on spirals.* In particular, this proposition shows that for every D connected on spirals, we can find a connected T^k-homogeneous domain that relates to D by (6.2.1.3).

Let us give some examples. The domain $D' = \mathbb{T}^2_P \cap \{|x - P/2| < P/4\}$ is not connected on spirals, whereas $D'' = \mathbb{T}^2_P \cap \{|y| < \pi/4\}$ is. It follows that $D' \cap D''$ is not connected on spirals whereas $D' \cup D''$ is.

The situation can be more complicated. Set

$$\begin{aligned}
x'(x, y, \alpha) &:= x\cos\alpha + y\sin\alpha;\\
y'(x, y, \alpha) &:= -x\sin\alpha + y\cos\alpha;\ 0 \le \alpha < \pi/4;\\
P_1 &:= (1/2)\ |x'(P, 2\pi, -\alpha)|;\\
P_2 &:= (1/2)\ |y'(P, 2\pi, -\alpha)|.
\end{aligned}$$

Then $R' = \{z' = x' + i'y' : -P_1 < x' < P_1;\ -P_2 < y' < P_2\}$ is a fundamental rectangle for $\mathbb{T}^2_P$ in the corresponding coordinates. Set $f(y') := (P_2 - y')^{-1} - (y' + P_2)^{-1}$ and $D_{0,0} := \{z' : -P_2 < y' < P_2;\ f(y') < x' < f(y') + d\}$ where $0 < d < P_1$. Then the domains $D_{l,m} := D_{0,0} + 2P_1 l + 2P_2 m i'$, $l, m \in \mathbb{Z}$ are disjoint, and their union D determines a domain $\hat{D} \subset \mathbb{T}^2_P$. This $\hat{D}$ is determined completely by the intersection of D with the rectangle $R = (0, P) \times (-\pi, \pi)$. The domain $\hat{D}$ is not connected on spirals.

One more example. Consider the family of lines $L_l := \{z = x + iy : y = \pi/(kP)x + l\pi/k,\ x \in \mathbb{R}\}$, $l \in \mathbb{Z}$. It determines a closed curve (spiral) γ on $\mathbb{T}^2_P$ with $n_1 = k$. The open set $D_k = \{z : |z - \zeta| < \epsilon, \zeta \in L_l,\ l \in \mathbb{Z}\}$, $0 < \epsilon < P/2\sqrt{\pi^2 + k^2}$, determines a domain $\hat{D}_k$ on $\mathbb{T}^2_P$ that is connected on spirals, and such that $\phi^{-1}(\hat{D}_k)$ consists of k components, every one of them T^k-homogeneous.

Since the function v in (6.2.1.1) is subharmonic, the function q of (6.2.1.2) is upper semicontinuous and in the D' topology on $\mathbb{T}^2_P$ satisfies the inequality $L_\rho q \geq 0$, where

$$L_\rho := \Delta + 2\rho\frac{\partial}{\partial x} + \rho^2. \tag{6.2.1.4}$$

Such functions q are called *subfunctions with respect to* L_ρ, or L_ρ*-subfunctions.*

$L_\rho q$ is a positive measure on $\mathbb{T}^2_P$.

The operator L_ρ arises naturally by changing variables $z \mapsto \log z$ in the Laplace operator Δ_ζ.

Exercise 6.2.1.1 Check this. Set $\zeta = e^z$.

Let us note that if q depends only on the variable y, it is a 2π-periodic ρ-trigonometric convex function because L_ρ turns into $T_\rho = (\bullet)'' + \rho^2(\bullet)$ (cf. Section 3.2.3).

6.2.2 Consider the solution of the homogeneous boundary problem

$$\begin{aligned} L_\rho q &= 0 \quad \text{in } D; \\ q\,\Big|_{\partial D} &= 0, \end{aligned} \tag{6.2.2.1}$$

where D is a domain in $\mathbb{T}^2_P$ and q is bounded in a neighborhood of ∂D with boundary value zero quasi-everywhere. This is a spectral problem for a *pencil* of differential operators ([Ma]).

A solution of this problem can be defined for an arbitrary domain $D \subset \mathbb{T}^2_P$ with a boundary of positive capacity.

The *spectrum* of the problem (6.2.2.1) consists of those (complex) ρ for which (6.2.2.1) holds for some function $q \not\equiv 0$. The minimal positive point of the spectrum $\rho(D)$ exists iff the spectrum exists. The spectrum exists iff the domain D is connected on spirals. In this case $\rho(D)$ is the order of the minimal harmonic function in every one of the domains G_i that corresponds to D by Proposition 6.2.1.1.

The quantity $\rho(D)$ is *strictly monotonic.* It means that if two domains D_1, $D_2 \in \mathbb{T}^2_P$ are such that $D_1 \subset D_2$ and the capacity of $D_2 \setminus D_1$ is positive, then $\rho(D_2) < \rho(D_1)$. For example, this is the case of $D_2 = \{|y| < d, d < 2\pi\}$ and D_1 is the same strip without the segment $\{it : 0 \leq t \leq d\}$.

In connection with the multiplicator problem we considered the maximal subharmonic minorant of a function $m = H - v$ where v is a T-automorphic function. From Theorem 6.1.1.2, 2 we can obtain that if v is a T-automorphic function, then $\mathcal{G}_H v$ is also T-automorphic.

Exercise 6.2.2.1 Check this.

Thus for this case, finding D_H in Theorem 6.1.1.1 is reduced to finding a maximal L_ρ-subfunction q that satisfies the inequality

$$q(z) \le m(z) := [H(e^z) - v(e^z)]e^{-\rho x}, \; z \in \mathbb{T}_P^2. \tag{6.2.2.2}$$

We say that $m(z)$ has an L_ρ-*subminorant.*

The idea of $\rho(D)$ gives a possibility for

Theorem 6.2.2.1 *If m has a non-zero L_ρ-subminorant, then $\rho(D) \le \rho$ for some component D of the open set $\mathcal{M}_+ := \{z : m(z) > 0\}$.*

Conversely, if $\rho(D) < \rho$ (strict inequality) for some component D of the set $\mathcal{M}_+$, and $m(z) \ge 0$ for all $z \in \mathbb{T}_P^2$, then m has a non-zero L_ρ-subminorant.

Exercise 6.2.2.2 Prove that $\mathcal{M}_+$ is open.

6.2.3 If $\rho \notin \mathbb{Z}$, the operator L_ρ has a fundamental solution $E_\rho(\bullet - \zeta)$ in $\mathbb{T}_P^2$, where ζ is a shift by the torus, i.e., by the modulus $P + i2\pi$. It means that

$$L_\rho E_\rho(\bullet - \zeta) = \delta_\zeta,$$

in $\mathcal{D}'(\mathbb{T}_P^2)$, where δ_ζ is the Dirac function, concentrated at ζ.

If $\rho \in \mathbb{Z}$, there exists, as for operator T_ρ and a spherical operator (see Theorem 3.2.4.2, Theorem 3.2.6.3), a generalized fundamental solution E'_ρ that satisfies the equation

$$L_\rho E'_\rho(\bullet - \zeta) = \delta_\zeta - \cos\rho(y - \eta), \; \zeta = \xi + i\eta$$

in $\mathcal{D}'(\mathbb{T}_P^2)$.

Theorem 6.2.3.1 *Let $\rho > 0$, $\rho \notin \mathbb{Z}$. Then every L_ρ-subfunction on $\mathbb{T}_P^2$ can be represented in the form*

$$q(z) = \int_{\mathbb{T}_P^2} E_\rho(z - \zeta)\nu(d\zeta), \tag{6.2.3.1}$$

where $\nu = L_\rho q$.

This theorem is the counterpart of Theorems 3.2.3.3, 3.2.6.2.

Theorem 6.2.3.2 *Let $\rho > 0$, $\rho \in \mathbb{Z}$. Then the mass distribution $\nu = L_\rho v$ satisfies the condition*

$$\int_{\mathbb{T}_P^2} e^{\pm i\rho y}\nu(dz) = 0, \tag{6.2.3.2}$$

and the representation

$$q(z) = \Re(Ce^{i\rho y}) + \int_{\mathbb{T}_P^2} E'_\rho(z - \zeta)\nu(d\zeta) \tag{6.2.3.3}$$

holds with C that is a complex scalar.

This theorem is the counterpart of Theorems 3.2.4.2, 3.2.6.2.

Let $D \subset \mathbb{T}_P^2$ and $\rho(D) > \rho$. Then the operator L_ρ has in D the Green function $-G_\rho(z, \zeta, D)$. Thus for every q that is an L_ρ-subfunction in D and bounded from above in $\overline{D}$, we have the representation

$$q(z) = g(z) - \int_D G_\rho(z, \zeta, D)\nu(d\zeta), \qquad (6.2.3.4)$$

in which $\nu = L_\rho q$ and g is the minimal majorant on ∂D of the function q, satisfying $L_\rho g = 0$ in D.

This is the counterpart of Theorem 2.6.4.3 (F. Riesz representation) and Theorem 3.2.5.1.

From (6.2.3.4) one can easily obtain

Theorem 6.2.3.3. (Maximum principle) *If $\rho(D) > \rho$ and $q(z)$ is an L_ρ-subfunction such that $q(z) \leq 0,\ z \in \partial D$, then $q(z) \leq 0,\ z \in D$.*

Exercise 6.2.3.1 Prove this.

Theorem 6.2.3.4 *An L_ρ-subfunction in $\mathbb{T}_P^2$ can not attain zero maximum if it is not zero identically.*

Exercise 6.2.3.1 Prove this by exploiting (6.2.1.2) and properties of subharmonic functions.

Theorem 6.2.3.5 *Let q be an L_ρ-subfunction in $\mathbb{T}_P^2$. If $q(z) \leq 0$ for $z \in \mathbb{T}_P^2$ then $q(z) \equiv 0$.*

Exercise 6.2.3.2 Prove this using Theorem 3.1.4.7 (**Liouville).

Proposition 6.2.3.6 *Let q_D be the solution of the problem* (6.2.2.1) *in a domain D with a smooth boundary, corresponding to $\rho = \rho(D)$. Suppose that $q_D(z_0) = 1$ for some $z_0 \in D$. Then*

$$\frac{\partial q_D}{\partial n} > 0, \forall z \in \partial D.$$

Exercise 6.2.3.3 Prove this, using properties of positive harmonic functions.

6.2.4 In the part devoted to completeness of an exponential system (Section 6.3) we will need the notion of minimality of a subharmonic function from $U[\rho]$. A function $v \in U[\rho]$ is called *minimal* if the function $v - \epsilon r^\rho$ has no subharmonic minorant for arbitrarily small $\epsilon > 0$. If v is T-automorphic, the corresponding L_ρ-subfunction q is called minimal if the function $q - \epsilon$ has no L_ρ-subminorant in $\mathbb{T}_P^2$. We formulate one sufficient condition for minimality and one sufficient condition for nonminimality.

Theorem 6.2.4.1 *Let $\mathcal{H}_\rho(q)$ be the maximal open set on which $L_\rho q = 0$. If there exists a connected component $M \subset \mathcal{H}_\rho(q)$ such that $\rho(M) < \rho$, then q is a minimal L_ρ-subfunction.*

For example, $q \equiv 0$ is minimal.

Proposition 6.2.4.2 *The function q is nonminimal if $q(z) \geq c$ or $L_\rho q - c > 0$ for some positive c for all $z \in \mathbb{T}^2_P$.*

For example, $q \equiv c > 0$ is nonminimal.

6.3 Completeness of exponential systems in convex domains

6.3.1 Let $\Lambda := \{\lambda_k\}$, $k = 1, 2, \ldots$ be a set of points in the complex plane $\mathbb{C}$, satisfying the condition $\lambda_k \neq 0$ and $\lambda_j \neq \lambda_k$, if $k \neq j$.

Consider the canonical product

$$\Phi_\Lambda(\lambda) := \prod_k (1 - \lambda/\lambda_k) \exp \lambda/\lambda_k. \tag{6.3.1.1}$$

We suppose in this section that $\Phi_\Lambda(\lambda)$ is an entire function of order 1 and normal type, i.e., a function of *exponential type* (see [Le, Ch. 1, § 20].

This fact can be expressed in terms of Λ by using the Brelot-Lindelöf Theorem 2.9.4.2.

Exercise 6.3.1.1 Formulate this theorem for entire functions of order 1 and normal type under assumption that $\rho(r) \equiv 1$.

We will suppose that the upper density of zeros (see Section 2.8, Section 5.1). $\overline{\Delta_\Lambda} > 0$.

6.3.2 Let $G \subset \mathbb{C}$ be a convex bounded domain containing zero. This last requirement does not restrict any of the further considerations connected to completeness, because $\exp \Lambda := \{e^{\lambda_j z} : \lambda_j \in \Lambda\}$ can be replaced by the system $\{e^{\lambda_j (z - z_0)} : \lambda_j \in \Lambda\}$ and $e^{\lambda_j(z-z_0)} = C_j e^{\lambda_j z}$. Let $A(G)$ be the space of holomorphic functions in G with the topology of uniform convergence on compact sets. We will study the completeness of the exponential systems

$$\exp \Lambda := \{e^{\lambda_j z} : \lambda_j \in \Lambda\} \tag{6.3.2.1}$$

in $A(G)$.

We will be interested in the following questions:

1. *completeness* of $\exp \Lambda$ in $A(G)$;
2. *maximality* of G for $\exp \Lambda$, which is complete in $A(G)$;
3. *extremal overcompleteness* of $\exp \Lambda$ in $A(G)$ for a maximal G.

Let us give precise definitions of maximality and extremal overcompleteness. The completeness means that every function $f \in A(G)$ can be approximated on every compact set $K \Subset G$ with arbitrary precision by linear combinations of functions from $\exp \Lambda$.

A convex domain G is called *maximal* for a system $\exp \Lambda$, which is complete in $A(G)$ if for every domain G_1 such that $G \Subset G_1$ $\exp \Lambda$ is not complete in $A(G_1)$.

A system $\exp \Lambda$ is called *extremely overcomplete* in $A(G)$ for a maximal G, if for every sequence $\Lambda_1 := \{\lambda_j^1\}$ such that $\Lambda_1 \cap \Lambda = \varnothing$ and $\overline{\Delta}_{\Lambda_1} > 0$ the domain G is not maximal for the system $\exp \Lambda \cup \Lambda_1$.

In other words, every essential enlargement of an extremely overcomplete system enlarges also the maximal domain of completeness.

6.3.3 Let

$$h_\Lambda(\phi) := \limsup_{r\to\infty} \log |\Phi_\Lambda(re^{i\phi})| r^{-1}$$

be the indicator of Φ_Λ. It is a 1-trigonometrically convex function or simply a *trigonometrically convex function* (t.c.f). Let G_Λ be the *conjugate indicator diagram* of Φ_Λ, i.e., a convex domain of the form

$$G_\Lambda := \{z : \max_{z\in G_\Lambda} \Re(ze^{i\phi}) \le h_\Lambda(\phi)\}.$$

Let us describe conditions for completeness, maximality and extremal overcompleteness when Λ is a *regular set* (see Section 5.6) and Φ_Λ is a CRG-function (see Section 5.6).

We say that G_Λ is *enclosed* in G if it can be enclosed in G by parallel translation, *enclosed with sliding*, if it can be moved after enclosing only in one direction, *enclosed rigidly* if it is impossible to move after enclosing, *freely enclosed* in every other case of enclosing.

Theorem 6.3.3.1 *Let Λ be a regular set. Then the following holds:*

1. $\{\exp \Lambda$ *is not complete in* $A(G)\} \Longleftrightarrow \{G_\Lambda$ *is freely enclosed in* $G\}$;
2. $\{G$ *is maximal for* $\exp \Lambda\} \Longleftrightarrow \{G_\Lambda$ *is not freely enclosed in* $G\}$;
3. $\{\exp \Lambda$ *is extremely overcomplete in* $A(G)\} \Longleftrightarrow \{G_\Lambda$ *is enclosed rigidly in* $G\}$.

Let us note that G is maximal for $\exp \Lambda$ but not extremely overcomplete if and only if G_Λ is enclosed with sliding in G.

This theorem is a corollary of the more general Theorem 6.3.4.1, but will be proved independently in Section 6.3.10.

6.3.4 If Λ is not regular, it is natural to exploit the notion of a limit set (see Section 3.1) to characterize $\exp \Lambda$.

Suppose the limit set of Φ_Λ has the form

$$\mathbf{Fr}[\Phi_\Lambda] := \{v(\lambda) = |\lambda|(ch_1 + (1-c)h_2)(\arg \lambda) : c \in [0;1]\}$$

where h_1, h_2 are t.c.f.

Such a limit set is a particular case of U_{ind} (6.1.2.2). It is called an *indicator limit set* and it is indeed a limit set of an entire function (see Exercise 6.1.2.4).

The asymptotic behavior of the set Λ (i.e., the limit set of the corresponding mass distribution) can be described completely using Theorem 3.1.5.2.

Exercise 6.3.4.2 Do that.

We will call such Λ an *indicator set.* Denote by G_1, G_2 the conjugate diagram of h_1, h_2. Since G_1, G_2 are convex, the set

$$\alpha G_1 + \beta G_2 := \{\alpha z_1 + \beta z_2 : z_1 \in G_1, z_2 \in G_2\},\ \alpha, \beta > 0$$

is also convex and is a conjugate diagram of the t.c.f. $h := \alpha h_1 + \beta h_2$.

Theorem 6.3.4.1 *Let a set Λ be an indicator set. Then the following holds:*

1. $\{\exp \Lambda$ *is not complete in* $A(G)\} \iff \{G_1$ *and* G_2 *are freely enclosed in* $G\}$*;*
2. $\{G$ *is maximal for* $\exp \Lambda\} \iff \{G_1$ *and* G_2 *are enclosed in* G *and at least one of them is not freely enclosed in* $G\}$*;*
3. $\{\exp \Lambda$ *is extremely overcomplete in* $A(G)\} \iff \{cG_1 + (1-c)G_2$ *is enclosed rigidly in* G $\forall c \in [0;1]\}$.

This theorem is proved in Section 6.3.10.

The equality holds:

$$h_\Lambda = \max(h_1, h_2). \tag{6.3.4.1}$$

Thus the conjugate diagram G_Λ of the function h_Λ is the convex hull of G_1 and G_2.

Let us note that the indicator h_Λ does not determine the completeness of the system $\exp \Lambda$ if Λ is not a regular set, as the following example shows.

Example 6.3.4.1 Let

$$\begin{aligned} G_1 &:= \{z = x + iy : x = 1; -1 \le y \le 1\},\\ G_2 &:= \{z = x + iy : x = -1; -1 \le y \le 1\},\\ \text{and} \quad G &= \{z : |z| < 1 + \epsilon\} \end{aligned}$$

with a small ϵ.

Exercise 6.3.4.3 Prove that G_1 and G_2 are freely enclosed in G and their convex hull is not enclosed.

Let Λ be a set such that the interior of G_Λ coincides with G. If Λ is a regular set, then $\exp \Lambda$ is complete in $A(G)$, G is maximal for $\exp \Lambda$ and $\exp \Lambda$ is extremely overcomplete in $A(G)$.

If Λ is an indicator set, then the first two assertions hold but $\exp \Lambda$ can be not extremely overcomplete:

Example 6.3.4.2 Set

$$G_1 := \{z = x + iy : -1 \leq x \leq 0; y = 0\};$$
$$G_2 := \{z = x + iy : x = 1; -1 \leq y \leq 1\}.$$

Here G_Λ is a triangle in which G_1 is freely enclosed and G_2 is rigidly enclosed, but $cG_1 + (1 - c)G_2$ is free enclosed for all $c : 0 < c < 1$.

Exercise 6.3.4.4 Check this.

Example 6.3.4.3 Set

$$G_1 := \{z = x + iy : x = -1; y \in [-1; 1]\};$$
$$G_2 := \{z = x + iy : x = 1; y \in [-1; 1]\}.$$

Exercise 6.3.4.5 Check that G_1 and G_2 are enclosed with sliding in G_Λ.

If G_1 and G_2 are rigidly enclosed in G it does not imply in general that $cG_1 + (1 - c)G_2$ are rigidly enclosed for all $c \in [0; 1]$.

Example 6.3.4.4 Let G_1 be an equilateral triangle inscribed in the circle $|z| = 1$, let G_2 be the same triangle rotated by the angle $\pi/6$, and let G be the unit disc.

Exercise 6.3.4.6 Show that $\frac{1}{2}(G_1 + G_2)$ is freely enclosed in G.

If $G_1, G_2 \subset G$ and $G_1 \cap G_2$ is rigidly enclosed in G, then $cG_1 + (1 - c)G_2$ is rigidly enclosed for $c \in [0; 1]$.

Exercise 6.3.4.7 Check this.

However this is not a necessary condition.

Example 6.3.4.5

$$G := \{z = x + iy : |x| < 1; |y| < 1\};$$
$$G_1 := \{z = x + iy : x \in (-1, 1); -x > y > -1\};$$
$$G_2 := \{z = x + iy : x \in (-1, 1); -1 < y < x\}.$$

Exercise 6.3.4.8 Check that every triangle $cG_1 + (1 - c)G_2$ is rigidly enclosed in G and $G_1 \cap G_2$ is freely enclosed.

6.3.5 Consider in more detail the conditions for extremal overcompleteness in the case when Λ is an indicator set and $G_\Lambda = G$ or, in other words, if

$$h_\Lambda = h_G. \tag{6.3.5.1}$$

We can suppose that h_1 and h_2 are linearly independent, otherwise we exploit Theorem 6.3.3.1. If, for example, the inequality $h_1(\phi) \leq h_2(\phi), \forall \phi$, holds, the extremal overcompleteness is in the case when G_1 is rigidly enclosed in G_2 because $G_1 \cap G_2 = G_1$, and this case was mentioned above (Exercise 6.3.4.7).

Consider the general case. Denote $g(\phi) := |h_1 - h_2|(\phi)$, and set $\Theta_\Lambda := \{\phi : g(\phi) > 0\}$. This is an open set on the unit circle. Denote as $I_\Lambda := (\alpha_1, \alpha_2)$ the maximal interval contained in Θ_Λ and denote by d_Λ its length. Since $g(\phi)$ is continuous,

$$g(\alpha_j) = 0, \ j = 1, 2. \tag{6.3.5.2}$$

If also at least one of the conditions

$$\liminf_{\phi \in I_\Lambda, \phi \to \alpha_j} \frac{g(\phi)}{\phi - \alpha_j} = 0, \ j = 1, 2,$$

is fulfilled, we say g is *zero with tangency* on ∂I_Λ.

Theorem 6.3.5.1 *Suppose Λ is an indicator set that satisfies* (6.3.5.1). *In order that* $\exp \Lambda$ *be extremely overcomplete in $A(G)$ it is necessary and sufficient that at least one of the following conditions holds:*

1. $d_\Lambda < \pi$;
2. $d_\Lambda = \pi$ *and g is zero with tangency on ∂I_Λ.*

This theorem is proved in Section 6.3.11.

6.3.6 We call Λ *periodic* if $\mathbf{Fr}[\Phi_\Lambda]$ is a periodic limit set (see Theorem 4.1.7.1). In such a case all the limit set is determined by one subharmonic function $v \in U[1]$ (see (4.1.3.1)). Let us characterize the system $\exp \Lambda$ for periodic Λ.

Set

$$h_G(\phi) := \max\{\Re(ze^{i\phi}) : z \in G\}, \tag{6.3.6.1}$$

$$m(\lambda, G, v) := |\lambda| h_G(\arg \lambda) - v(\lambda). \tag{6.3.6.2}$$

Denote by $\mathcal{G}_G v$ the maximal subharmonic minorant of the function $m(\lambda, G, v)$. A function $w \in U[1]$ is called *minimal* if the function $w - \epsilon|\lambda|$ has no subharmonic minorant in $U[1]$ for every small $\epsilon > 0$. The harmonic function of the form

$$H(\lambda) := |\lambda|(A \cos(\arg \lambda) + B \sin(\arg \lambda)), \tag{6.3.6.3}$$

for example, is minimal.

We will denote as HARM the set of functions of the form (6.3.6.3).

Theorem 6.3.6.1 *Let Λ be a periodic set. The following holds:*

1. $\{\exp \Lambda$ *is not complete in* $A(G)\} \iff \{\mathcal{G}_G v$ *exists and is non-minimal*$\}$;
2. $\{G$ *is maximal for* $\exp \Lambda\} \iff \{\mathcal{G}_G v$ *exists and is minimal*$\}$;
3. $\{\exp \Lambda$ *is extremely overcomplete in* $A(G)\} \iff \{\mathcal{G}_G v \in \mathrm{HARM}\}$.

This theorem is proved in Section 6.3.12.

6.3.7 Let us characterize the completeness of $\exp_\Lambda$ for periodic Λ in other terms. For this we need the information that was presented in Section 6.2. We will take $\rho = 1$. Denote

$$q_\Lambda(z) := v_\Lambda(e^z)e^{-x} \tag{6.3.7.1}$$

(compare with (6.2.1.2)). As it was explained in Section 6.2 this function is an L_1-subfunction on the torus $\mathbb{T}^2_P$. Set

$$m(z, G, q_\Lambda) = h_G(y) - q_\Lambda, \ D(G, \Lambda) := \{z : m(z, G, q_\Lambda) > 0\} \subset \mathbb{T}^2_P.$$

The set $D(G, \Lambda)$ is open because $-m$ is an upper semicontinuous function (see Theorem 2.1.2.4), denote

$$\rho(\Lambda, G) := \min \rho(M)$$

where the minimum is taken over all components M of $D(G, \Lambda)$, and it is attained on one of the components because they are not intersecting and $\mathbb{T}^2_P$ is compact.

Exercise 6.3.7.1 Explain this in detail, using properties of $\rho(D)$ (Section 6.2).

Theorem 6.3.7.1 *If*

$$\rho(\Lambda, G) \geq 1, \tag{6.3.7.2}$$

then $\exp \Lambda$ *is complete in* G.

This theorem is proved in Section 6.3.12.

Let $w := g_G q_\Lambda(z)$ be the maximal L_1-subminorant of $m(z, G, q_\Lambda)$. Denote by $\mathcal{H}_\Lambda$ the open set in $\mathbb{T}^2_P$ where $L_1 w = 0$.

Theorem 6.3.7.2 *If there exists a component M of $\mathcal{H}_\Lambda$ such that $\rho(M) < 1$, then w is minimal, and, hence, G is maximal for* $\exp \Lambda$.

This theorem follows directly from Theorem 6.2.4.1. It is not known if the condition (6.3.7.2) is necessary.

Consider in detail the situation in which the domain G coincides with G_Λ, the conjugated indicator diagram of h_Λ, i.e., we suppose that

$$h_G(\phi) = h_\Lambda(\phi), \ \forall \phi. \tag{6.3.7.3}$$

In this case $m(z, G, q_\Lambda) \geq 0$ and we obtain the following criterion:

Theorem 6.3.7.3 *In order that* $\exp \Lambda$ *be complete in* G_Λ *it is necessary and sufficient that*

$$\rho(\Lambda, G_\Lambda) \geq 1. \tag{6.3.7.4}$$

This theorem is proved in Section 6.3.12. The condition (6.3.7.3) automatically implies maximality if there is completeness.

Since

$$h_\Lambda(y) = \max\{q_\Lambda(x + iy) : x \in [0; P]\}, \tag{6.3.7.5}$$

the function $m(z, G_\Lambda, q_\Lambda)$ has a zero in x for every fixed y.

Thus the set $D(G, \Lambda)$ does not contain any curve $y = \text{const}$ on the torus.

Theorem 6.3.7.4 *Let G_0 be a strictly convex domain and let $D_0 \subset \mathbb{T}_P^2$ be such that $\mathbb{T}_P^2 \setminus D_0$ intersect every line $\{y = y_0\}$, $y_0 \in [0, 2\pi]$.*

Then there exists a periodic Λ such that

$$G_\Lambda = G_0, \ D(G_\Lambda, \Lambda) = D_0. \tag{6.3.7.6}$$

This theorem is proved in Section 6.3.13.

Example 6.3.7.1 Let D_0 be the complement in $\mathbb{T}_P^2$ to the set

$$M := \{z = x + iy : x = f(y), y \in [0; 2\pi]\} \tag{6.3.7.7}$$

where $f(y)$ is a continuous 2π-periodic function satisfying the condition

$$0 < f(y) < P.$$

Then $\rho(D_0) = \infty$, because this domain is not connected on spirals (see Section 6.2.). It means that for every strictly convex G_0 there exists a periodic Λ such that $G_\Lambda = G_0$ and $\exp \Lambda$ is extremely overcomplete in G_0.

Example 6.3.7.2 Let D_0 be the complement to the set

$$M := \left\{ z = x + iy : x = \frac{P}{2\pi} y, \ 0 \leq y \leq 2\pi \right\}.$$

Then

$$\rho(D_0) = \frac{1}{2}\left(1 + (2\pi/P)^2\right) \tag{6.3.7.8}$$

(see Section 6.3.13).

Thus, choosing P, and using Theorem 6.3.7.4, it is possible make $\exp \Lambda$ complete or non-complete in $G_0 (= G_\Lambda)$ for every strictly convex domain G_0.

6.3.8 Now pass to generalizations. Denote by D_G the natural domain of definition of the operation $\mathcal{G}_G$, i.e., the set of $v \in U[1]$ for which $m(\lambda, G, v)$ (see (6.3.6.2)) has a subharmonic minorant belonging to $U[1]$.

Let Φ_Λ be defined by the equality (6.3.1.1). The condition that for every $v \in \mathbf{Fr}[\Phi_\Lambda]$ the function $m(\lambda, G, v)$ has a subharmonic minorant belonging to $U[1]$ is possible to express by the relation

$$\mathbf{Fr}[\Phi_\Lambda] \subset D_G \tag{6.3.8.1}$$

(compare with (6.1.1.1)).

We call the set $U \subset U[1]$ *minimal* ($U \in$ MIN) if for arbitrarily small $\epsilon > 0$ there exists $w = w_\epsilon \in U$ such that the function $w_\epsilon - \epsilon|\lambda|$ has no subharmonic minorant, belonging to $U[1]$.

Let us note that if U contains a minimal function (in the sense of Section 6.3.6), then $U \in$ MIN. We denote the image of $\mathbf{Fr}[\Phi_\Lambda]$ under the mapping by the operator $\mathcal{G}_G$ as $J_G(\Lambda)$.

Theorem 6.3.8.1 *The following holds:*

1. $\{\exp_\Lambda$ *is not complete in* $A(G)\} \Longleftrightarrow \{((6.3.8.1)$ *holds*$) \wedge (J_G(\Lambda) \notin \mathrm{MIN})\}$;
2. $\{G$ *is maximal for* $\exp_\Lambda\} \Longleftrightarrow \{((6.3.8.1)$ *holds*$) \wedge (J_G(\Lambda) \in \mathrm{MIN})\}$;
3. $\{\exp\Lambda$ *is extremely overcomplete for maximal* $G\} \Longleftrightarrow \{((6.3.8.1)$ *holds*$) \wedge (J_G(\Lambda) \in \mathrm{HARM})\}$;

6.3.9 In the proof of Theorem 6.3.8.1 that we are going to prove now we exploit

Theorem 6.3.9.1. (A.I. Markushevich) see [Le, Ch. 4, § 7] *Let* $A(\mathbb{C}\setminus\overline{G})$ *be a class of functions* ψ *which are holomorphic in* $\mathbb{C}\setminus\overline{G}$ *and equal to zero in infinity. In order that the system* $\exp\Lambda$ *be complete in* $A(G)$*, it is necessary and sufficient that the function*

$$\Phi(\lambda) := \int_{L_\psi} e^{\lambda z}\psi(z)dz, \tag{6.3.9.1}$$

where $\psi \in A(\mathbb{C}\setminus\overline{G})$*, and* $L_\psi \Subset G$ *is a rectifiable closed curve, has the following property: the condition*

$$\Phi(\lambda_k) = 0, \ \forall\lambda_k \in \Lambda \tag{6.3.9.2}$$

implies $\Phi(\lambda) \equiv 0$.

Proof of Theorem 6.3.8.1, 1. Necessity. Let $\exp\Lambda$ be not complete . By Theorem 6.3.9.1 $\Phi(\lambda_k) = 0$, but $\Phi(\lambda) \not\equiv 0$. The function $g(\lambda) := \Phi(\lambda)/\Phi_\Lambda(\lambda)$, where Φ_Λ is from (6.3.1.1), is an entire function and it has order one and normal or minimal type by Theorem 2.9.3.1. Set

$$u^g := \log|g(\lambda)|; \ u^\Phi(\lambda) := \log|\Phi(\lambda)|; \ u^\Lambda(\lambda) := \log|\Phi(\lambda)|.$$

We have from (6.3.9.1) $u^\Phi(\lambda) \le \max\{\Re(\lambda z) : z \in L_\psi\} + C_\psi\}$, where C_ψ is a constant, depending possibly on ψ.

This implies that

$$u^\Phi(\lambda) \le h_{G_1}(\phi)r + C_\psi, \ \lambda = re^{i\phi}, \tag{6.3.9.3}$$

for some convex domain $G_1 \Subset G$.

Let $v \in \mathbf{Fr}[\Phi_\Lambda]$. Choose a sequence $t_j \to \infty$ for which $(u^\Lambda)_{t_j} \to v$, and the sequences $(u^\Phi)_{t_j}$ and $(u^g)_{t_j}$ also converge to v^Φ and v^g respectively. From the equality $u^g(\lambda) = u^\Phi(\lambda) - u^\Lambda(\lambda)$ we obtain $v^g(\lambda) = v^\Phi(\lambda) - v(\lambda)$ where $v^g \in \mathbf{Fr}[g]$, $v^\Phi \in \mathbf{Fr}[\Phi]$.

Since (6.3.9.3) implies $v^\Phi(\lambda) \le h_{G_1}(\phi)r$,

$$v^g(\lambda) \le h_{G_1}(\phi)r - v(\lambda) \tag{6.3.9.4}$$

and it means that for every $v \in \mathbf{Fr}[\Phi_\Lambda]$ $\mathcal{G}_{G_1}v$ and hence $\mathcal{G}_G v$ exist, i.e., the condition (6.3.8.1) holds.

Let us show that the condition $J_G(\Lambda) \notin \text{MIN}$ is satisfied. We have for some $\delta > 0$ the relation

$$h_{G_1}(\phi) - h_G(\phi) \le -\delta.$$

From (6.3.9.4) we obtain

$$v^g(\lambda) + \delta r \le m(\lambda, G, v). \tag{6.3.9.5}$$

The left-hand side of the inequality (6.3.9.5) belongs to $U[1]$. Thus $w_v := \mathcal{G}_{G_1} v$ satisfies the condition $v^g(\lambda) + \delta r \le w_v(\lambda)$ for every $v \in \mathbf{Fr}[\Phi_\Lambda]$. It means that $J_G(\Lambda) \notin \text{MIN}$.

Necessity is proved. □

For proving sufficiency we exploit the following assertion.

Theorem 6.3.9.2 (I.F. Krasichkov-Ternovskii) *Suppose there exists an entire function g such that*

$$h_{g\Phi_\Lambda}(\phi) < h_G(\phi), \ \forall \phi. \tag{6.3.9.6}$$

Then the system $\exp \Lambda$ *is not complete for some convex domain* $G_1 \Subset G$.

This theorem connects the problem of completeness to the multiplicator problem.

Proof of Theorem 6.3.9.2. Let $g(\lambda)$ satisfy (6.3.9.6). Denote by $\psi(z)$ the Borel transformation for $\Phi(\lambda) := g(\lambda)\Phi_\Lambda(\lambda)$. By the Pólya Theorem (see, for example, [Le, Ch. 1, § 20]), all the singularities of ψ are contained in a convex domain G_Φ which is the conjugate diagram of the indicator $h_\Phi(\phi)$. Thus the representation (6.3.9.1) holds with L_ψ that embraces G_Φ. It follows from (6.3.9.6) that $G_\Phi \Subset G$. Thus it is possible to choose L_ψ between ∂G_Φ and ∂G. Since (6.3.9.2) for Φ is fulfilled and $\Phi(\lambda) \not\equiv 0$, $\exp \Lambda$ is non-complete in some convex $G_1 \Subset G$ such that $L_\psi \Subset G_1$ by Theorem 6.3.9.1. □

Now we can prove sufficiency in Theorem 6.3.8.1, 1. From the condition $J_G(\Lambda) \notin \text{MIN}$ it follows that one can choose $\delta > 0$ such that $\forall v \in \mathbf{Fr}[\Phi_\Lambda]$ the functions $w_v - \delta r$ where $w_v := \mathcal{G}_G$, have subharmonic minorants. As we already said in Section 6.3.2, completeness does not depend on shift by any fixed z_0. Thus we can suppose that $0 \in G$ and, hence, $h_G(\phi) > 0$ for all ϕ. Let $\gamma < 2\delta$ be such that $h_G(\phi) - \gamma > 0$ and $G_1 \Subset G$ satisfy

$$h_{G_1}(\phi) - \gamma/3 > 0, \ h_G(\phi) - h_{G_1}(\phi) \le \gamma/2. \tag{6.3.9.7}$$

Let us check that

$$D_{G_1} \supset \mathbf{Fr}[\Phi_\Lambda], \tag{6.3.9.8}$$

Indeed, for $v \in \mathbf{Fr}[\Phi_\Lambda]$ we have

$$\begin{aligned} m(\lambda, G_1, v) := h_{G_1}(\phi)r - v(\lambda) &\ge h_G(\phi)r - (\gamma/2)r - v(\lambda) \\ &\ge h_G(\phi)r - v(\lambda) - \delta r \\ &\ge w_v - \delta r. \end{aligned} \tag{6.3.9.9}$$

Since the right-hand side of (6.3.9.9) has a subharmonic minorant from $U[1]$, then (6.3.9.8) is proved. By Theorem 6.1.1.1 there exists a multiplicator $g(z) \in A(1)$ such that

$$h_{g\Phi}(\phi) \leq h_{G_1}(\phi) < h_G(\phi). \tag{6.3.9.10}$$

From Theorem 6.3.9.2 we obtain that $\exp \Lambda$ is non-complete in G.

Proof of Theorem 6.3.8.1, 2. Necessity. Let G_j, $j = 1, 2, \dots$ be a sequence of convex domains, satisfying the conditions $G_j \Supset G$, $G_j \downarrow G$. Since $\exp \Lambda$ is non-complete in every $A(G_j)$, $D_{G_j} \supset \mathbf{Fr}[\Phi_\Lambda]$ by Theorem 6.3.8.1, 1.

The sequence $w_j := \mathcal{G}_{G_j} v$ satisfies

$$w_j(\lambda) \leq h_{G_j}(\phi)r - v(\lambda), \ \lambda \in \mathbb{C}.$$

Since $\{w_j\}$ is compact and $h_{G_j} \to h_G$, one can find a subsequence with the limit $w \in U[1]$. Then $w(\lambda) \leq h_G(\phi)r - v(\lambda)$. Hence $\mathcal{G}_G v$ exists.

If $J_G \in$ MIN would not hold, then, by Theorem 6.3.8.1, 1, $\exp \Lambda$ is non-complete in $A(G)$, which contradicts maximality.

Necessity is proved. Let us prove sufficiency.

Completeness of $\exp \Lambda$ in $A(G)$ follows from Theorem 6.3.8.1, 1. We will prove that $\exp_\Lambda$ is non-complete in $A(G_1)$ for every $G_1 \Supset G$ under the condition $D_G \supset \mathbf{Fr}[\Phi_\Lambda]$. Set

$$\delta := \min_\phi [h_{G_1}(\phi) - h_G(\phi)] > 0.$$

Then $\forall v \in \mathbf{Fr}[\Phi_\Lambda]$,

$$\mathcal{G}_G v + \delta r \leq h_{G_1}(\phi)r - v(\lambda), \ \lambda \in \mathbb{C}.$$

This means that $\mathcal{G}_{G_1} v \geq \mathcal{G}_G v + \delta r$. Hence $J_{G_1}(\Lambda) \notin$ MIN and, by Theorem 6.3.8.1, 1, $\exp \Lambda$ is non-complete in $A(G_1)$. □

Proof of Theorem 6.3.8.1, 3. Necessity. By Theorem 6.3.8.1, 2 from maximality G (6.3.8.1) follows. We will prove that $\mathcal{G}_G v \in$ HARM $\forall v \in \mathbf{Fr}[\Phi_\Lambda]$. Suppose it is not fulfilled, i.e., there exists $v_0 \in \mathbf{Fr}[\Phi_\Lambda]$ such that the mass distribution ν_0 of the function $w_0 = \mathcal{G}_G v_0$ is not zero. By Proposition 6.1.1.3 there exists a multiplicator g such that $v_0 + w_0 \in \mathbf{Fr}[g\Phi_\Lambda]$. Let Λ_0 be the set of zeros of g. Since $\nu_0 \in \mathbf{Fr}\Lambda_0$, $\underline{\Delta}(\Lambda_0) > 0$, because $\nu_0 \neq 0$ and by the definitions in Section 3.3.1.

We can shift a little zeros of g and suppose without lack of generality that they are simple and $\Lambda_0 \cap \Lambda = \varnothing$.

The condition for a multiplicator gives the inequality:

$$h_{g\Phi_\Lambda}(\phi) \leq h_G(\phi), \ \forall \phi.$$

It implies

$$m(\lambda, G, v_\Pi) = r h_G(\phi) - v_\Pi \geq 0$$

for all $v_\Pi \in \mathbf{Fr}[g\Phi_\Lambda]$. It means that $m(\lambda, G, v_\Pi)$ has zero as a minorant $\forall v_\Pi \in \mathbf{Fr}[g\Phi_\Lambda]$, i.e., $D_G \supset \mathbf{Fr}[g\Phi_\Lambda]$. So the domain G is maximal although the system $\exp\Lambda$ is replaced with the system $\exp(\Lambda \cup \Lambda_0)$. This contradicts the extremal overcompleteness. Hence, $\nu_0 \equiv 0$ and $w_0 = \mathcal{G}_G v_0 \in$ HARM.

Necessity is proved. Let us prove sufficiency.

Let the condition $\mathcal{G}_G v \in$ HARM $\forall v \in \mathbf{Fr}[\Phi_\Lambda]$ hold. Suppose that there exists Λ_0 such that $\overline{\Delta}_{\Lambda_0} > 0$ and G is maximal for the system $\exp(\Lambda \cup \Lambda_0)$.

Theorem 6.3.8.1, 2 implies

$$D_G \supset \mathbf{Fr}[\Phi_{\Lambda_1}], \tag{6.3.9.11}$$

where $\Lambda_1 = \Lambda \cup \Lambda_0$.

For every $v_0 \in \mathbf{Fr}[\Phi_{\Lambda_0}]$ one can find $v \in \mathbf{Fr}[\Phi_\Lambda]$ such that

$$v_1 := v_0 + v \in \mathbf{Fr}[\Phi_{\Lambda_1}].$$

The condition $\overline{\Delta}_{\Lambda_0} > 0$ implies that one can choose v_0 for which the Riesz measure $\nu_0 \not\equiv 0$. For $w_1 = \mathcal{G}_G v_1$ one has the inequality $w_1 \leq rh_G - v_1$ by (6.3.9.11), so $w_1 + v_0 \leq rh_G - v$ holds. Hence $w_v := \mathcal{G}_G v$ satisfies the inequality

$$(w_1 + v_0)(\lambda) \leq w_v(\lambda), \ \forall \lambda \in \mathbb{C}. \tag{6.3.9.12}$$

Let us show that (6.3.9.12) is impossible.Indeed, since $w_v \in$ HARM $w := w_1 + v_0 - w_v \leq 0$ and $w \in U[\rho]$. Thus $w \equiv 0$. However the Riesz measure $\nu_w \geq \nu_0 \not\equiv 0$, hence $w \not\equiv 0$. This contradiction proves sufficiency. □

6.3.10 Now we prove Theorems 6.3.3.1, 6.3.4.1 and 6.3.5.1. We need some auxiliary assertions.

Lemma 6.3.10.1 *Let $v := rh_1(\phi)$ and G_1 be the conjugated diagram of h_1. Then the following holds:*

1. $\{G_1$ *is freely enclosed in* $G\} \iff \{\mathcal{G}_G v$ *is non-minimal*$\}$;
2. $\{G_1$ *is enclosed to* G *but not free enclosed*$\} \iff \{\mathcal{G}_G v$ *is minimal*$\}$;
3. $\{G_1$ *is rigidly enclosed in* $G\} \iff \{\mathcal{G}_G v \in$ HARM$\}$;
4. $\{G_1$ *is not enclosed in* $G\} \iff \{\mathcal{G}_G v$ *does not exist*$\}$.

To prove this lemma we need the following two lemmas.

Lemma 6.3.10.2 *Let $v := rh(\phi)$. Then $\mathcal{G}_G v = rh_1(\phi)$ where h_1 is the maximal trigonometrically convex minorant of the function*

$$m(\phi, G, h) := h_G(\phi) - h(\phi).$$

Proof. Let $v_1 = \mathcal{G}_G v$. Since $v_{[t]} = v$ for all $t > 0$,

$$(v_1)_{[t]} = \mathcal{G}_G v_{[t]} = \mathcal{G}_G v$$

by Theorem 6.1.1.2, 2.

Thus the function

$$\hat{v}_1 := \left(\sup_t (v_1)_{[t]}\right)^* (\lambda) \geq v_1(\lambda)$$

and is also a subharmonic minorant belonging to $U[1]$. Thus $v_1 = \hat{v}_1$. However, the function $\hat{v}_1$ is invariant with respect to the transformation $(\bullet)_{[t]}$. Hence it has the form $rh_1(\phi)$. The maximality of $h_1(\phi)$ follows from the maximality v_1. □

Lemma 6.3.10.3 *In order that $v := rh_1$ be a minimal function, it is necessary and sufficient that G_1, the conjugate diagram of h_1, be a segment (in particular, a point).*

Proof. Let $v = rh_1$ be minimal and let G_1 be the conjugate diagram of h_1. If G_1 is not the segment, then it contains some disc of radius $\delta > 0$. Hence there exists a trigonometric function $A\cos(\phi - \phi_0)$ such that

$$\delta + A\cos(\phi - \phi_0) \leq h_1(\phi).$$

Multiplying this inequality by r, we obtain that $v - \delta r$ has a harmonic (and hence subharmonic) minorant. This contradicts minimality.

Inversely, suppose v is not minimal. Then there exists $\delta > 0$ and t.c.f. $h_2(\phi)$ such that

$$h_2(\phi) \leq h_1(\phi) - \delta. \tag{6.3.10.1}$$

For every t.c.f. h_2 there exists a trigonometric function $A\cos(\phi - \phi_0)$ such that

$$h_2(\phi) + A\cos(\phi - \phi_0) \geq 0. \tag{6.3.10.2}$$

This corresponds to a shift of the diagram which contains zero. From (6.3.10.1) and (6.3.10.2) we obtain

$$\delta - A\cos(\phi - \phi_0) \leq h_1(\phi),$$

which means that G_1 contains some disc of radius $\delta > 0$. So it is not a segment. □

Proof of Lemma 6.3.10.1. G is freely enclosed iff the following assertion holds: there exists $\delta > 0$ and a trigonometrical function $A\cos(\phi - \phi_0)$ such that the inequality

$$h_1(\phi) + \delta - A\cos(\phi - \phi_0) \leq h_G(\phi) \tag{6.3.10.3}$$

holds.

Exercise 6.3.10.1 Prove this.

Let $\mathcal{G}_G v$ be non-minimal. By Lemma 6.3.10.2 it has the form $w_2 = rh_2$, where h_2 is the maximal trigonometrically convex minorant of $m(\phi, G, h_1)$. There exists $\delta > 0$ such that the function $w_2 - \delta r$ has the maximal subharmonic minorant $v_3 = rh_3(\phi)$. Let $A\cos(\phi - \phi_0)$ be a trigonometric function for which

$$h_3(\phi) + A\cos(\phi - \phi_0) \geq 0.$$

In addition,

$$h_3(\phi) \leq h_2(\phi) - \delta,\ h_2(\phi) \leq h_G - h_1(\phi).$$

From this we obtain (6.3.10.3) and hence that G_1 is free enclosed.

Inversely, let G_1 be freely enclosed in G. From (6.3.10.3) it follows that

$$\delta - A\cos(\phi - \phi_0) \leq h_G(\phi) - h_1(\phi). \tag{6.3.10.4}$$

Multiplying (6.3.10.4) by r, we obtain that $m(\lambda, G, v)$ has a minorant $v_0 = r(\delta - A\cos(\phi - \phi_0))$ which obviously is non-minimal. Hence, $\mathcal{G}_G v$ is non-minimal.

G_1 is enclosed in G with sliding, hence there does not exist $\delta > 0$ such that (6.3.10.3) is fulfilled, but there exists a segment with support function

$$E(\phi) = B|\sin\phi| + A\cos(\phi - \phi_0),$$

such that the inequality

$$h_1(\phi) + E(\phi) \leq h_g(\phi) \tag{6.3.10.5}$$

holds.

Exercise 6.3.10.2 Prove this.

Let $\mathcal{G}_G v$ be minimal. By Lemma 6.3.10.2 it has the form $w_2 = rh_2$ and by Lemma 6.3.10.3, $h_2 = E(\phi)$. Thus $E(\phi) \leq (h_G - h_1)(\phi)$, which is equivalent to (6.3.10.5).

Prove 2, suppose G is not freely enclosed and hence only (6.3.10.5) is possible. If $\mathcal{G}_G v$ were non-minimal, (6.3.10.3) would follow, as it was proved above. This contradicts the supposition.

The rigid enclosure is equivalent only to the inequality of the form

$$h(\phi) - A\cos(\phi - \phi_0) \leq h_G(\phi)\ \forall\phi,$$

and impossibility of enclosure is equivalent to the impossibility of even such an inequality. Thus all other assertions of the lemma can be proved analogously.

Exercise 6.3.10.3 Do this in detail. □

Proof of Theorem 6.3.3.1. Regularity of Λ means that $\mathbf{Fr}[\Phi_\Lambda] = \{v_0\}$ where $v_0 = rh_\Lambda$. Thus $J_G(\Lambda) = \{\mathcal{G}_G v_0\}$ and all the assertions of Theorem 6.3.3.1 follows from Theorem 6.3.8.1 and Lemma 6.3.10.1. □

Exercise 6.3.10.4 Check this in detail.

For proving Theorem 6.3.4.1 we need an additional

Lemma 6.3.10.4 *Let* Λ *be an indicator set,* $v_1 = rh_1,\ v_2 = rh_2$. *Then*

$$\{J_G(\Lambda) \notin \text{MIN}\} \Longleftrightarrow \{\mathcal{G}_G v_1 \quad \textit{and} \quad \mathcal{G}_G v_2 \quad \textit{are non-minimal}\}.$$

Proof. Suppose $w_1 := \mathcal{G}_G v_1$ and $w_2 := \mathcal{G}_G v_2$ are not minimal, i.e., $w_1 - \delta r$ and $w_1 - \delta r$ have subharmonic minorants g_1 and g_2.

Then $cg_1 + (1-c)g_2$ is a minorant of the function $cw_1 + (1-c)w_2 - \delta r$, i.e., $J_G(\Lambda) \notin$ MIN. The inverse implication is trivial. □

Exercise 6.3.10.5 Prove this.

Proof of Theorem 6.3.4.1. Suppose $\exp\Lambda$ is not complete. By Theorem 6.3.8.1 $J_G \notin$ MIN. By Lemma 6.3.10.4, $\mathcal{G}_G v_1$ and $\mathcal{G}_G v_2$ are not minimal. Hence G_1 and G_2 are freely enclosed in G by Lemma 6.3.10.1. Since every one of these assertions is reversible, the inverse implication also holds. Analogously the other cases are proved. □

Exercise 6.3.10.6 Prove all this in detail.

6.3.11 To prove Theorem 6.3.5.1 we need some auxiliary assertions.

Lemma 6.3.11.1 *Let* ϕ_0 *be a maximum point of t.c.f.* $h(\phi)$ *and* $h(\phi_0) \geq 0$. *Then*

$$h(\phi) \geq h(\phi_0)\cos(\phi - \phi_0),\ |\phi - \phi_0| \leq \pi/2. \tag{6.3.11.1}$$

Proof. We write $y(\phi) := h(\phi_0)\cos(\phi - \phi_0)$. We have $y(\phi_0) = h(\phi_0)$ and $y(\phi)$ is a trigonometric function. If $y(\phi_1) = h(\phi_1)$ for some ϕ_1 such that $|\phi_1 - \phi_0| < \pi/2$ this contradicts Theorem 3.2.5.2. If $y(\phi)$ does not intersect $h(\phi)$, this contradicts Theorem 6.2.3.4 applied to the function $h(\phi) - y(\phi)$, which is an L_ρ-subfunction with $\rho = 1$. □

Lemma 6.3.11.2 *Let* $H(\phi)$ *be a trigonometric function on the interval* $I = (\alpha, \beta)$ *of length* $\leq \pi$, *such that* $H(\phi) = 0$ *at one of the ends of* I. *Then every one of the conditions*

1. $H(\phi_0) = 0,\ \phi_0 \in (\alpha;\beta)$;
2. $H(\phi)$ *is zero on* ∂I *with tangency;*

implies $H(\phi) \equiv 0,\ \phi \in I$.

Exercise 6.3.11.1 Prove this.

Lemma 6.3.11.3 *Let* $g \geq 0$ *be a continuous periodic function, and let* $\Theta_\Lambda, I_\Lambda, d_\Lambda$ *be defined as in Theorem* 6.3.5.1. *In order that its maximal t.c.minorant be a trigonometrical function, it is necessary and sufficient satisfying at least one of the conditions:*

1. $d_\Lambda < \pi$;
2. $d_\Lambda = \pi$ *and* $g(\phi)$ *is zero with tangency on* ∂I.

Proof. Necessity. Suppose $d_\Lambda > \pi$. Without loss of generality we can suppose that $I_\Lambda = (\alpha; -\alpha)$, where $\alpha > \pi/2$.

Set $\cos^+ \phi := \max(\cos\phi, 0)$,

$$a = \inf\left(\frac{g(\phi)}{\cos^+ \phi} : \phi \in (-\alpha; \alpha)\right). \tag{6.3.11.2}$$

We have $a > 0$. Set

$$h(\phi) := \begin{cases} a_1 \cos\phi, & |\phi| \leq \pi/2 \\ 0 & |\phi| > \pi/2, \end{cases} \tag{6.3.11.3}$$

where $a_1 \leq a$.

The function $h(\phi)$ is a t.c.minorant of $g(\phi)$ and it is not a trigonometric function, which contradicts the supposition. Thus $d_\Lambda \leq \pi$.

Suppose $d_\Lambda = \pi$ and the condition to be zero with tangency on ∂I does not hold. Then for a defined by (6.3.11.2) the condition $a > 0$ holds and $h(\phi)$ defined by (6.3.11.3) is a non-trigonometric minorant of g that contradicts the supposition.

Sufficiency. Let the first condition hold and let $I = (\alpha; \beta)$ be an arbitrary interval belonging to Θ_Λ; let $h(\phi)$ be the maximal t.c.minorant of $g(\phi)$.

Set

$$H(\phi) := h(\phi_0)\cos(\phi - \phi_0),$$

where ϕ_0 is the maximum point of $h(\phi)$ on I. From inequality (6.3.11.1) and the conditions $g(\alpha) = g(\beta) = 0$ follows $H(\alpha) = H(\beta) = 0$. Then, by Lemma 6.3.11.2, we obtain $H(\phi) \equiv 0$. Thus $h(\phi_0) = 0$ and $h(\phi) \equiv 0$ for $\phi \in (\alpha; \beta)$, i.e., $h(\phi)$ is trigonometric.

Let the second condition be fulfilled. Lemma 6.3.11.1 implies that $H(\phi)$ is zero with tangency on ∂I. By Lemma 6.3.11.2 we obtain that $h(\phi) \equiv 0$. □

Proof of Theorem 6.3.5.1. Necessity. Let us note that if $v \in \mathbf{Fr}[\Phi_\Lambda]$, then for $c \in [0; 1]$ we have the equality

$$m(\lambda, G, v) = r\left(c(h_2 - h_1)^+ + (1-c)(h_2 - h_1)^-\right)(\phi) := rm(\phi, c). \tag{6.3.11.4}$$

Let $\exp\Lambda$ be extremely overcomplete in $A(G)$. By Theorem 6.3.8.1 $J_G \subset$ HARM, i.e., for every $c \in [0; 1]$ the maximal t.c.minorant of the function $m(\phi, c)$ is trigonometric. Since

$$\forall c \in [0; 1],\ \Theta_\Lambda = \{\phi : m(\phi, c) > 0\}$$

the necessity follows from Lemma 6.3.11.3.

Sufficiency. It follows directly from Lemma 6.3.11.3. □

Exercise 6.3.11.2 Explain this.

6.3.12 To prove Theorem 6.3.6.1 we need

Lemma 6.3.12.1 *If $w \in U[1]$ is non-minimal, then*

$$\mathbb{C}(w) := \{w_{[t]} : 1 \le t \le e^P\} \notin \text{MIN}.$$

It follows from Theorem 6.1.1.2, 2

Exercise 6.3.12.1 Explain this in detail.

Proof of Theorem 6.3.6.1. By Theorem 6.1.1.2, 2. $J_G(\Lambda) = \mathbb{C}(\mathcal{G}_G v)$. Thus Lemma 6.3.12.1 implies that $J_G(\Lambda) \notin$ MIN if and only if $\mathcal{G}_G v$ is not minimal. Thus Theorem 6.3.8.1 implies Theorem 6.3.6.1, 1 and 2.

Suppose $J_G(\Lambda) \subset$ HARM. Hence, $\mathcal{G}_G v = rH_0(\phi)$, where H_0 is trigonometric. Inversely, Lemma 6.3.12.1 implies $J_G(\Lambda) = \{rH_0(\phi)\}$. □

Proof of Theorem 6.3.7.1. Let $\rho(\Lambda, G) > 1$. Suppose $w_q := g_G q$ exists. By definition of $\rho(\Lambda, G)$ we have $w_q(z) \le 0$ for $z \in \partial D(\Lambda, G)$. By Theorem 6.2.3.3 (Maximum principle) $w_q \le 0$ for $z \in D(G, \Lambda)$. Also $w_q \le 0$ for $z \in \mathbb{T}^2_P \setminus D(G, \Lambda)$ by definition of $D(\Lambda, G)$. By Theorem 6.2.3.4 $w_q \equiv 0$ and hence is minimal. So $\exp \Lambda$ is complete by Theorem 6.3.6.1. If $\rho(\Lambda, G) = 1$, then the system $\exp \Lambda$ is complete for every $G_n \Supset G$, because of strict monotonicity of $\rho(\bullet)$ (see Section 6.2.2) so G is the maximal domain. □

For the proof of Theorem 6.3.7.3 we need an auxiliary assertion. We suppose that D is an image on $\mathbb{T}^2_P$ by the map (6.2.1.3) of the domain G with a smooth boundary.

Theorem 6.3.12.2 *Let $D \subset \mathbb{T}^2_P$ and $\rho(D) \le 1$. Then $\rho(\mathbb{T}^2_P \setminus \overline{D}) > 1$, if $D_\Lambda \ne \{\Re z > 0\}$.*

For the proof we need the following assertion which was proved originally by A. Eremenko and M. Sodin:

Theorem 6.3.12.3 (Eremenko, Sodin) *Let Γ be a Jordan curve, connecting 0 and ∞, $T\Gamma = \Gamma$ for some $T > 1$. Let D_+, D_- be domains, into which Γ divides the plane, and let ρ_1, ρ_2 be the orders of the minimal harmonic functions in D_+ and D_- respectively. Then*

$$\frac{1}{\rho_1} + \frac{1}{\rho_2} \le 2,$$

and equality is attained only if Γ consists of two rays.

We will give prove this theorem in Section 6.3.14.

Proof of Theorem 6.3.12.2. Let $\rho_1 = \rho(D)$, and suppose $q_1(z)$ is a solution of boundary problem (6.2.2.1), $\rho_2 = \rho(\mathbb{T}^2_P \setminus \overline{D})$, $q_2(z)$ is a solution of the corresponding boundary problem. Then the image of the boundary under the map $\lambda = e^z$ (we denote it as Γ) satisfies the conditions of Theorem 6.3.12.3 and the functions

$v_1(\lambda) := q_1(\log\lambda)|\lambda|^{\rho_1}$ and $v_2(\lambda) := q_2(\log\lambda)|\lambda|^{\rho_2}$ are positive harmonic functions in D_+, D_- with orders ρ_1 and ρ_2 respectively. By Theorem 6.3.12.3 we obtain

$$1/\rho(D) + 1/\rho(\mathbb{T}^2_P \setminus \overline{D}) \le 2.$$

and equality holds only if Γ is a pair of rays, i.e., $D_\Lambda = \{\Re z > 0\}$. □

Proof of Theorem 6.3.7.3. Necessity. Suppose $\rho(\Lambda, G_\Lambda) < 1$. Let us prove that $\exp\Lambda$ is not complete. To this end we construct an L_1-minorant of $m(z, G_\Lambda, \Lambda)$ and prove that it is not minimal.

Let $D_0 \Subset D(G_\Lambda, \Lambda)$ be a domain with smooth boundary for which $\rho(D_0) = 1$. This is possible because of strict monotonicity $\rho(D)$ (Section 6.2.2). Let q_0 be a solution of the problem (6.2.2.1) satisfying the condition

$$0 < \max\{q_0(z) : z \in D_0\} \le \min\{m(z, G_\Lambda, \Lambda) : z \in D_0\} - 2\epsilon$$

for sufficiently small ϵ. By Theorem 6.3.12.2, $\rho(\mathbb{T}^2_P \setminus \overline{D}_0) > 1$. Thus the potential

$$\Pi(z) = -\int_D G_\rho(z, \zeta, D)\nu(d\zeta)$$

exists and ν can be chosen in such way that $\operatorname{supp}\nu \Subset \mathbb{T}^2_P \setminus \overline{D}_0$. By Proposition 6.2.3.6,

$$\frac{\partial q_0}{\partial n} > 0, \; z \in D_0.$$

Thus ν can be chosen in such a way that

$$-\frac{\partial \Pi}{\partial n} < \min \frac{\partial q_0}{\partial n}, \; z \in \partial D_0.$$

Then the function

$$q(z) = \begin{cases} q_0(z), & z \in D_0, \\ \Pi(z), & z \in \mathbb{T}^2_P \setminus D_0, \end{cases}$$

is an L_1-subfunction on $\mathbb{T}^2_P$.

Exercise 6.3.12.2 Explain this in detail, exploiting Theorem 2.7.2.1.

The function $q(z)$ satisfies the condition

$$q(z) \le m(z, G_\Lambda, \Lambda) - 2\epsilon, \; \forall\, z \in \mathbb{T}^2_P,$$

because of negative potential. Hence,

$$q_1(z) := q(z) + \eta$$

for some $\eta > 0$ also is a minorant of $m(z, G_\Lambda, \Lambda)$ and it is not minimal. Necessity is proved. Sufficiency follows from Theorem 6.3.7.1. □

6.3.13 Now we pass to the proof of Theorem 6.3.7.4 and construction of Example 6.3.7.2.

Proof of Theorem 6.3.7.4. The set $\mathbb{T}^2_P \setminus D_0$ is closed. Let $\phi(z)$ be an infinitely differentiable function equal to zero on $\mathbb{T}^2_P \setminus D_0$ and positive on D_0. Set

$$q(z) := h_0(y) - \epsilon\phi(z),$$

where $h_0(y)$ is a t.c.f., corresponding to G_0, and let ϵ be small enough to satisfy $L_1 q(z) > 0,\ z \in \mathbb{T}^2_P$. It is possible, because $L_1 h_0(y) > 0$ by the condition of the theorem.

Then we have

$$m(z, G_0, q) = \epsilon\phi(z), \quad \text{hence} \quad \{z : m(z, G_0, q) > 0\} = D_0.$$

Take

$$v(\lambda) := |\lambda| q(\log \lambda)$$

and construct an entire function Φ_Λ for which

$$\mathbf{Fr}[\Phi_\Lambda] = \{v_{[t]} : 1 \le t \le e^P\}.$$

It is easy to check that the zero distribution of this function has all the properties demanded by Theorem 6.3.7.4. □

Exercise 6.3.13.1 Check this.

Proof of (6.3.7.8). Consider the problem

$$L_1 q(z) = 0, \ q|_{x=(2\pi/P)y} = 0. \tag{6.3.13.1}$$

Let us pass in the equation to new coordinates

$$\begin{cases} \xi = x\cos\alpha + y\sin\alpha, \\ \eta = -x\sin\alpha + y\cos\alpha, \ \tan\alpha = 2\pi/P. \end{cases}$$

Then the equation takes the form:

$$\left[\frac{\partial^2}{\partial\xi^2} + \frac{\partial^2}{\partial\eta^2} + 2\rho\left(\cos\alpha\frac{\partial}{\partial\xi} - \sin\alpha\frac{\partial}{\partial\eta}\right) r^2\right] R(\xi, \eta) = 0.$$

The condition of being zero on D_0 is

$$R(\xi, 2\pi l\cos\alpha) = 0. \ l \in \mathbb{Z}.$$

The condition of periodicity gives

$$R(\xi + (P/\cos\alpha)k, \eta) = R_1(\xi, \eta), \ k \in \mathbb{Z}.$$

We search for a solution that does not depend on ξ. We have

$$R''(\eta) - 2\rho \sin \alpha R'(\eta) + \rho^2 R(\eta) = 0, \; R(0) = R(2\pi \cos \alpha) = 0.$$

Further,

$$R(\eta) = C_1 e^{(\rho \sin \alpha)\eta} \cos((\rho \cos \alpha)\eta) + C_1 e^{(\rho \sin \alpha)\eta} \sin((\rho \cos \alpha)\eta).$$

Exploiting the boundary condition, we have

$$\rho_{\min} = (2\cos^2 \alpha)^{-1} = \frac{1}{2}\left[1 + \left(\frac{2\pi}{P}\right)^2\right].$$

The corresponding eigenfunction is

$$R = \exp\left(\rho_{\min} \sin \alpha\right)\eta \sin((\rho_{\min} \cos \alpha)\eta).$$

It is zero on $\mathbb{T}_P^2 \setminus D_0$ and positive in D_0, so it is determined up to a constant multiple. □

6.3.14 We are going to prove Theorem 6.3.12.3. Actually we prove

Theorem 6.3.14.1 *Let $\Gamma_1, \Gamma_2, \dots, \Gamma_n$ be Jordan curves, such that*

1. *Γ_i, $i = 1, 2, \dots, n$ connect 0 and ∞;*
2. *there exists a number T, $|T| > 1$ (not necessarily real) for which $T\Gamma_i = \Gamma_i$, $i = 1, 2, \dots, n$.*

Let D_i, $i = 1, 2, \dots, n$ be domains into which the plane is divided, and let ρ_i be the order of the minimal harmonic function in D_i. Then

$$\sum_i 1/\rho_i \leq 2 \tag{6.3.14.1}$$

and equality holds if and only if Γ_i are a logarithmic spirals (or rays, when $T \in \mathbb{R}_+$).

Proof. Denote by H_i the minimal harmonic function in D_i. Then $H_i = \Im \phi_i$ where $\phi_i : D_i \mapsto \Pi^+$ is a conformal map of D_i to the upper half-plane, $\phi(0) = 0$. The maps $g_i := \phi_i(T\phi_i^{-1}) : \Pi^+ \mapsto \Pi^+$ are continued by isomorphism to $\mathbb{C}$, and $g_i(0) = 0$. Thus $g_i(z) = \sigma_i z$, where $\sigma_i > 1$. Hence, $\phi_i(Tz) = \sigma_i \phi_i(z)$ or

$$Th_i(z) = h_i(\sigma_i z), \; h_i := \phi_i^{-1} : \Pi^+ \mapsto D_i.$$

Now we exploit the following inequality from [Lev]

$$\sum_{i=1}^{n} \frac{1}{\log \sigma_i} \leq \frac{2\log T}{|\log T|^2} \leq \frac{2}{\log T}. \tag{6.3.14.2}$$

The equalities in (6.3.14.2) are attained only when Γ_i are logarithmic spirals or rays.

Since $\rho_i = \log \sigma_i / \log |T|$ (6.3.14.2) implies (6.3.14.1). □

Notation

2.1 $\mathbb{R}^m$, $M(f,x,\varepsilon)$, $f^*(x)$, $C^+(E)$, $C^-(E), \chi_G$, χ_F, Γ_A, F^A, K, $M(f,K)$, K_n, $K_{\max}$.

2.2 $\sigma(G)$, μ, $G_0(\mu)$, $\operatorname{supp}\mu$, $\mathcal{M}(G)$, $\mu_F(E)$, ν, $\mathcal{M}^d$, ν^+, ν^-, $|\nu|$, $\operatorname{supp}\phi$, $\xrightarrow{*}$, $\overset{\circ}{E}$, $\overline{E}$, $\sigma(\mathbb{R}^{m_1}\times\mathbb{R}^{m_2})$, $\Phi_1\otimes\Phi_2$, $\mu_1\otimes\mu_2$.

2.3 $\varphi_n \xrightarrow{\mathcal{D}} \varphi$, $\alpha(t)$, $\alpha_\varepsilon(x)$, $\psi_\varepsilon(x)$, $\langle f,\varphi\rangle$, $\langle\delta_x,\varphi\rangle$, $\langle\delta_x^{(n)},\varphi\rangle$, $\langle\mu,\varphi\rangle$, $\langle\alpha f,\varphi\rangle$, $\langle f_1+f_2,\varphi\rangle$, $\langle\frac{\partial}{\partial x_k}f,\varphi\rangle$, $f_\epsilon(x)$, $f\,|_{G_1}$, $\widetilde{\cos\rho}(\phi)$.

2.4 $\Delta_{\boldsymbol{x}^0}$, $\mathcal{E}_m(x)$, θ_m, $G(x,y,\Omega)$, $G(x,y,K_{a,R})$.

2.5 $\Pi(x,\mu,D)$, $G_N(x,y)$, $\Pi_N(x,\mu,D)$, $\Pi(x,\mu)$, $\Pi(z,\mu)$, $\mathbf{cap}_G(K,D)$, $\mathbf{cap}_m(K)$, $\mathbf{cap}_m(D)$, $\overline{\mathbf{cap}}_m(E)$, $\underline{\mathbf{cap}}_m(E)$, $\mathbf{cap}_l(K)$.

2.6 $\mathcal{M}(x,r,u)$, $\mathcal{N}(x,r,u)$, E^ϵ, $D^{-\epsilon}$, $u_\epsilon(x)$, K_R, $M(r,u)$, $\mu(r,u)$, $\mathcal{M}(r,u)$, $N(r,u)$, $M(z)$.

2.7 $\tilde{u}(x)$, $\mu_x(t)$, $E(\alpha,\alpha',\epsilon,\mu)$, E_{n,δ_0}.

2.8 $a(r)$, $\rho[a]$, $\sigma[a]$, $\rho(r)$, $\sigma[a,\rho(r)]$, $V(r)$, $L(r)$, $\delta SH(\mathbb{R}^m)$, $T(r,u)$, $\rho_T[u]$, $\sigma_T[u]$, $\sigma_T[u,\rho(r)]$, $\rho_M[u]$, $\sigma_M[u]$, $\sigma_M[u,\rho(r)]$, $\rho[\mu]$, $\bar{\Delta}[\mu]$, $\bar{\Delta}[\mu,\rho(r)]$, $N(r,\mu)$, $\rho_N[\mu]$, $\delta\mathcal{M}(\mathbb{R}^m)$.

2.9 $H(z,\cos\gamma,m)$, $G(x,y,\mathbb{R}^m)$, $D_k(x,y)$, $H(z,\cos\gamma,m,p)$, $G_p(x,y,m)$, $G_p(z,\zeta,2)$, $\Pi(x,\mu,p)$, $\delta SH(\rho)$, $\Pi^R_<(x,\nu,\rho-1)$, $\Pi^R_>(x,\nu,\rho)$, $\delta_R(x,\nu,\rho)$, $\delta_R(z,\nu,\rho)$, $\delta_R(x,u,\rho)$, $M(r,\delta)$, $\bar{\Delta}_\delta[u,\rho]$, $\Omega[u,\rho(r)]$, $T(r,\lambda,>)$, $T(r,\lambda,<)$.

3.1 V_t, P_t, $SH(\mathbb{R}^m,\rho,\rho(r))$, $SH(\rho(r))$, $u_t(x)$, $\mathbf{Fr}[u,\rho(r),V_\bullet,\mathbb{R}^m]$, $U[\rho,\sigma]$, $U[\rho]$, $v_{[t]}$, $\mathcal{M}(\mathbb{R}^m,\rho(r))$, $\mu\in\mathcal{M}(\rho(r))$, $\mathbf{Fr}[\mu,\rho(r),V_\bullet,\mathbb{R}^m]$, $\mathbf{Fr}[\mu]$, $\mathcal{M}[\rho,\Delta]$, $\mathcal{M}[\rho]$, $\nu_{[t]}$.

3.2 $h(x,u)$, $\underline{h}(x,u)$, $l_{\boldsymbol{x}^0}$, $x^0(x)$, T_ρ, $G_I(\phi,\psi)$, $\Pi_I(\phi,ds)$, TC_ρ, Co_Ω.

3.3 $\overline{\Delta}(G,\mu)$, $\overline{\Delta}(E,\mu)$, $\underline{\Delta}(K,\mu)$, $\underline{\Delta}(E,\mu)$, $Co_\Omega(I)$, $\overline{\Delta}^{\mathrm{cl}}(E)$, $\underline{\Delta}^{\mathrm{cl}}(E)$, $\Omega^G(\epsilon)$, $\Omega^K(\epsilon)$.

4.1 T^t, $(T^\bullet,M)$, $d(\bullet,\bullet)$, $\Omega(T^\bullet)$, $\mathbb{C}(m)$, $\Omega(m)$, $A(m)$, $T_t v$.

4.2 U_0, $\beta(x)$, b_0, $k(s)$, $R_\epsilon v(x)$, $Str(\delta)$, $v(x|t)$, $v(\bullet, t)$

4.3 $w(\bullet|t)$, $w(\bullet|\bullet)$.

4.4 $\boldsymbol{u}$, $(\boldsymbol{u})_t$, $\mathbf{Fr}[\boldsymbol{u}]$, $\boldsymbol{U}[\rho]$.

5.1 $M(r,f)$, $T(r,f)$, $\rho_T[f]$, $\rho_M[f]$, $\sigma_T[f,\rho(r)]$, $\sigma_M[f,\rho(r)]$, $n(K_r)$, $n(r)$, $\rho[n]$, $\overline{\Delta}[n]$, $N(r,n)$, $\rho_N[n]$, $\overline{\Delta}_N[n]$, $p[n]$, $\mathbf{Fr}[f]$, $\mathbf{Fr}[n]$, $\mathrm{Mer}(\rho,\rho(r))$, $T(r,f)$, $\rho_T[f]$, $\sigma_T[f,\rho(r)]$.

5.2 $\alpha - \overline{\mathrm{mes}} C$, C_0^α, C_0^0.

5.3 $\|g\|_p$.

5.4 $\underline{h}_1(\phi,f)$, $\underline{h}_2(\phi,f)$, $\underline{h}(\phi,f)$.

5.5 $N(\delta,\mathcal{X})$, $(\mathcal{X})\int f d\delta$, $(\mathcal{X})\int_E f d\delta$, $\delta(\Theta^F)$, $D_{r,\Theta}$, $\delta_z(D_{r,\Theta})$, $A^{cl}(\delta,\chi_\Theta)$.

5.7 $\mathcal{F}(u)$, $H_\phi(u)$, $T(u)$, $M_\alpha(u)$, $M(u)$, $I_{\alpha\beta}(u)$, $I(u,g)$, $\overline{\mathcal{F}}[f]$, $\underline{\mathcal{F}}[f]$, χ_H, χ_I, χ_{Fo}.

5.8 K_{S_1}, S_1, $G(t,\gamma,\rho)$, $\hat{\mathbb{G}}(s,S_1-S)$, $(\mathcal{F}\nu)(s)$.

6.1 $H(z)$, $m(z,v,H)$, $\mathcal{G}_H$, D_H, U_{ind}, $\hat{U}_{\mathrm{ind}}$.

6.2 $\mathbb{T}_P^2$, $\mathcal{D}'(\mathbb{T}_P^2)$, $q(z)$, L_ρ, $E_\rho(\bullet - \zeta)$, $E'_\rho(\bullet - \zeta)$, q_D, $G_\rho(z,\zeta,D)$, $\mathcal{H}_\rho(q)$.

6.3 Λ, $\Phi_\Lambda(\lambda)$, $\exp\Lambda$, $A(G)$, $h_\Lambda(\phi)$, G_Λ, $\alpha G_1 + \beta G_2$, Θ_Λ, I_Λ, d_Λ, $h_G(\phi)$, $m(\lambda,G,v)$, $H(\lambda)$, $q_\Lambda(z)$, $D(G,\Lambda)$, $\rho(\Lambda,G)$, $g_G q$, $\mathcal{G}_G$, D_G, MIN, $J_G(\Lambda)$, HARM, $m(\phi,G,h)$, $E(\phi)$.

List of Terms

2.1 upper semicontinuous regularization
upper semicontinuous function
lower semicontinuous function

2.2 measure
mass distribution
support of μ
μ is *concentrated* on $E \in \sigma(G)$
restriction of μ onto $F \in \sigma(G)$
charge
positive and *negative*, respectively, *variations* of ν
full variation of ν
variation
Borel function
restriction of μ on the set E
product of measures

2.3 linear space
topological space
linear continuous functional on $\mathcal{D}$
Schwartz *distribution*
Dirac delta-function
the nth derivative of the Dirac delta-function
regular distribution
positive distribution
product of a distribution f by
an *infinitely differentiable* function $\alpha(x)$
sum of distributions f_1 and f_2
partial derivative of distribution
sequence of distributions f_n *converges* to a distribution f
regularization of the distribution f
restriction of distribution $f \in \mathcal{D}'(G)$ to $G_1 \subset G$

fundamental solution of L at the point y
spherical operator

2.4 *harmonic* distribution
Lipschitz boundary, *Lipschitz* domain
harmonic measure
spherical function of a *degree* ρ
Green potential of μ relative to D
Newton potential
logarithmic potential

2.5 *balayage, sweeping*
Green capacity of the compact set K relative to the domain D
Wiener capacity
external and *inner* capacity of any set E
capacible set
logarithmic capacity
irregular point
equilibrium mass distribution
h-Hausdorff measure
Carleson measure

2.6 *mean value* of $u(x)$ on the sphere $S_{x,r} := \{y : |y - x| = r\}$
subharmonic function
the least harmonic majorant of u in K
Riesz measure of the subharmonic function u

2.7 *precompact* family of functions
a sequence f_n of locally summable functions *converges in* L_{loc}
quasi-everywhere convergence
a sequence of functions u_n *converges* to a function u *relative* to α-Carleson measure
a point $x \in \mathbb{R}^m$ $(\alpha, \alpha', \epsilon)$-*normal* with respect to the measure μ

2.8 *order* of $a(r)$
type number of $a(r)$
$a(r)$ of *minimal type*
$a(r)$ of *normal type*
$a(r)$ of *maximal type*
convergence exponent for the sequence $\{r_j\}$
a *proximate order* with respect to order ρ
equivalent proximate orders
type number with respect to a proximate order
proper proximate order

Nevanlinna characteristic
order of $u(x)$ with respect to $T(r)$
characteristics $\rho_M[u]$, $\sigma_M[u]$, $\sigma_M[u, \rho(r)]$
convergence exponent of μ
upper density of μ
genus of μ
N-order of μ
N-type of μ

2.9 *Gegenbauer* polynomials
Chebyshev polynomials
primary kernel
canonical potential
zero distribution
canonical Weierstrass product

3.1 *limit set* of the function $u(x)$
limit set of the mass distribution μ

3.2 *indicator* of growth of u
lower indicator
ρ-subspherical function
ρ-trigonometrically convex (ρ-t.c.)
fundamental relation of indicator

3.3 *upper (lower) density* of μ
subadditivity of $\overline{\Delta}(E, \bullet)$
superadditivity of $\underline{\Delta}(E, \bullet)$
semi-additivity
generalized semi-additivity
monotonic function of $E \in \mathbb{R}^m$
t-extension of E
to be *dense in*
angular densities

4.1 *dynamical system*
(ϵ, s)-chain from m to m'
chain recurrent dynamical system
non-wandering point
attractor
completely regular growth
polygonally connected set
periodic dynamical system

4.2 *partition of unit*

4.3 *pseudo-trajectory*
asymptotically dynamical pseudo-trajectory
with *dynamical asymptotics* $T_\bullet$ (a.d.p.t.)
piecewise continuous pseudo-trajectory $w(\bullet|\bullet)$
ω-dense pseudo-trajectory

4.4 *subharmonic curve*

5.1 *entire function of order ρ and normal type*
with respect to proximate order $\rho(r)$
entire function with prescribed limit set
meromorphic function of order ρ and normal type
with respect to a proximate order $\rho(r)$

5.2 *relative* Carleson α-measure

5.3 *lower indicator* of entire function

5.4 *maximal interval of ρ-trigonometricity*
strictly ρ-t.c.f.
concordant h and g

5.5 *upper density of zeros* of entire function
$(\mathcal{X})$-integral with respect to a *nonnegative measure* δ

5.6 *completely regular growth* function
CRG-function
regular zero distribution
regular zero distribution with integer ρ
completely regular growth functions
along curves of regular rotation
curve of regular rotation

5.7 *growth characteristic*
continuity, positive homogeneity
asymptotic characteristics of growth
total family of growth characteristics
non-rarefied set
rarefied set
thinly closed set
independent family of characteristic

6.1 *ideally complementing* H-multiplicator
entire function is of *minimal type*
with respect to a proximate order $\rho(r), \rho(r) \to \rho$
limit set of indicators
the maximum principle for $U[\rho]$ *is valid in the domain* G

6.2 *automorphic*
connected on spirals
spectrum
strictly monotonic
minimal $v \in U[\rho]$

6.3 function of *exponential type*
completeness
maximality
extremal overcompleteness
maximal domain of completeness
extremely overcomplete system $\exp \Lambda$
trigonometrically convex function (t.c.f)
conjugate indicator diagram
regular set
G_Λ is *enclosed* in G
enclosed with sliding
enclosed hardly
enclosed freely
indicator limit set
indicator set
zero with tangency
Λ is *periodic*
$w \in U[1]$ is *minimal*
$U \subset U[1]$ is *minimal*

Bibliography

[An] Anosov, D.V. et al., *Ordinary differential equations and smooth dynamical systems*, Dynamical systems I, Ency. Math. Sci, 1, Springer-Verlag, 1997.

[Ar] Arakeljan, N.U., *Uniform approximation by entire functions on unbounded continua; an estimate of the rate of their growth*, Acad. N. Armjan. SSR Dokl. **34** (1962), 145–149, (Russian).

[Az(1969)] Azarin, V.S., *The rays of completely regular growth of an entire function*, Math. USSR Sb. **8** (1969), 437–450.

[Az(1979)] Azarin, V.S., *Asymptotic behavior of subharmonic functions of finite order*, Math. USSR Sb **36** no. 2 (1979), 135–154.

[Az(1998)] Azarin, V.S., *Completely regular growth on a prescribed set of rays and the limit sets of entire functions*, Compl. Var. **37** (1998), 53–66.

[Az(2007)] Azarin, V.S., *On the polynomial asymptotics of subharmonic functions of finite order and their mass distributions*, J. Math. Ph., Anal. Geom. **3**, No 1 (2007), 5–12.

[Az(2008)] Azarin, V.S., *Limit sets and a problem in dynamical systems*, arXiv: 0804.0348v2

[AG(1982)] Azarin, V.S., and Giner, V.B., *On a structure of limit sets of entire and subharmonic functions*, Teor. Funkts. Anal. Prilozh. **38** (1982), 2–12, (Russian).

[AG(1992)] Azarin, V.S., and Giner, V.B, *Limit Sets and Multiplicators of Entire Functions*, Advances in Soviet Mathematics **11** (1992), 251–275.

[AG(1994)] Azarin, V.S., and Giner, V.B., *Limit Sets of Entire Functions and Completeness of Exponent Systems*, Matematicheskaia Fizika, Analiz, Geometria, vol 1, 1994, 3–30; (Russian – Predel'nye mnozhestva zelych funkcij i polnota sistem exponent).

[AGL] Azarin, V.S, Giner, V.B. and Lyubich, M.Yu., *Dinamicheskie Systemy i Kompleksnyi Analiz*, Naukova Dumka, Kiev, 1992, pp. 3–17, (Russian).

[ADP] Azarin, V., Drasin D., and Poggi-Corradini, P., *A generalization of trigonometric convexity and its relation to positive harmonic functions in homogeneous domains*, J. d'Analyse Math. **95** (2005), 173–220.

[AD] Azarin, V., and Drasin, D., *A Generalization of Completely Regular Growth*, Israel Mathematical Conference Proceedings **15** (2001), 21–30.

[AP] Azarin, V., and Podoshev, L., *Limit sets and indicators of entire functions*, Sib. Mat. J. **XXV** no. 6 (1984), 3–16, (Russian).

[Ax] Axler, Sh., et al. *Harmonic Function Theory*, Springer, 1992.

[BM] Beurling, A., and Malliavin, P., *On Fourier transforms of measures with compact support*, Acta Math. **107** (1962), 291–309.

[BP] Bessaga, C., and Pelczynski, A., *Selected topics in infinite-dimensional topology*, PWN, Warsaw, 1975.

[Bo] Bourbaki, N., *Integration I. Elements of Mathematics*, Springer-Verlag, Berlin, 2004.

[Br] Brelo, M., *Foundations of Classic Potential Theory.*

[Bai] Bailette, J., *Fonctions approchables par des sommes d'exponentielles*, J. d'Analyse Math. **10** (1962–1963), 91–114.

[Bal(1973)] Balashov, S.K., *On entire functions of finite order with zeros on curves of regular rotation*, Math. USSR. Izv. **7** (1973), 601–627.

[Bal(1976)] Balashov, S.K., *On entire functions of completely regular growth along curves of regular rotation*, Math. USSR. Izv. **10** (1976), 321–328.

[Ca] Carleson, L., *Selected problem on Exceptional Sets*, D. Van Nostrand Company, Inc., Princeton, New Jersey, 1967,

[De] Delange, H.Y., *Un théorème sur les fonctions entières à zeros réeles et négatifs*, J. Math. Pures Appl. **(9)** 31 (1952), 55–78.

[Ev] Evgrafov, M.A., *Asymptoticheskie ocenki i celye funkcii*, FM, "Nauka" L, 1979 (Russian).

[Fa] Fainberg, E.D., *Integral with respect to nonadditive measure and estimates of indicators of entire functions*, Siberian Math. J. **24** (1984), 143–153.

[GG] Girnyk, M., and Gol′dberg, A., *Approximation of subharmonic functions by logarithms of moduli of entire functions in integral metrics*, Isr. Math. Conf. Proc. **15** (2001), 117–135.

[Gi] Giner, V.B., *On approximation limit sets of subharmonic and entire functions in $\mathbb{C}$ by periodic limit sets*, Manuscript No. 1033-Ukr87 deposed at the UkrNIINTI (1987), (Russian).

[GPS] Giner, V.B., Podoshev, L.R., and Sodin, M.L., *On summing lower indicators of entire functions*, TFFA **42** (1984), 27–36.

[GLO] Gol′dberg, A.A., Levin, B.Ja., and Ostrovskii, I.V., *Entire and meromorphic functions*, Encycl. Math. Sci., vol. 85, Springer, 1997, 4–172.

[Go(1967)] Gol′dberg, A.A., *Estimates of indicators of entire functions and integral on non additive measures*, Contemporary problems of Analytic Function Theory, 1967, 88–93; (Russian – Sovremennye prolemy teorii analiticheskikh funkcij, M.Nauka, 1967).

[Go(1962)] Gol′dberg, A.A., *Integral with respect to a semi-additive measure and its application to the theory of entire functions*, AMS Transl. **88** (1970), 105–289.

[Gr] Grishin, A.F., *On sets of regular growth of entire functions, I*, TFFA **40** (1983), 36–47, (Russian).

[Ha] Halmos, P., *Measure Theory*, D. Van Nost. Com., NY, 1954.

[HN] Havin, V.P., and, Nikolskii, N.K., (eds.) *Linear and Complex Analysis Problem Book* 3, (Lecture Notes in Mathematics No. 1574) Springer, 1994.

[HK] Hayman, W.K., and Kennedy, P.B., *Subharmonic functions, I*, Academic Press, 1976.

[He] Helms, L.L., *Introduction to Potential Theory*, Wiley-Interscience, 1969.

[Hö] Hörmander, L., *The Analysis of Linear Partial Differential Operators I*, Springer, 1983.

[HS] Hörmander, L., and Sigurdsson, R., *Limit sets of plurisubharmonic functions*, Math. Scand. **65** (1989), 308–320.

[Ho] Hopf, E., *A remark on linear elliptic differential equations of second order*, Proc. Amer. Math. Soc. **3** (1952), 791–793.

[Ke] Kellogg, O.D., *Foundations of Potential Theory*, Dover Publ., 1953.

[Kj] Kjellberg, B., *On certain integral and harmonic functions: a study in minimum modulus, Thesis*, Uppsala, 1948.

[Ko] Koosis, P., *La plus petite majorante surharmonic*, Ann. Inst. Fourier **33** (1983), 67–107.

[Kr] Krasichkov-Ternovskii, I.F., *Lower estimates for entire functions of finite order*, Sib. Mat. J. **6** No. 4 (1956), 840–861, (Russian).

[Kon] Kondratyuk, A.A., *Entire functions with finite maximal density of zeros*, TFFA **No. 11** (1970), 35–40, (Russian).

[KF] Kondratyuk, A.A., and Fridman, A.A., *Limit value of lower indicator and lower estimates of entire functions with positive zeros*, Ukr. Math. J. **24** No. 4 (1972), 488–494, (Russian).

[La] Landkoff, N.S., *Foundation of modern potential theory, Die Grundlehren Mathematischer Wissenchaften. Band 180*, Springer-Verlag, New-York-Heidelberg, 1972.

[Le] Levin, B.Ya., *Distribution of zeros of entire functions*, AMS, Providence, Rhode Island, 1980, (English).

[Li] Lindelof, E., *Mémoire sur la théorie des fonctions entières de gendre fini*, Acta Soc. Sci. Fenn. **31** No. 1 (1902).

[Lev] Levin, G.M., *Boundaries for the multipliers of periodic points of holomorphic mappings*, Sib. Math. J. **31** No. 2 (1990), 273–278.

[LM] Lyubarskii, Yu., and Malinnikova, E., *On approximation of subharmonic functions*, J. Anal. Math. **83** (2001), 121–149.

[LS] Lyubarskii, Yu, and Sodin, M., *Preprint No.* 17, Institute for Low Temperatures Ukr. Acad. Sci, Kharkov, 1986, (Russian).

[Ma] Markus, A., *Introduction to the Spectral Theory of Polynomial Operator Pencils*, AMS, 1988.

[Oz] Ozawa, M., *On an estimate for* $\int_0^\infty m(t, E(-z,q))t{-}1-\beta dt$, Kodai Math. J. **8** (1985), 33–35.

[Pf(1938)] Pfluger, A., *Die Wertverteilung und das Verhalten von Betrag und Argument einer speciellen Klasse analytischer Funktionen I*, Comment. Math. Helv. **11** (1938), 180–214.

[Pf(1939)] Pfluger, A. *Die Wertverteilung und das Verhalten von Betrag und Argument einer speciellen Klasse analytischer Funktionen II*, Comment. Math. Helv. **12** (1939), 25–65.

[Po(1985)] Podoshev, L.R., *On summing of indicators, and Fourier coefficients of logarithm of the modulus of an entire functions*, TFFA **No. 43** (1985), 100–107, (Russian); English transl. J. Sov. Math. **48** No. 2 (1990), 203–209.

[Po(1992)] Podoshev, L.R., *Complete description of the pair indicator – lower indicator of an entire function*, Advances Soviet Mathematics **11** (1992), 75–105.

[PS] Pólya, G., and Szegő, G., *Aufgaben und Lehrsätze aus der Analysis I*, Springer-Verlag, Berlin-Göttingen-Heidelberg-New York, 1964.

[PW] Protter, H.H, and Weinberger, H.F, *Maximum principles in differential equations*, Prentice-Hall, Englewood Cliffs, N.J., 1967.

[Ro] Ronkin, L., *Introduction to the theory of entire functions of several variables*, Nauka, Moscow, 1971, (Russian).

[Si] Sigurdsson, R., *Growth properties of analytic and plurisubharmonic functions of finite order*, Math. Scand. **59** (1986), 235–304.

[So] Sodin, M.L., *On growth in L_p metrics of entire functions of finite low order*, UkrNIINTI No. 420 (1983), 2–20, (Russian).

[Ti] Tichmarsh, E.C., *On integral functions with real negative zeros*, Proc. London Math. Soc. **26** (1927), 185–200.

[TT] Timan, A., and Trofimov, V., *Introduction to the harmonic function theory*, FML, Moscow, 1966, (Russian)

[Va] Valiron, G., *Sur les fonctions entières d'ordre fini*, Ann. Fac. Sci. Univ. Toulouse (3) **5** (1913), 117–257.

[Vl] Vladimirov, V.S., *Equations of Mathematical Physics*, FML, "Nauka", 1971, (Russian).

[Yu(1982)] Yulmukhametov, R.S., *Approximation of subharmonic functions*, Sib. Math. J. **26** no. 3 (1985), 603–618.

[Yu(1985)] Yulmukhametov, R.S., *Approximation of subharmonic functions*, Anal. Math. **11** no. 3 (1985), 257–282.

[Yu(1996)] Yulmukhametov, R.S., *Entire functions of several variables with given behavior at infinity*, Izv. Math. **60** no. 4 (1996), 857–879.

Birkhäuser Advanced Texts (BAT)

Edited by
Herbert Amann, Zürich University, Switzerland
Steven G. Krantz, Washington University, St. Louis, USA
Shrawan Kumar, University of North Carolina, Chapel Hill, USA
Jan Nekovář, Université Pierre et Marie Curie, Paris, France

BIRKHÄUSER

This series presents, at an advanced level, introductions to some of the fields of current interest in mathematics. Starting with basic concepts, fundamental results and techniques are covered, and important applications and new developments discussed. The textbooks are suitable as an introduction for students and non–specialists, and they can also be used as background material for advanced courses and seminars.

Azarin, V.
Growth Theory of Subharmonic Functions (2008).
ISBN 978-3-7643-8885-0

Quittner, P. / Souplet, P.
Superlinear Parabolic Problems. Blow-up, Global Existence and Steady States (2007).
This book is devoted to the qualitative study of solutions of superlinear elliptic and parabolic partial differential equations and systems. This class of problems contains, in particular, a number of reaction-diffusion systems which arise in various mathematical models, especially in chemistry, physics and biology.
The book is self-contained and up-to-date, it has a high didactic quality. It is devoted to problems that are intensively studied but have not been treated so far in depth in the book literature. The intended audience includes graduate and postgraduate students and researchers working in the field of partial differential equations and applied mathematics.
ISBN 978-3-7643-8441-8

Drábek, P. / Milota, J.
Methods of Nonlinear Analysis. Applications to Differential Equations (2007).
In this book, fundamental methods of nonlinear analysis are introduced, discussed and illustrated in straightforward examples. Every method considered is motivated and explained in its general form, but presented in an abstract framework as comprehensively as possible. Applications and generalizations are shown. In particular, a large number of methods is applied to boundary value problems for partial differential equations.
The text is structured in two levels: a self-contained basic level and an advanced level – organized in appendices – for the more experienced reader. It thus serves as both a textbook for graduate-level courses and a reference book for mathematicians, engineers and applied scientists.
ISBN 978-3-7643-8146-2

Krantz, S.G. / Parks, H.R.
A Primer of Real Analytic Functions (2002)
ISBN 978-0-8176-4264-8

DiBenedetto, E.
Real Analysis (2002).
ISBN 978-0-8176-4231-0

Estrada, R. / Kanwal, R.P.
A Distributional Approach to Asymptotics. Theory and Applications (2002).
ISBN 978-0-8176-4142-9

Chipot, M.
ℓ goes to plus Infinity (2001).
ISBN 978-3-7643-6646-9

Sohr, H.
The Navier–Stokes Equations. An Elementary Functional Analytic Approach (2001).
ISBN 978-3-7643-6545-5

Conlon, L.
Differentiable Manifolds (2001).
ISBN 978-0-8176-4134-4

Chipot, M.
Elements of Nonlinear Analysis (2000).
ISBN 978-3-7643-6406-9

Gracia-Bondia, J.M. / Varilly, J.C. / Figueroa,H.
Elements of Noncommutative Geometry (2000).
ISBN 978-0-8176-4124-5